BATS OF THE WEST INDIES

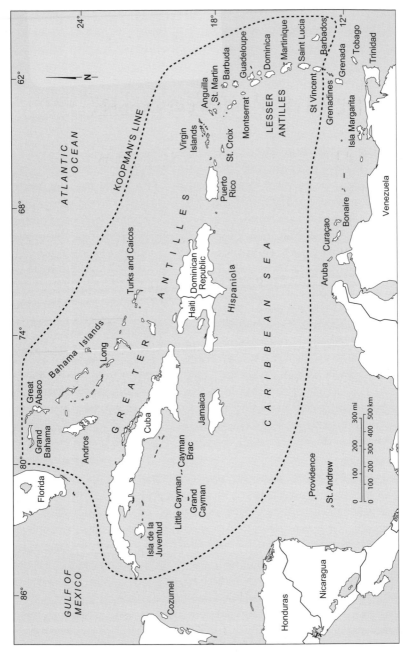

The West Indies. Map by Bill Nelson.

Bats of the West Indies

A NATURAL HISTORY AND FIELD GUIDE

EDITED BY Allen Kurta & Armando Rodríguez-Durán

DRAWINGS BY ASHLEY K. WILSON

COMSTOCK PUBLISHING ASSOCIATES

AN IMPRINT OF

CORNELL UNIVERSITY PRESS

ITHACA AND LONDON

Publication of this book was made possible by generous grants from Eastern Michigan University, Para la Naturaleza, and Universidad Interamericana de Puerto Rico.

First published 2023 by Cornell University Press

Black-and-white line drawings © Ashley K. Wilson, used by permission

Design and composition by Julie Allred, BW&A Books, Inc.
Printed in South Korea

Library of Congress Cataloging-in-Publication Data
Names: Kurta, Allen, 1952- editor. | Rodríguez Durán, Armando, editor. | Wilson, Ashley K., illustrator.
Title: Bats of the West Indies : a natural history and field guide / edited by Allen Kurta and Armando Rodríguez-Durán; drawings by Ashley K. Wilson.
Description: Ithaca [New York] : Comstock Publishing Associates, an imprint of Cornell University Press, 2023. | Includes bibliographical references and index.
Identifiers: LCCN 2022022666 | ISBN 9781501768934 (paperback)
Subjects: LCSH: Bats—West Indies. | Bats—West Indies—Identification.
Classification: LCC QL737.C5 B379 2023 | DDC 599.409729—dc23/ eng/20220528
LC record available at https://lccn.loc.gov/2022022666

To our early mentors,
Rollin H. Baker, Michigan State University, and
Allen R. Lewis and Juan A. Rivero,
University of Puerto Rico, Mayagüez,
for providing us with the basic tools,

To our academic father,
Thomas H. Kunz, Boston University,
for bringing us to Boston and initiating nearly four decades
of collaboration and friendship,

and

To our students,
for continuing our education

CONTENTS

FAMILY AND SPECIES ACCOUNTS

PREFACE

Bats are the most important terrestrial mammals in the West Indies, both in number of species and in how they influence natural ecosystems. Although numerous scientific publications have described various traits and behaviors of these bats over the last century, most observations are limited to single islands and are published in a myriad of technical outlets. Our goal in creating *Bats of the West Indies* is to synthesize, in a readable fashion, that diverse information concerning the history, structure, distribution, ecology, behavior, and reproduction of the sixty-one species currently living throughout these islands. This book is not meant to be a compendium of geographic variation, an in-depth analysis of phylogenetic relationships, or a detailed diary of short-term expeditions to each island. Instead, *Bats of the West Indies* is intended as a concise reference for students, wildlife managers, research biologists, and interested laypersons. The content is solid biology, but with minimal jargon and writing typically geared toward an educated nonspecialist. To assemble this book, we relied on twenty-seven other authors, from ten countries, ranging from Canada and Brazil to Cuba and the Dominican Republic; each author has extensive field experience in the Neotropics and usually firsthand knowledge of their species. Although each chapter or account was written as a stand-alone article, all were rigorously edited so that they flowed smoothly and appeared, as much as possible, as if written in "one voice."

But what are the West Indies? Geographers loosely define the West Indies as an arc of islands separating the Gulf of Mexico and Caribbean Sea from the Atlantic Ocean, and together, these islands and waters often are simply referred to as the Caribbean region. Various authors, though, have delineated the exact boundaries of the West Indies based on different parameters, such as depth of water between an island and the continent, number of endemic species of birds or reptiles that are present, or even human population density (Hedges et al. 2019). For this book, we generally follow Genoways et al. (1998, 2010; Durocher et al. 2021), who defined the West Indian faunal region based on the distribution of insular mammals.

Genoways et al. (1998, 2010) included the Cayman Islands, Bahama Islands, Greater Antilles, and most of the Lesser Antilles within the region, and labeled its encompassing boundary as "Koopman's Line," in honor of the great twentieth-century taxonomist Karl F. Koopman. In the Lesser Antilles, Koopman's Line lies between Saint Vincent and the Grenadine isle of Bequia. Consequently, continental islands in the Caribbean Sea, such as Trinidad, Aruba, and Curaçao, are outside the West Indies, as are the Grenadine Islands and Grenada itself. This placement is reasonable for bats. Grenada, for example, hosts two species (only 16% of its bat fauna) that are endemic to the West Indies, and the rest are species associated with mainland South America. In contrast, the endemic faunas of Saint Vincent and Barbados, located immediately north of the line, are markedly higher at 40 and 50%, respectively, which is more in line with the overall level of endemism in the Lesser Antilles as a whole (62%). Furthermore, Grenada harbors a single species from the Emballonuridae, a family that is absent from all islands inside Koopman's Line. Similarly, we excluded the Colombian islands of Saint Andrew and Providence, because up to six species of bats are known from those tiny islands but only one is a West Indian endemic that also lives on Jamaica and Hispaniola (Instituto de Investigación de Recursos Biológicos Alexander von Humboldt 2021).

Organization

The book begins with an Introduction and five chapters that set the stage for descriptive accounts of individual families and species. The Introduction familiarizes the nonspecialist with these airborne mammals and describes their structure, flight characteristics, and elementary echolocation strategies. Chapter 1 presents, in a more technical fashion, the biotic, abiotic, and historical factors that affect the modern-day ecosystems of which bats are a part, whereas Chapter 2 combines these concepts with the theory of island biogeography to illustrate how communities of bats formed on the islands over thousands of years. Chapter 3 moves into modern times, examines how contemporary species partition roosting and aerial space, and points out distinct differences between present-day insular and continental assemblages. Chapter 4, in contrast, summarizes interactions by these animals with indigenous peoples, early colonists, and modern residents of the archipelago, emphasizing the beneficial services that bats provide and making a case for their preservation in our modern world. In the final chapter, the authors depict the challenges faced by bats today as a result of direct and indirect human impacts and suggest that successful conservation of these beneficial creatures requires local knowledge of bat populations that is generated by resident researchers, managers, and actively involved citizens. All chapters, of course, indicate what biologists know about a particular topic but also frequently point out information that is lacking and offer suggestions for future research.

The six family and sixty-one species accounts, which are the heart of the book, appear after brief identification keys. For ease of use, families are arranged alphabetically, as are the species within each family. All family accounts summarize the physical characteristics and diversity of each group, both worldwide and within the West Indies, and include representative skull drawings, whereas every species account features a range map and color photo.

Whenever possible, we use photos of individual animals from the West Indies but occasionally rely on pictures taken of the same species on the mainland. However, three endemic species—the Cuban pallid bat (*Antrozous koopmani*), Cuban yellow bat (*Lasiurus insularis*), and Jamaican red bat (*Lasiurus degelidus*)—are so uncommon that no photos are available. In addition, we could not locate a quality photo showing the naked back of Davy's naked-backed bat (*Pteronotus davyi*). Consequently, to aid in identification of each of these four species, we substituted an image of a close continental relative with essentially identical external appearance.

On each map, dark brown designates the islands on which biologists have documented a species as a resident, whereas white indicates land where the bat does not reside; a cream color, in contrast, is used for peninsular Florida, to show that it is outside the West Indies. The maps are suggestions as to which species likely are resident on each island, but an important caveat is that many islands, especially small ones, have not been investigated in detail, if at all, and ranges can change as hurricanes eliminate species from some places and introduce the same animal to other locations (e.g., Mathis and Reed 2021). Similarly, the maps are meant to provide a quick visual representation of the animal's overall range within the West Indies, and obviously, it is not possible to portray every capture locality or even every island on maps of such small size. To prepare our maps, we greatly relied on the detailed compilation of occupied islands in Hoffman et al. (2019), as well as the input of our authors.

The text of each species account begins with the scientific name, followed by common names in English, Spanish, and French, which are the official languages on most islands in the region. Scientific names generally follow Wilson and Mittermeier (2019), as do most common names. However, common names concocted in North America and then translated into Spanish or French often ignore vernacular names already in use or that actually might have meaning in Spanish- or French-speaking countries, such as Cuba or Guadeloupe, respectively. Consequently, we occasionally deviate from Wilson and Mittermeier (2019) and list common names that make sense within local cultural contexts. The text that follows the names integrates information from throughout the islands and occasionally the mainland to describe the life of each mammal and is organized into six sections dealing with the animal's name, distribution, measurements, physical description, natural history, and conservation status.

Latinized names are often mysterious to the nonbiologist, so every account begins with a paragraph that explains the meaning of the binomial, based on the original description of the species,

Latin and Greek dictionaries, or published compendia (e.g., Braun and Mares 1995; Álvarez-Castañeda and Álvarez Solórzano 1996; Gotch 1996; Beolens et al. 2009). After a brief description of the geographic range, a list of standard measurements and a physical description are presented, and these should allow identification of the bat in the field, especially when combined with the accompanying photo. The largest portion of each account, though, is devoted to the natural history of the animal, and the length of this section is directly proportional to what naturalists know about a particular species; hence, this part also consistently indicates what information is still needed for a thorough understanding of the life of each mammal. The next segment concerning conservation status indicates how the animal is classified by the International Union for the Conservation of Nature (see Chapter 5) and often provides additional information that is meant to stimulate awareness, promote conservation efforts, or highlight research needs. Each account concludes with several key references, with a preference to articles published in the last 10 to 15 years that provide an introduction to the literature on each species; additional information for each account was extracted from the references tabulated in Appendix 5.

After the accounts, we include appendices that summarize useful information in mostly tabular form, thus facilitating comparison among species without having to thumb through every account. The first three appendices include a summary of simple body and skull measurements, an explanation and presentation of dental formulas to aid in identification, and a tabulation of echolocation call parameters and studies. This information is followed by a list of major islands occupied by each species and a table of references arranged by island, thus making it easy for tourists and researchers alike to prepare for an upcoming visit or to facilitate research. Following the appendices, we present a glossary of technical terms to aid the nonbiologist and detail all references. The book concludes with an alphabetical roster of contributing authors and their affiliations, a list of photo credits, and the index.

ACKNOWLEDGMENTS

The authors of our introductory chapters are indebted to a number of people. Ariel Lugo, who wrote Chapter 1 on climate and vegetation, is grateful to Tania López Marrero, for allowing use of Figure 1.1; Mariela Ortiz, for assistance in preparation of Figure 1.2; Helen Nunci, for help with producing the manuscript; and Victor Cuevas, William Gould, and Ernesto Medina, for comments that improved the chapter. Chapter 1 was developed in collaboration with the University of Puerto Rico. J. Angel Soto-Centeno thanks Bruna da Silva Fonseca for support and conversations that improved Chapters 2 (biogeography) and 5 (conservation), Nancy B. Simmons and David W. Steadman for friendship and mentoring, and the U.S. National Science Foundation for financial support (grant DEB-2135257).

In putting together such a comprehensive work, the editors accrued a number of debts. Both Eastern Michigan University and Universidad Interamericana de Puerto Rico, Bayamón Campus, have been most gracious, providing released time from teaching and financial support that made this book possible. Additional funding came from the Puerto Rican environmental organization Para la Naturaleza, which has been an important partner throughout the years, both sponsoring research and disseminating information to the layperson. This project was conceived at Mata de Plátano, a field station at the heart of Puerto Rico's karst country, where a good deal of information about the bats of the West Indies has been gained. The station is the result of joint efforts between the Bayamón Campus of Universidad Interamericana and Ciudadanos del Karso, a local environmental organization, with support from the National Science Foundation.

We thank the twenty-seven authors, who submitted their work promptly and graciously and put up with our nagging and editing, and we deeply appreciate the willingness of sixteen photographers to share their images; all authors and photographers are individually listed at the end of the book. Béatrice Ibéné in Guadeloupe, Susan Koenig in Jamaica, Arnaud Lenoble in Bordeaux, France, and Carlos Mancina in Cuba, were especially helpful in finding photos and completing other aspects of this book. The archaeologists Reniel Rodríguez Ramos and Carlos Ayes Suarez contributed to our understanding of pre-Columbian mythology. Burton Lim commented on the keys, and he and Gary Kwiecinski provided other authors with unpublished measurements of various species. We are grateful to Cara Rogers for proofreading, Brooke Daly for assistance with creating distribution maps (when she should have been working on her thesis), Bill Nelson for preparing the frontispiece, and especially Ashley K. Wilson for producing such wonderful drawings. We also thank Giselle Delgado and Sandra Rosa for administrative support and Rafael Ortíz for helping with figures in Chapter 3. In addition, AK thanks Scott Pedersen for the opportunity to experience the bats and culture of Montserrat, and ARD recognizes the continuing cooperation of the Programa de Conservación de Murciélagos de Puerto Rico. Finally, ARD and AK acknowledge our better halves, Zulma Ayes and Robin Slider, respectively, who have learned to love these secretive nocturnal creatures and have become our partners in their study and conservation.

Allen Kurta and Armando Rodríguez-Durán

BATS OF THE WEST INDIES

Bats: The Basics

Allen Kurta

What are bats? Bats, of course, are mammals—vertebrates that produce milk and wear a hairy coat. Although bats are numerous and very diverse, they have small bodies compared with zebras, bears, wolves, squirrels, and even most mice; half of all bats have a mass less than 20 g (Jones and Purvis 1997). The bumblebee bat (*Craseonycteris thonglongyai*) of Thailand weighs as little as 1.5 g, is only 3 cm long from nose to rump, and is arguably the smallest mammal in the world. The largest bats, in contrast, are the golden-crowned flying fox (*Acerodon jubatus*) of the Philippines and the large flying fox (*Pteropus vampyrus*) of Asia that weigh only 1100 to 1600 g, perhaps as much as a small rabbit. Nevertheless, these Old World bats do have a substantial wingspan of up to 1.7 m, which is greater than the average height of human females in Canada and the United States (1.6 m).

Adaptations for Flight

Among vertebrates, powered flight has developed three times, twice in reptiles (pterosaurs and birds) and once in mammals (bats), and each time, it required considerable modification of the standard body plan of a four-legged animal (Alexander 2015). Although these alterations concerned many aspects of physiology and anatomy, the changes are most obvious in the forelimb, which had to be transformed into a wing that could provide lift against gravity and thrust to propel the animal through the air. Birds used stiff feathers to create the wing surface, whereas pterosaurs and bats developed a living skinlike membrane that stretched outward from the body, from the shoulder to the leg.

Unlike pterosaurs, though, that relied primarily on one large metacarpal and a single finger to buttress the outer half of the wing's surface, the forearm and most bones of the hand in bats became greatly lengthened to provide support (Fig. 0.1). This structure is reflected in the technical name for bats, Chiroptera, which is based on Greek words meaning "hand wing." The slender fingers have moveable joints that are controlled by muscles, just as in other mammals. However, these muscles and joints, when combined with bands of muscle and stretchy connective tissue within the wing membrane, provide exceptional control over the shape of the flight surface and allow bats to perform acrobatic maneuvers unlike those of any bird (Swartz and Allen 2020).

Over evolutionary time, four fingers and the associated metacarpals of bats became elongated, but the first finger, the thumb, remained short and free of the wing membrane (see Fig. 0.1). The thumb is also the only digit in New World bats that retains a claw, and it is used mostly as a hook to help the mammal climb the rough wall of a cave, scurry up a tree trunk, or manipulate a small fruit (Castillo-Figueroa 2022). The claw is quite sharp and can accidently puncture the skin of humans who casually contact bats without gloved hands.

In general, the leg bones of bats, the femur and tibia, tend to be thin and somewhat long compared with those of other small terrestrial mammals. The legs, though, typically terminate in standard-sized feet, each bearing five clawed toes. The bat uses the hind claws to groom the fur, to scratch an itch, and, of course, to suspend itself from a tree branch, house rafter, or cave ceiling. A human dangling from a perch by her fingers (or toes!) would quickly tire and fall to the ground, but many bats hang for 12 to 15 hours each day, and in temperate species that hibernate, the animals remain suspended for weeks at a time. Bats can stay in this position because their claws are anatomically locked by ligaments into the curved, grasping position, and the muscles are not continually contracting (Simmons

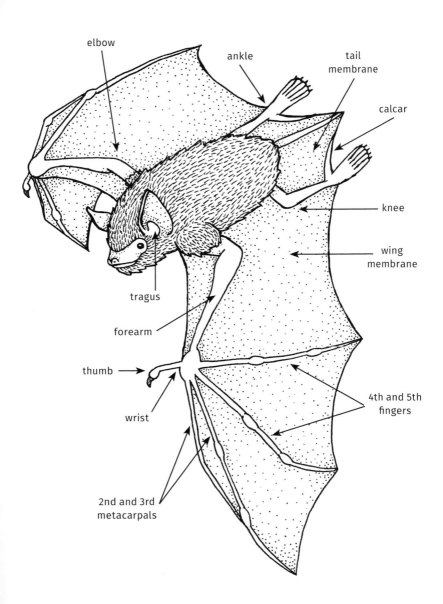

Figure 0.1. Anatomy of a bat.

and Quinn 1994); a small amount of energy is actually required for the bat to release its grip. Many songbirds (Passeriformes) have a similar passive-locking mechanism that allows them to sleep on skinny tree branches and not topple over in the middle of the night.

In flight, the long legs of a bat extend backward, rather than to the side, and support a tail membrane, which is also called the uropatagium (see Fig. 0.1). This membrane is a complex double-layered piece of skin, similar to the wing but thicker, that stretches between the right and left legs and encases some or all of the tail vertebrae. Many bats also have a calcar, which is a thin, moveable, cartilaginous spur that protrudes from the inner ankle; this projection supports the trailing edge and provides tension that helps stiffen the membrane during flight (Stanchak et al. 2019).

The shape of the uropatagium and its relationship to the tail are often useful in identifying the family to which a particular bat belongs (Fig. 0.2). For example, the rear edge of the membrane forms a backward-pointing V in vesper bats (Vespertilionidae) and funnel-eared bats (Natalidae), creates a frontward-pointing U or V in most leaf-nosed bats (Phyllostomidae), but is perpendicular to the body axis in mustached bats (Mormoopidae). Similarly, the tail protrudes from the center of the membrane in mustached bats and stretches to the back of the uropatagium in funnel-eared bats and vespers, but half the tail continues past the rear margin in free-tailed bats (Molossidae). Some of this structural variation corresponds to lifestyle, with frugivorous and nectarivorous species that scamper through trees having the least-developed uropatagia, perhaps to minimize snagging on branches, whereas aerial insectivores have the most expansive membranes.

The original function of the uropatagium probably was to provide an additional surface to supplement the lift created by the wings, and indeed, many pterosaurs independently evolved a similar structure (Alexander 2015). However, over time, the uropatagium has assumed additional roles for some bats. Well-developed tail membranes, for instance, act as a rudder or brake, helping with quick aerial maneuvers, or are used as a basket to catch the neonate at birth. Unlike wings that are largely hairless, uropatagia are occasionally 50 to 100% furred on their upper surface, and

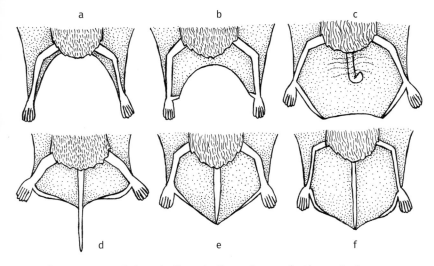

Figure 0.2. Dorsal view of tails and tail membranes. (a–b) Most leaf-nosed bats (Phyllostomidae). (c) Ghost-faced and mustached bats (Mormoopidae). (d) Free-tailed bats (Molossidae). (e) Vesper bats (Vespertilionidae). (f) Funnel-eared bats (Natalidae).

some species, such as red and hoary bats (*Lasiurus*), use these hairy membranes as blankets during cool weather, folding the membrane forward to cover the abdomen and chest. In addition, the uropatagium commonly functions as a food-gathering device; most insectivorous bats do not initially grab prey with their teeth but instead scoop the flying insect out of the air with the tail membrane, or sometimes the wing, and then pass the morsel to the mouth.

The hind limbs of bats are unusual in two other ways. In most mammals, the legs are positioned to the side of the animal, and the socket in the hip for attachment of the leg, called the acetabulum, is pointed somewhat laterally. However, bats hang and fly with their legs pointed behind the body, and the location of the acetabulum has shifted to the rear of the pelvis to accommodate this orientation. In addition, bat hind legs have rotated 180° so that the knees point backward (dorsally), thus allowing the animal to bring the legs and tail membrane, as well as any insect or pup that it might hold, forward toward the head. Both these modifications, though, make terrestrial locomotion by bats generally more awkward than movement by other small mammals, such as rodents and shrews (Riskin et al. 2016).

Variability in Wings

Different species of bats do different things. Some maneuver through dense woods, many fly high overhead searching for insects, a few hover above flowers, and others hunt along a forest-field edge (see Chapter 3). In general, though, bats are limited by the shape and size of their wings, and no single individual or species is physically capable of doing all these things. Two parameters concerning wing structure—loading and aspect ratio—are particularly useful in illustrating how body shape might determine how and where a bat flies.

An engineer calculates loading by dividing the body mass of the airplane, bird, or bat by the area of its wing surface. A flying mammal must generate sufficient lift to counteract the pull of gravity on the animal's body, and lift, to a large degree, results from air rushing over the wings. The faster an animal flaps, the more air surges over the wing surface, and the greater the resulting lift. Consequently, bats with high wing loading often fly fast in order to generate sufficient lift to keep their bodies in the air. A bat with low wing loading, in contrast, does not require as much lift, so the animal can reduce the flapping rate and still remain airborne—in other words, it can fly more slowly.

Aspect ratio is the ratio of the wing length (from wingtip to wingtip) to the average breadth (from front to back), so a short broad wing has low aspect ratio and a long skinny one has high aspect ratio (Fig. 0.3). A wing with a high aspect ratio experiences less of something called induced drag and creates more lift than one with similar area but a lower aspect ratio. Although wings of high aspect ratio produce greater lift, there is a drawback. Think of the other kind of bat—a piece of wood to hit a baseball or cricket ball. Would it be easier to control the bat and hit the ball if the bat were 3 m long or 1 m long? Obviously, the shorter stick is easier to maneuver, and the same is true of wings. Wings of high as-

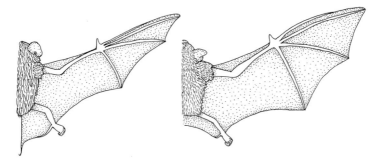

Figure 0.3. Wings of high (*left*) and low (*right*) aspect ratio.

pect ratio are less maneuverable and require a larger turning radius than those of low aspect ratio. A bat foraging within the understory of a crowded forest must be able to fly slowly, but the animal must also have a small turning circle, to avoid obstacles, as well as short wings that do not bang into every branch and leaf; consequently, these mammals tend to have low aspect ratio.

Nonetheless, the details of bat flight are complex (Swartz et al. 2012), and aspect ratio and loading are just two wing traits that are modified over evolutionary time to determine the flying style of a particular species (Norberg and Rayner 1987). Other parameters that contribute include the shape of the wing tip (rounded or pointed) and the total area of the wing, as well as the relative proportions of the inner (plagiopatagium) and outer (dactylopatagium) sections of the wing. The plagiopatagium, which is proximal to the wrist, provides most of the lift as a bat flies, whereas the dactylopatagium, which is supported by the metacarpals and fingers, is responsible for forward thrust. Any flight apparatus, though, has to coordinate with the navigation system, and how and where bats fly and forage is further constrained by the limits of their echolocation system.

Echolocation: Visualizing with Sound

To varying degrees, different bats integrate sensory information provided by vision, hearing, and sometimes olfaction to navigate, orient, and interact in a dark or dimly lit world (Boonman et al. 2013; Gorresen et al. 2015; Thiagavel et al. 2020). Bats, though, depend on sound more than any other group of terrestrial vertebrates, because most of these airborne mammals use echolocation to detect their food and find their way in the darkness. Although representatives of a few other terrestrial groups, including birds (Brinkløv and Warrant 2017), shrews (Siemers et al. 2009), rodents (Panyutina et al. 2017), and even humans (Norman et al. 2021), sometimes echolocate, none is as sophisticated or dependent on the technique as are bats. Echolocation is integral to the lives of all bats, except the 190 or so species of Old World flying foxes (Pteropodidae) that rely primarily on vision instead.

Echolocation by bats, at its simplest, is analogous to the sonar (*so*und *na*vigation and *r*anging) systems used by submarines to avoid collisions and by fishermen to locate their quarry. Both operate on the same principle. An electronic instrument emits a sound, which travels outward and is reflected back to the device by an underwater obstacle or school of fish. The CPU (central processing unit) of the instrument records the time between release of the sound and receipt of the echo and then computes the distance to the object, based on that elapsed time and the speed of sound. A bat hunting insects follows a similar procedure. The mammal produces a sound in its larynx (voice box) that bounces off an insect and returns to the bat's ear, thus allowing its CPU (brain) to measure elapsed time and calculate distance to the prey.

Human conversation typically involves sounds with frequencies of less than 3 kHz, and the upper limit of hearing for most people is 20 kHz. Echolocating bats, in contrast, typically emit sounds with frequencies between 20 and 100 kHz, although some species, such as the spotted bat (*Euderma maculatum*) of North America, use calls as low as 11 kHz, and those of Percival's trident bat (*Cloeotis percivali*) from Africa may be as high as 212 kHz (Jones and Holderied 2007). Because most sounds produced by most bats are above 20 kHz, they generally appear silent to us as they fly by; nevertheless, the calls produced by a bat trapped in your house may drive the family dog or cat bonkers, because their ears respond to sounds with frequencies up to 45 kHz or more. Even though people cannot hear these pulses, they can be quite intense (loud), often exceeding 120 dB at a distance of 10 cm, which is about as loud as a police siren sitting next to your ear; such high energy helps broadcast the sound over a greater distance (Jakobsen et al. 2013). Note that bats do not continuously produce sound when echolocating; instead they emit a sequence of pulses or calls—sound on, sound off, sound on, etc.—thus providing most bats with time to listen for the soft echoes between the booming pulses.

Use of high frequencies is a necessity in most situations. In general, the strongest (loudest) echo results if the wavelength of an emitted sound matches the size of an item that the animal is trying to detect. In the West Indies, foraging bats must identify twigs, thorns, beetles, moths, flies, and other tiny objects, and to do so, these mammals must use sounds with correspondingly small wavelengths to produce echoes of sufficient strength. Wavelength and frequency, though, are inversely related, so sounds having the small wavelengths needed by a bat necessarily have high frequencies. To illustrate, sounds with frequencies of 10, 30, 50, and 100 kHz correspond to wavelengths of 34, 11.4, 6.8, and 3.4 mm, respectively. Obviously, any bat would have difficulty flying through a dense woodland or "looking" for a mosquito while using a frequency of 10 kHz and a gigantic wavelength of 34 mm.

A serious drawback, however, to high frequencies (and a reason for the loudness of echolocation pulses) is that sound energy is more rapidly absorbed by air molecules at high rather than low frequencies, thus reducing the intensity of the outgoing pulse and incoming echo. Because of this attenuation problem, all echolocating bats are capable of perceiving small objects only at short distances, often just a few meters, and insectivorous species typically detect their prey only 0.5 second or less before capture (Jung et al. 2014). Echolocation inevitably involves tradeoffs—high-frequency sound attenuates rapidly but provides sufficient echoes from small objects, whereas lower-frequency sound travels farther but cannot be used to distinguish tiny items. Hence, a slow-flying bat searching for small fruits amid the twigs and leaves of a tangled forest typically relies on calls of very high frequency, often 50 to 100 kHz, whereas a fast-flying bat hunting for large moths in open space high above a forest might use frequencies of only 15 to 25 kHz (Luo et al. 2019).

Regardless of whether the sounds are at the high or low end of the frequency spectrum, biologists classify echolocation pulses into two broad types that are at the ends of a continuum—frequency modulated and constant frequency (Schnitzler 2009; Fig 0.4). Most bats in the West Indies rely on frequency-modulated calls that change or "sweep" through a range of frequencies over a short period, whereas a few species, such as the Puerto Rican mustached bat (*Pteronotus portoricensis*), consistently emit long pulses, mostly at a single frequency. In general, constant-frequency sounds are best for simple detection of a distant object, but frequency-modulated calls are better at localizing an item in space and providing details about its surface, potentially allowing a bat, for instance, to determine whether an insect is a beetle or a fly.

Not surprisingly, the typical calls of many species combine elements of both types (see Fig. 0.4). For example, sooty mustached bats (*Pteronotus quadridens*) begin each call with a short segment of quasi-constant frequency that transitions into a frequency-modulated sweep of longer duration, whereas Cuban evening bats (*Nycticeius cubanus*) do the reverse, starting with a steep sweep but finishing with a component of quasi-constant frequency. Furthermore, many bats emit calls containing harmonics (overtones) that are integral multiples of the basic or fundamental frequency; this is an additional way of expanding the range of frequencies to investigate an object and provide more detail. Fruit-eating leaf-nosed bats generally use harmonics as part of their normal calling strategy, whereas many insectivorous species from other families introduce harmonics only when closing in on a target.

All but a few bats create echolocation sounds using their vocal chords, and by contracting tiny laryngeal muscles, the animals change the tension on the cords, which results in the production of different frequencies. Most species in the New World project these sounds through their mouth, just as a human does when speak-

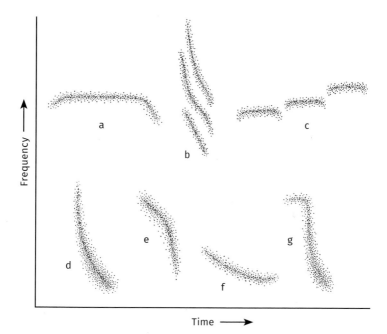

Figure 0.4. The design of echolocation calls, illustrating variety in use of frequency-modulated (FM) and constant-frequency (CF) components. (a) Predominantly CF, as in Parnell's mustached bat (*Pteronotus parnellii*). (b) Harmonics, typical of many leaf-nosed bats (Phyllostomidae). (c) Triplet of quasi-CF sounds, as in the Pallas's mastiff bat (*Molossus molossus*). (d) Steep, slightly concave, FM call, typical of many vesper bats (Vespertilionidae). (e) Slightly convex, FM pulse, as in the Greater Antillean ghost-faced bat (*Mormoops blainvillei*). (f) Shallow, concave, FM call, as in the Brazilian free-tailed bat (*Tadarida brasiliensis*). (g) Short quasi-CF component, followed by a steep FM sweep, as in the sooty mustached bat (*Pteronotus quadridens*).

ing, and the bat controls directionality of the beam by adjusting the gape of the mouth (Surlykke et al. 2009). Most leaf-nosed bats, in contrast, emit some or all their pulses through the nasal openings (Gessinger et al. 2021). These bats have a spear- or leaf-shaped appendage of variable size, sitting above the nostrils (Fig. 0.5); this structure, the nose leaf, is moveable and apparently helps control

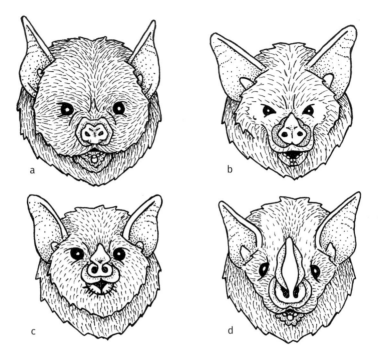

Figure 0.5. Variety in the structure of nose leafs, from small to substantial. (a) Antillean fruit-eating bat (*Brachyphylla cavernarum*). (b) Brown flower bat (*Erophylla bombifrons*). (c) Greater Antillean long-tongued bat (*Monophyllus redmani*). (d) Jamaican fruit-eating bat (*Artibeus jamaicensis*).

the directionality of the sound beam, particularly in the vertical plane (Surlykke et al. 2013).

To receive the faint echoes and channel the sound to the eardrums, bats, of course, rely on their external ears, the pinnae (López-González and Ocampo-Ramírez 2021). Most bats also have a well-developed tragus, and some, such as the free-tailed bats, have a prominent antitragus; these fleshy projections are associated with the front and rear margins, respectively, of the ear canal (Fig. 0.6). These structures interact with the returning sound waves and help the animal localize the source of the echo in terms of ele-

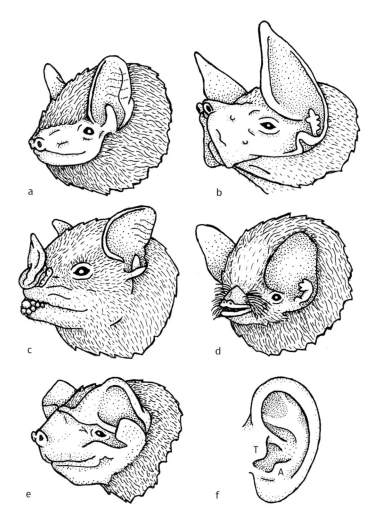

Figure 0.6. Variety in the structure of ears. (a) Simple tragus of a vesper bat (Vespertilionidae). (b) Branched tragus of a bulldog bat (Noctilionidae). (c) Simple tragus of a leaf-nosed bat (Phyllostomidae). (d) Twisted tragus of a funnel-eared bat (Natalidae). (e) Antitragus (A) in a freetailed bat (Molossidae). (f) Tragus (T) and antitragus (A) in a human for comparison.

vation. A bat with a deformed tragus, for instance, has difficulty in gauging its position (height) relative to a surface, such as that of a pond (Hoffmann et al. 2015).

Flying bats are constantly vocalizing, and scientists who study these mammals often depend on those very sounds to answer questions such as, where does this bat forage, at what time of night is it most active, and which species are present here versus there? To detect echolocation calls, field biologists rely on "bat detectors" (Fraser et al. 2020; Runkel et al. 2021). These handheld electronic instruments have a microphone that responds to the ultrasonic frequencies produced by bats. After sensing a high-frequency call, the device emits a sound of lower frequency that humans can hear, and many bat detectors also provide an immediate visual display of the call's structure (i.e., a graph of frequency against time). Thus, these instruments allow a person to know when bats are flying by, even though they are not visible, and sometimes identify the animals to species, based on the structure of the calls, their duration, and the time between pulses. Unlike birds, though, that repeat the same notes over and over and over again in precisely the same boring pattern, the timing and structure of echolocation calls by bats is dynamic, varying with the habitat and the task at hand (e.g., searching for prey at a distance vs. imminent capture). Hence, identifying individual bats to species based on sound can be tricky, even for experienced biologists (Russo et al. 2017). Nevertheless, the cost of these instruments has declined as their sophistication has increased, and some low-priced models now work with a smartphone or tablet, making bat detectors available to curious people everywhere.

Evolution, Classification, and Distribution

For some types of animals, the fossil record is extraordinary in portraying evolutionary experiments over time. In horses, for example, fossils reveal a complex transition from a fox-sized animal with five toes on each foot to multiple descendants, including the modern long-legged mammals with one toe per foot. But how did the wing of a bat develop? What were the intermediate steps toward a mammal capable of flight? Some scientists estimate that bats first evolved perhaps 65 to 70 million years ago, long before humans appeared, although the oldest complete fossils, called *Icaronycteris* and *Onychonycteris*, are only about 50 million years old (Gunnell and Simmons 2005; Brown et al. 2019). Nevertheless, these ancient animals were already bats, possessing elongated fingers and forearms, and not some creatures halfway on the evolutionary road toward a flying mammal. Consequently, the specific ancestors of bats and the transitional steps toward flight are unknown.

Icaronycteris and *Onychonycteris* are bats, but they differ from modern forms. *Onychonycteris*, for example, retains a claw on all five fingers, not just the thumb, and its hind limbs are robust (Simmons et al. 2008). The two characters together suggest that these animals were agile climbers and possibly "under-branch hangers" that used both hands and feet when suspended from branches or clambering through them. In addition, the small size of the inner ear, the cochlea, strongly indicates that *Onychonycteris* could not echolocate.

Although biologists do not know precisely when or how bats evolved, it is clear that these animals began diversifying in the Eocene, and over time, they became one of the most successful groups of mammals. Today, there are about 1450 living species in the order Chiroptera, which represent 22.5% of all mammals (www.mammal diversity.org); in other words, one of every four or five mammalian species is a bat. Only the order Rodentia contains more species, and the two orders combined encompass almost two-thirds of all mammals. Modern bats have diversified into twenty-two different families and inhabit every continent except Antarctica. Although absent from polar regions, extreme deserts, and the highest of mountains, these intriguing animals dwell in virtually every corner of the globe. The ability to fly has allowed bats to colonize even remote oceanic islands, and after centuries of isolation, many insular populations have evolved into separate and unique species (Barreto et al. 2021).

Bats and the West Indies

In the West Indies, sixty-one species of bats, representing six different families, currently live in three archipelagos—the Greater Antilles, Lesser Antilles, and the Lucayan Archipelago, which consists of the Bahama Islands and the Turks and Caicos Islands. Today, bats are the only native terrestrial mammals in the region, except on three of the Greater Antilles that still harbor a few hutias and solenodons (Turvey et al. 2017). The number of bat species per island varies considerably and is influenced by numerous factors, including size of the island, distance to the mainland, and availability of diverse habitats, which, in turn, is greatly affected by differences in elevation and human activity (Barreto et al. 2021; see Chapters 2–3). Cuba, for example, is the largest island in the Caribbean at 110,000 km², and it harbors twenty-seven species. However, Saint Eustatius in the Lesser Antilles, located 750 km from South America, is tiny (21 km²) and suffers from deforestation and overgrazing (Pedersen et al. 2018); consequently, only five species of bats are present.

Forty-five of the sixty-one species of bats in the West Indies are endemic to the region, that is, they inhabit only one or a few islands and are not found anywhere else in the world. The little goblin bat (*Mormopterus minutus*), for instance, inhabits only a small portion of Cuba, and the Saint Vincent big-eared bat (*Micronycteris buriri*) and Guadeloupean big brown bat (*Eptesicus guadeloupensis*) are restricted to their namesake islands. Indeed, this uniqueness of the bat fauna helps define the West Indies as a distinct biogeographical region (Genoways et al. 1998, 2010; Durocher et al. 2021). Which species actually are present on each island, though, is the result of complex interactions among geologic events, climatic factors, and vegetative influences, as well as human intervention (see Chapters 1–3).

CHAPTER 1

Climate, People, and Vegetation
of the West Indies

Ariel E. Lugo

The definition of the West Indies varies depending on the group of organisms that one considers, and in this book, the region is delineated in the south by "Koopman's Line," which passes between Saint Vincent and Bequia, the northernmost of the Grenadine Islands (Genoways et al. 2010). Islands north of this boundary support at least some species of bats that are endemic to the region, whereas islands to the south are primarily inhabited by species that also live on mainland South America. Hence, the West Indies includes three island chains: the Lesser Antilles from Saint Vincent to the Virgin Islands, the Greater Antilles, and the Bahamas and Turks and Caicos Islands. The total number of islands probably exceeds 7000, because many smaller islets are associated with each major land mass. For example, the governmental entity known as Puerto Rico is really an archipelago of 660 individual islands, including cays as small as 0.001 ha (Lugo et al. 2016). The region spans a latitudinal range of 14°, from Saint Vincent (13° N) to Little Abaco in the Bahamas (27° N). Geographers consider all these islands "tropical," except the northern half of the Bahamas, which lie above the Tropic of Cancer (23.4° N).

The diversity of animals in any part of the West Indies is strongly dependent on the type of plant community that is present. The goal of this chapter is to present a realistic overview of contemporary vegetation in these diverse archipelagos and describe how it is shaped by abiotic factors and human activity. The interrelationships among plants, climate, and anthropic change are central to understanding the following introductory chapters that concern the historical biogeography of bats, their modern ecology, and current conservation needs, as well as the details of the species accounts that follow.

The Importance of Elevation on Temperature and Rainfall

Elevation plays an important role in the climate of these islands, which consist of two types of landmasses—high and low. High islands have elevations from near sea level up to 3.1 km at Pico Duarte on Hispaniola, the highest point in the West Indies. Different sites on high islands typically experience several distinct climatic regimes, ranging from dry woodlands (500–1000 mm of annual rainfall) to rain forests (>4000 mm). Low islands, in contrast, such as Anguilla, Antigua, and Barbuda, are usually small in area and are typically dry.

Prevailing winds that bring moisture to each island also influence climate. In the Northern Hemisphere, the trade winds blow across the Atlantic Ocean from the northeast and strike the Lesser Antilles from Dominica southward. Those southern islands are described as windward, whereas islands to the north and not in the direct flow are leeward. Thus, islands can be low or high, based on topography, and windward or leeward, depending on their position relative to the trade winds. For example, Barbados is a low windward island, whereas Dominica, a high island, is leeward.

The classification of entire islands as leeward or windward is applied mostly to the Lesser Antilles, but the Greater Antilles and other high islands have windward and leeward aspects. Elevation, combined with a persistent direction of moisture-laden winds, greatly affects the amount and distribution of rain on these landmasses. Moist air blowing over the ocean strikes the uplands of any high island and rises upward, cooling the air, condensing the water vapor, and resulting in rain. Thus, the windward side of high islands receives more rainfall and supports wet forests, while a rain shadow occurs on the leeward side, which sustains only dry to moist forests. Overall, annual rainfall is greater on high windward islands and lower on low leeward islands. Most islands experience a wet and a dry season, with annual rainfall typically ranging from 800 to 2000 mm but reaching more than 5000 mm on the highest peaks.

Hurricanes, with strong winds and intense rainfall, occur anytime between May and December, although the "official" Atlantic hurricane season is between 1 June and 30 November (Neumann et al. 1978). Hurricanes have passed over the West Indies for millions of years and have affected the vegetation by limiting tree height and resulting in high species dominance in plant communities (Lugo 1991). On average, daily maximum air temperature on the islands ranges from the upper 20s °C in winter to the lower 30s °C in spring and summer; nighttime temperatures are about 6 °C cooler. Frost, which determines the types of plants that can grow, does not occur in the lowlands of the West Indies, and the resulting life zones range from tropical to subtropical (Holdridge 1967). Nevertheless, frost is a regular occurrence above 2200 m in elevation in the pine forests of Hispaniola and is a factor in the reduced diversity of plants, as well as bats, high in the mountains of that island (Martin et al. 2011; Núñez-Novas et al. 2021).

The Effect of Humans, History, and Economics on the Landscape

> Three sea-crossing technologies—canoe, caravel, and container ships—serve as symbols of the main periods of Caribbean history. . . . Each of these vessels carried with them whole cultures, representing an increasingly global cargo.
>
> —Higman (2011, 327)

Human activity in the West Indies dates to about seven millennia before present and involves periods of distinct activities, such as hunting and gathering, subsistence agriculture, plantations, industry, and political dominance, including colonialism, rebellions, democracies, and dictatorships (Higman 2011). The canoe of indigenous peoples represents 5000 years of history, whereas the European caravel spanned 500 years; the modern container ship, in contrast, involves only a short period of 50 years. Each technology had consequences to the people, the economy, and the ecosystems of the region. From its beginning, the progression of human history in the West Indies is characterized by the changing diversity of peoples and cultures inhabiting the various isles, with each island evolving along different trajectories than the others, even if distances among them were short (Horowitz 1971), resulting in contrasting levels of economic development and environmental effects.

The population of the region, as of 2010, is forty-one million people, with 90% living in the Greater Antilles. Of this population, 66% live in urban environments, with the greatest densities occurring in Anguilla, the Cayman Islands, and Puerto Rico (Fig. 1.1). On most inhabited islands more than half the population is classified as urban, and such populations increased by 81% between 1980 and 2010. Overall, average population density is about 180 people/ km^2, ranging from 664 in Barbados and 449 in Puerto Rico to 31, 46, and 50 people/km^2 for the Bahamas, Turks and Caicos, and Montserrat, respectively. The human population is not continuing to grow on all islands, as Puerto Rico and many of the Lesser Antilles lost population between 1980 and 2010, and this decline accelerated in multiple locations after Hurricane Maria in 2017 (Lugo 2019).

The landscapes of the Caribbean islands reflect the development of their local economies (Lugo 1996). Human economic activity influences the kinds of ecosystems that prevail on each island, as well as their species composition. Indeed, all components of the social-ecological-technological systems of the West Indies function in synchrony, responding and adapting to extreme events originating either in the anthropogenic realm, such as wars, political actions, and economic turmoil, or the nonanthropogenic realm, such as hurricanes and volcanoes.

The effects of severe events on landscapes are epitomized by Puerto Rico over the last 125 years, as changes in political regimes, economic upheavals, and differing agricultural practices combined with periodic high-intensity hurricanes to alter the landscape of

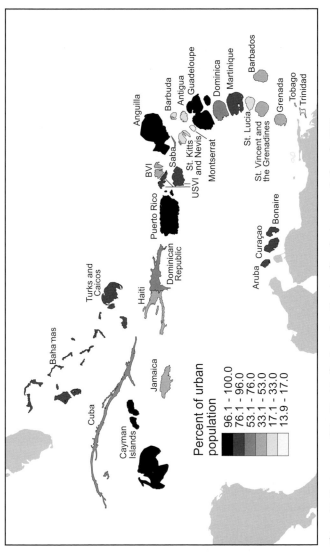

Figure 1.1. Cartogram of the West Indies based on the density of urban populations in 2010 (from López Marrero et al. 2012). Courtesy of Tania López Marrero.

the island and the composition of the bat community. Native plant species once maintained high concentrations of soil organic matter in historic forests, but by the early 1900s, deforestation and widespread agriculture had degraded soils and greatly reduced soil organic matter. After the collapse of agriculture following Hurricane San Ciprián in 1932, the reforestation that followed involved introduced species that are now contributing to increased forest cover and soil organic matter. Both the agricultural activity and the prevailing condition of the landscape, including forests, changed after the passage of each intense storm. By 2020, the local vegetation had shifted from the pre-twentieth-century indigenous combinations of species to maturing, novel plant communities dominated by introduced species (Lugo 2019, 2020); consequently the plant communities that feed and shelter bats today are not the same as those 200 years ago. All West Indian islands share the key events in the Puerto Rican story, but they occurred at different times, with varying intensity, and with different long-term effects, because each island has its own unique culture, level of economic activity, governance, historical context, and ecosystems. Regardless of inter-island differences, human activity is linked with the state of the landscape and its ecological systems, and it is impossible to study one without considering the other.

The Vegetation of the West Indies

> This climax [plant community] is that which develops not
> only in the entire absence of anthropic influences, but in the
> absence of all natural accidents, catastrophes and "abnormal"
> disturbances.
>
> —Egler (1952, 74)

When thinking about the vegetation of the West Indies, one usually imagines the green fringe of mangroves lining low-energy coastal and estuarine waters, beautiful dry forest shrublands, magnificent lower montane forests, impressive windward mountain slopes dominated by palm forests with even-height canopies, or cloud-shrouded montane forests and woodlands. These are all iconic examples of Caribbean flora. However, modern tourists flock to the West Indies to spend time on beaches lined with coconut palms (*Cocos nucifera*) that originated somewhere in Indomalaya. As visitors travel along island roads, people marvel at the beautiful flowers of flamboyan (*Delonix regia*) from Madagascar and those of the African tulip tree (*Spathodea campanulata*) from Ghana and Kenya, or observe the large fruits of the breadfruit tree (*Artocarpus altilis*) from New Guinea via Tahiti. In fact, tourists typically encounter vegetation quite different from the climax communities described in the classic botanical literature of the West Indies (Stehlé 1945; Beard 1949).

Egler (1952) observed that the writings and classification systems of Beard (1949) and Stehlé (1945) reflected the different experiences and worldviews of those authors, resulting in an incomplete picture of reality, because of what they did not consider when developing their respective frameworks. Egler (1952) believed that the frequent description of plant associations by Beard (1949) and Stehlé (1945) as ruinate, reduced, defective, or degraded suggested a bias that supported their idealistic model of a vegetation adapted to climatic conditions. Moreover, he noted that most plant studies focused on the mountains, where human effects were less apparent, and less so in the lowlands, where humans were more prevalent. In the spirit of Egler (1952), the following discussion concerning the modern flora of the West Indies takes a broad view and examines the importance of non-native species of plants; explores the interactions among climate, topography, soil, and extreme events; and concludes with a description of the role that animals play in structuring plant communities in the islands.

The Present-Day Flora Consists of Native and Novel Species Assemblages

Technically, botanists classify the flora of the West Indies as part of the Antillean subregion of the Caribbean province of the Neotropical kingdom (Good 1953; Borhidi 1991). Within the subregion, there are six distinct floras associated with different islands or groups of islands. These six include the Lesser Antilles, Cuba,

Jamaica, Hispaniola, Puerto Rico plus the Virgin Islands, and the combination of south Florida, the Bahamas, and Bermuda.

The West Indies is a global hot spot of endemism, with at least 2% of the world's endemic plant and vertebrate species on only 0.4% of the earth's land surface (Myers et al. 2000). The number of native plant species, which reflects development over millennia, is 10,948 (Acevedo Rodríguez and Strong 2008). Furthermore, there are 1899 introduced species that have become naturalized over the last 500 years, for a total of 12,847 species, with introduced plants representing 14.8% of this total. The number of introduced forms on an island is positively related to the amount of urban cover, which is an index of the level of human activity; such activity, in turn, creates opportunities for the importation and ultimate naturalization of more species (Lugo et al. 2012b). Because naturalized species are recent additions to the flora, their presence has created novel assemblages of species that, unlike native plants, can colonize deforested and degraded lands and rehabilitate their productivity and biodiversity.

Effects of Topography, Climate, Soils, and Extreme Events on the Flora

Depicting a realistic scenario for the development of vegetation in the West Indies requires an examination of four principal conditions—topography, climate, soils, and extreme events—that, together, determine the ecological space available for the growth and dispersal of plants and their assembly into interacting communities. This space is characterized in Figure 1.2 by four quadrants, one for each condition. The totality of the flora is represented by a central circle at the intercept of the four quadrants, and components of the vegetation (species) spread outward along environmental gradients, according to their life-history traits and physiological tolerances. The diagram illustrates eighteen such continuums, because empiric data support the distribution of species along those gradients within the West Indies. Although the diagram depicts the gradients and quadrants as discrete entities, they interact with each other in determining which plants occupy an area.

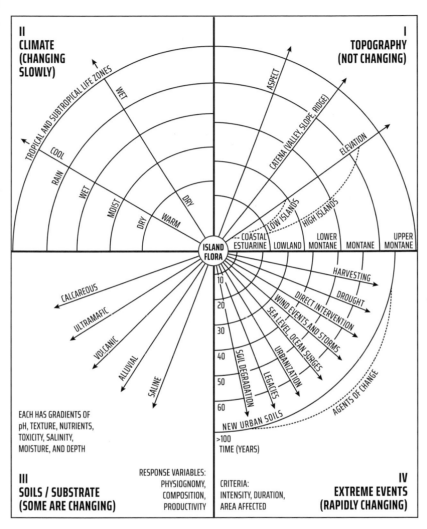

Figure 1.2. Four-quadrant diagram depicting the environmental conditions or ecological space within which the vegetation of the West Indies grows and functions. Each quadrant represents a major group of environmental conditions to which vegetation responds. The eighteen radiating lines represent environmental gradients along which the insular flora (including naturalized species) is distributed. Prepared by Ariel E. Lugo.

Quadrant I represents topography. At the scale of the West Indies, the topography of the islands is the most stable of the four quadrants and provides the basal conditions for classifying vegetation according to elevation (Beard 1949). However, at smaller scales, forces of change from within this quadrant and from other quadrants can significantly affect the vegetation of a particular area. For example, tectonic forces slowly move the geologic plates of the Caribbean, inducing earthquakes and volcanic activity that can alter surface features in either the short or the long term. Landslides, because of the interaction of climate and topography (quadrants I and II), also change surface conditions, and some geologic substrates in quadrant III are significantly more susceptible to sliding than others. In addition, use of heavy machinery and explosives (quadrant IV) modifies significant portions of the topography of the West Indies, thus affecting the cover of vegetation on montane (quarries), coastal (construction over sand dunes), wetland (drainage and filling), and lowland (urbanization) locations.

Within quadrant I, there are three principal gradients— elevation, catena, and aspect. The high islands of the West Indies contain the range of vegetation types described by Beard (1949), including coastal, lowland, lower montane, montane, and cloud forests. Each elevational belt is associated with a particular group of species distributed along the catena and aspect. The catena (a gradient of location—from a valley, along a slope, to the top of a ridge) represents contrasting conditions for the establishment of vegetation, as well as community organization and functioning (Scatena and Lugo 1995). Aspect (direction that a site faces), especially in relation to rainfall (quadrant II) and wind exposure (quadrant IV), is another critical component regulating community composition and its physical appearance.

Quadrant II represents the climate of the islands, which both Beard (1949) and Stehlé (1945) used to classify climax plant communities. However, the tropical climate of the West Indies is slowly but measurably changing (Jennings et al. 2014) because of human activity at the global scale (quadrant IV). For instance, climate change in the Luquillo Mountains of Puerto Rico is affecting some avian species, such as the endemic lizard cuckoo (*Coccyzus vieilloti*), which has shifted its range upslope because of environmental warming (Campos-Cerqueira et al. 2017). Although global climate models also predict future reductions in rainfall (Henareh Khalyani et al. 2016), so far there is no evidence that such a decrease is happening (but see Hernández Ayala and Heslar 2019), perhaps because the climate of islands is too complicated and difficult to predict with global-scale models.

There are two major climatic gradients affecting vegetation in the West Indies. One is the moisture gradient caused by differences in rainfall. This gradient is closely linked to the topographic conditions in quadrant I, with more precipitation at higher elevations (García Martinó et al. 1996). The other gradient, temperature, defines the West Indies as tropical and subtropical because of the absence of frost in the lowlands, as indicated earlier. Furthermore, air temperature differs with elevation, time of day, and season of the year, with more variation at higher latitudes. Plant and ecosystem processes respond to all of these variations.

Quadrant III represents the rocky substrates and soils that are ultimately derived from the bedrock. These gradients represent the simplest aggregation of substrates possible, as geologists recognize many others. This quadrant is complicated because of the diversity of soils and complexity of the substrates that occur in the West Indies. Even in the absence of human activity, the conditions in this quadrant result from forces associated with time, topography, and elevation. For example, karst substrates occur in areas submerged by the ocean earlier in geologic time, but the soils that are present may originate on volcanic mountain slopes; in addition, the climate in karst areas may be dry, moist, or wet, depending on elevation and aspect. Although the geology of the islands is well known, as are the locations and identities of the soils, ecologists understand less about how these substrates and soils affect vegetation than they do about the effects of climate.

Nevertheless, vegetation does change dramatically along the five gradients of quadrant III. Changes in vegetative structure and functioning along gradients of soil moisture and salt concentration were detected long ago (Gleason and Cook 1926). Mangroves, as an illustration, become smaller and less productive as salinity increases (Cintrón et al. 1978; Lugo and Medina 2014), and tree height decreases with elevation on high islands, culminating in elfin woodlands on mountaintops (Odum 1970).

Quadrant IV comprises the forces of anthropogenic change acting on the West Indies. There are eight gradients depicted, each associated with events that have a measurable intensity and duration (Lugo 2020). In contrast to older historic vegetation in quadrants I and II, the vegetation that predominates in quadrant IV is novel and generally less than 100 years old. When historic vegetation is subjected to extreme events, it recovers through pathways of succession that evolved over thousands of years, but historic vegetation yields to novel vegetation when site conditions change so that the older species assemblages cannot compete with novel assemblages (Álvarez Ruiz and Lugo 2012). For example, the native tree that normally grows at the beginning of succession, the trumpet tree (*Cecropia schreberiana*), cannot successfully colonize lowland pastures, because soil moisture is too low, and its seeds are consumed by predators (Silander 1979). Instead, non-native trees such as guava (*Psidium guajava*) and African tulip now fulfill the role as colonizers (Aide et al. 1995, 1996).

In the upper montane forests of Hispaniola, the complexity of plant distribution makes it difficult to anticipate changes in response to global warming and increased frequency of hurricanes (Martin et al. 2004). However, anthropogenic effects on vegetation are measurable and already reflect that introduced species account for 29% of the "importance value" (an index derived from the frequency, density, and basal area) of all tree species. Despite the prevalence of introduced species on Hispaniola, the structure (e.g., height and canopy closure) of woody vegetation in secondary riparian forests approaches the structure of mature forests within a 40-year span, which is similar to results from the mountains of Puerto Rico (Aide et al. 1995, 1996).

The Role That Animals Play in Structuring Plant Communities

The list of plants used as food by bats (Gannon et al. 2005) contains numerous introduced species, indicating that these mammals are providing pollination and seed dispersal for the non-native as well as the native flora. This dietary behavior of bats is consistent with their frequent reliance on remnant blocks of forest, which are mostly dominated by naturalized species of plants (Rodríguez-Durán and Feliciano Robles 2016). In fact, there are many examples of symbiotic relationships between native and non-native plants and animals other than bats. These include the role of Africanized honey bees in pollinating the endangered endemic beautiful goetza (*Goetzea elegans*) (Caraballo Ortiz 2007) and synchrony between the swarming behavior of European honey bees (*Apis mellifera*) and the massive flowering of Sierra palms (*Prestoea montana*) (Snyder et al. 1987).

Naturalized animal species are ubiquitous in the West Indies and include representatives of most animal groups (Kairo et al. 2003). The number of introduced species varies with the size and location of each land mass, with the larger islands of the Greater Antilles having more than the smaller isles of the Bahamas or Lesser Antilles. Like introduced plants, non-native animals assemble into novel combinations of indigenous and introduced species that function within historic and novel environmental conditions on each island (Lugo et al. 2012a).

In so doing, animals become part of the functioning of these new ecological systems. One of the key ecological roles of birds, for example, is bringing tree seeds into pastures, whereby the animals contribute to the establishment of novel plant communities that include both native and non-native species (Carlo and Morales 2016). Bats play an analogous role by accelerating the transition of tree plantations to secondary forests, and these flying mammals also influence the dispersal of both native and introduced plants in novel forests, such as those dominated by the African tulip tree (Abelleira Martínez 2008; Abelleira Martínez et al. 2015).

Species diversity of bats changes with alterations in land use (Mendes and Srbek Araujo 2020) and also with floral succession (Martínez-Ferreira et al. 2020). Populations of bats respond to the overall physical structure of plant communities, such as the height of the canopy and the density of the understory, as well as the ability of plants to produce food suitable for the animals.

In general, the vegetation of the West Indies consists of a diverse matrix of communities within a small geographic space; this distribution likely helps maintain a diverse bat community more effectively than in continental regions, where individual vegetation types span a much larger area, necessitating a greater flying distance to access a similar variety of plants. Moreover, the development of novel communities may not be as detrimental to bats as expected by preservationists, who deny a positive role to introduced species, because the structure of these novel forests resembles that of mature historical woodlands (Aide et al. 1995, 1996), thus not affecting connectivity among communities in the West Indies. Habitat heterogeneity is one of the most important predictors of bat species richness in the Caribbean (Hoffman et al. 2019; see Chapter 2), and the current trend toward novel communities contributes to this heterogeneity, as evidenced by the use of remnant habitats by many species of bats in an urban matrix on Puerto Rico (Rodríguez-Durán and Feliciano Robles 2016).

The Past, Present, and Future of Bat Biogeography in the Caribbean

J. Angel Soto-Centeno and Camilo A. Calderón-Acevedo

Islands are excellent places to study biodiversity, with many of them allowing examination of entire communities of plants and animals, to determine the different factors that shape these assemblages. Archipelagoes are dynamically affected by climatic, ecological, and geological processes; show a range of habitat and topographic variability; and, in more recent times, are differentially influenced by human actions (Fernández-Palacios et al. 2021; see Chapter 1). These factors act simultaneously to create discrete and naturally replicated systems that are manageable in size and in which biologists can measure the responses of species. Because of this replication, biogeographers often regard the West Indies as a "natural laboratory," where studies can help determine the patterns and processes that shape the diversity of life on earth (Ricklefs and Bermingham 2008).

Biologists studying insular biotas have identified broad factors affecting the diversity of organisms and assembled these concepts into the so-called classical theory of island biogeography (MacArthur and Wilson 1963, 1967). Two main inferences stem from this theory. First, a positive relationship exists between the area of an island and the number of species present (i.e., species richness), and second, the species richness of an island decreases as its distance from a source of colonizers increases. Species richness of remote islands also depends on the time available for colonization, and a dynamic balancing act occurs between colonization and the extinction of species on each landmass. These basic patterns have been widely examined using West Indian vertebrates, including bats (Koopman 1959, 1989; Baker and Genoways 1978; Griffiths and Klingener 1998; Rodríguez-Durán and Kunz 2001; Willig et al. 2009; Hoffman et al. 2019). In this chapter, we provide an overview of the diversity of West Indian bats in space and time and emphasize the processes that help shape these insular communities, rather than discussing details of the theoretical underpinnings that led to the formation of this biodiversity.

The Past: How Did Bat Communities Form?

The West Indian archipelago is geologically complex and ancient. Most of the region sits on the Caribbean Plate, with some geologic units dating to the Cretaceous period between 100 and 75 million years ago (Robertson 2009). Similarly, the deepwater channels and basins of the Bahamas, which lie north of the Caribbean Plate, have likely existed in their present positions since the late Cretaceous (Carew and Mylroie 1997). By the Miocene epoch, about 23 million years ago, the islands of the Greater and Lesser Antilles began approaching their current geographic position. Although uplifted landmasses that contribute to present-day Cuba, Hispaniola, and Puerto Rico existed in some form during that time, Jamaica spent a longer period underwater, emerging only about 10 million years ago. The Lesser Antilles, in contrast, formed as a volcanic arc at the eastern edge of the West Indies; the first rocks began breaching the surface as early as the Eocene or Oligocene, perhaps 40 million years ago, with uplift continuing until about 6 to 2.5 million years ago (Robertson 2009).

Throughout geologic history, multiple glacial and interglacial cycles prompted the fall and rise of sea level, which exposed and submerged different landmasses, thus creating variation in the size and shape of different islands and their proximity to each other. These fluctuations also contributed to the appearance and subsidence of temporary land bridges and smaller terrestrial stepping-stones between islands, as well as linking some islands to the mainland (Iturralde-Vinent and MacPhee 1999; Lambeck and Chappell 2001; Pindell and Kennan 2009; Bellard et al. 2014). Over time, these numerous geologic and climatic changes produced opportunities for ancient bats to colonize the West Indies, occupy various habitats on each island, and ultimately differentiate into the modern forms.

How did bats arrive in the newly formed West Indies? The geographic position of the archipelago suggests three main routes of colonization (Baker and Genoways 1978; Rodríguez-Durán and Kunz 2001). Some species may have come from the north, reaching the Bahamas and Greater Antilles via Florida, whereas other bats may have followed a southern track, gradually moving from South America northward through the Lesser Antilles. A third possibility is a western origin, with colonizers coming from Central America or the Yucatán Peninsula. Based on the bat faunas existing on continental areas along these possible paths of dispersal, biologists believe that each route likely was exploited by different types of bats to access the islands (Baker and Genoways 1978).

Some studies investigating how species arrived in the West Indies propose a vicariant origin for the region's terrestrial mammals; that is, most West Indian mammals originally crossed an ancient land bridge that connected South America to the eastern Greater Antilles around the time of the Eocene-Oligocene boundary, about 34 million years ago (Iturralde-Vinent and MacPhee 1999). After subsidence of the land bridge, populations became geographically separated, leading to the formation of new species in isolation. However, other biologists believe that the time of subsidence of the proposed land bridge, the geologic age of the West Indian islands, their changing size and position across time, and their historical isolation (i.e., many islands were never fully connected to their neighbors) suggest that overwater dispersal was required for bats to reach the islands (Baker and Genoways 1978; Dávalos 2004).

The distribution of bats in the West Indies understandably mirrors the patterns described in the general theory of island biogeography. Island area, for example, helps explain why the largest island (Cuba) has the greatest richness (twenty-seven species) in the region. Additionally, West Indian islands closer to a potential mainland source of colonizers typically possess higher species richness, compared with islands that are more isolated. This pattern of increased richness with proximity to the mainland is evident even when comparing the tiny (389 km^2) Lesser Antillean isle of Saint Vincent, which is only 275 kilometers from South America, with the much larger (9100 km^2) but more isolated Puerto Rico that lies at least 750 kilometers from any continent. Saint Vincent harbors twelve species, whereas Puerto Rico supports only thirteen. Furthermore, several genera of bats that form "core members" of the community, namely *Artibeus*, *Molossus*, *Noctilio*, and *Tadarida*, have extensive distributions across the West Indies. Mammalogists typically consider these animals to be strong fliers with a good ability to disperse long distances, which could explain the prevalence of these kinds of bats in the archipelago. Such patterns of species richness also support overwater dispersal as the primary way in which bats reached the West Indies.

As early bats colonized and became established on the West Indies, some of these animals occupied novel habitats, which promoted diversification and resulted in the high proportion of endemic species that occurs today. Estimates of the time of genetic divergence suggest that endemic species within the family Mormoopidae and the phyllostomid tribes Brachyphyllini, Glossophagini, and Stendodermatini, originated at times of high sea level during the Miocene epoch, at least 19 million years ago (Dávalos 2009). The family Natalidae is similar. Although fossils suggest that natalids originated in North America or Europe during the Eocene about 50 million years ago (Morgan and Czaplewski 2003), the group did not radiate into the eight species that live on the islands today until the early Miocene, about 22 million years ago, which, again, was a time of high ocean levels that would have covered any land bridges. Therefore, phylogenetic and fossil evidence independently supports the concept that the ancestors of these West Indian groups first arrived by overwater dispersal.

Although the ancestral origin of most West Indian bats points to colonization from mainland America, the evolutionary history of some groups suggests that the West Indies themselves acted as an additional source of diversity. Biogeographic patterns are typically

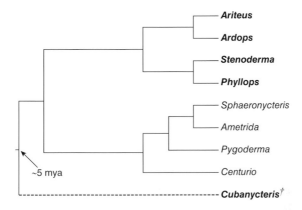

Figure 2.1. Phylogenetic relationships of the genera of short-faced bats (subtribe Stenodermatina: family Phyllostomidae). West Indian members of the subtribe are in bold. The extinct Silva's Cuban bat (*Cubanycteris silvai*) is indicated by a dagger, and the dashed line indicates that only morphological data are available for this species, compared with the other taxa that have both morphological and molecular data. All genera contain only one species, except the Cuban fig-eating bats (*Phyllops*), which include three species, two of which are extinct and not shown. The arrow indicates the estimated time of divergence from the common ancestor of the subtribe. Note that, in the absence of Silva's Cuban bat, the ancestral origin of Stenodermatina could not be resolved. Drawing is based on data extracted from Tavarez et al. (2018). Prepared by J. Angel Soto-Centeno and Camilo A. Calderón-Acevedo.

examined under the assumption that the extant fauna of a region is representative of that community of species since its founding. For example, studies of existing members of the phyllostomid subtribe Stenodermatina (short-faced bats), which includes eight genera of West Indian endemic and mainland bats, produced ambiguous explanations for the origin of species diversity in this subtribe. Evolutionary biologists classically interpreted the relationships of living short-faced bats as the product of two different colonization events from the mainland to the West Indies. However, the discovery of an extinct ancestor from Cuba, Silva's Cuban bat (*Cubanycteris silvai*), provided concrete evidence that short-faced bats actually originated in the West Indies and, from there, colonized the continent (Fig. 2.1; Tavares et al. 2018). This so-called reverse colonization from the West Indies to the mainland has also been demonstrated for some frogs (*Eleutherodactylus*; Heinicke et al. 2007), lizards (*Anolis*; Nicholson et al. 2005), and birds (*Amazona*; Russello and Amato 2004). Existence of reverse colonization in West Indian bats emphasizes the importance of these airborne mammals to the broader understanding of regional biogeographic patterns and highlights this archipelago as a key source of chiropteran diversity in the Neotropics.

Past events added to the diversity of bats, but historical factors, such as previous climate change and human colonization, also could affect the composition of communities by causing the loss of species from islands. Fossils indicate that at least twelve species of West Indian bats became extinct in the late Quaternary period (Dávalos and Turvey 2012; Soto-Centeno and Steadman 2015). Changes in climate and sea level that occurred since the Pleistocene-Holocene transition, 15,000 to 9000 years ago, altered the size of islands across the archipelago (Soto-Centeno and Steadman 2015). As glaciers melted, seas rose 80 m to present levels, inundating ancient coastal zones and fragmenting large expanses of low-lying islands; for example, total area of land in the Bahamian Archipelago decreased by 90%, from 130,000 km^2 to just 13,000 km^2.

This well-documented reduction in available land that occurred millennia ago led to the notion that past climate change caused the extinction of many species on the islands (Dávalos and Russell 2012). However, a comparison of fossil and extant communities of bats, followed by an assessment of the amount and kind of habitat available over time, reveals that the loss of species in the West Indies is more complex and cannot be explained by Pleistocene sea levels alone (Soto-Centeno and Steadman 2015; Orihuela et al. 2020b). The dating of fossils indicates that most bats in the West Indies that are now extinct survived climate change during the Pleistocene-Holocene transition, only to vanish in the late Holocene, 5000 to 500 years ago, during a time when climate and island areas were broadly similar to those of today. The temporal and geographic patterns of species loss observed in the West Indies suggest that modern communities of bats likely resulted from multiple factors, including changes in climate and habitat, as well as anthropogenic forces. Documenting these time-scaled changes is important to understand the factors that affected bat communities within islands over time, particularly because humans continue to modify these fragile insular ecosystems.

The Present: How Are Communities of Bats Maintained?

Ecological and geological factors influence the diversity and richness of biological communities on islands. In turn, these communities are maintained by deterministic (predictable) processes, such as competition for resources, and stochastic (random) processes, such as fluctuations in population size or chance colonization. The interaction among these processes also helps shape other population dynamics, such as the spatial organization of genetic variation (i.e., population structure), resulting in nonrandom patterns of community assembly.

According to classical biogeographic theory, the species richness of islands is lower than that of continental regions with a similar areal extent (MacArthur and Wilson 1967). This characteristic of insular faunas results in part from the large distances that bats must overcome to colonize an island, as well as differences in vagility among species. Typically, biologists have considered bats more likely than other terrestrial mammals to travel long distances, because of the animals' ability to fly. Long-distance movements, though, are either migration, which occurs seasonally and entails movement from and to a particular place of origin, or dispersal, which, in a strict sense, is a one-way trip.

Migration by bats across large distances is well known. For example, Mexican long-nosed bats (*Leptonycteris yerbabuenae*) fly over 1700 km during spring, on a migratory path that tracks the blooming of several tropical plants (Wilkinson and Fleming 1996). Although the only way a bat may reach an isolated archipelago is via long-distance dispersal, scientists have documented few such cases, and perhaps the best-known example is the northern hoary bat (*Lasiurus cinereus*). Two ancestral stocks of this mammal dispersed across 3665 km of the Pacific Ocean from North America to the Hawaiian archipelago, founding an insular population about

10,000 years ago (Russell et al. 2015). These two examples, however, represent some of the extremes in terms of movement by flying mammals, and most bats are arguably poor dispersers, as exemplified by the high proportion of West Indian species (41%) that are restricted to just one island.

Genetic data obtained from individual populations, either within or among islands, can shed light on the likelihood of chance colonization, dispersal, or migration of insular bats, thus providing insight into how communities assemble and are maintained. Most insular populations of bats are assumed to be highly structured, with individuals of a species from one island being genetically more similar to each other than to individuals of the same species living on different islands, and this pattern is attributed to the broad oceanic straits that isolate the populations and act as barriers to interisland movement (Koopman 1977). Studies of population structure in West Indian bats are sparse but suggest that the frequency with which bats cross these aquatic boundaries varies depending on the origin and ecology of each species.

First, consider three widespread species that occur in the West Indies and in continental America. The Jamaican fruit-eating bat (*Artibeus jamaicensis*) inhabits the southern Bahamas, Greater Antilles, and northern Lesser Antilles, but it displays low levels of population structure, meaning that there is little genetic variation among islands (Carstens et al. 2004; Fleming et al. 2009). This species appears to have colonized the West Indies in the late Pleistocene, and insular populations have had little time to differentiate genetically (Larsen et al. 2017). In contrast, Waterhouse's leaf-nosed bat (*Macrotus waterhousii*) also occupies many islands of the Bahamas and the Greater Antilles, but it appears more genetically structured within islands (Fleming et al. 2009; Muscarella et al. 2011), which may be related to the animal's ecology and morphology. This bat has short, broad wings that allow it to maneuver within vegetative clutter and glean insects from the ground or tree branches, but such wings are not designed for prolonged flight and presumably limit the movement of these animals and their genes among islands. Mammalogists assume that some species that are widespread in the archipelago, such as the Brazilian free-tailed bat

(*Tadarida brasiliensis*), have good dispersal ability, because mainland populations of the same animal undergo extensive annual migrations. Nevertheless, individual insular populations of this bat display high levels of genetic structure, which suggests limited movement among islands (Speer et al. 2017).

Second, the complexity in patterns of population structure is also exemplified by species that do not occur on the continents but are endemic to the West Indies. For instance, genetic structure is low in the Antillean fruit-eating bat (*Brachyphylla cavernarum*), despite having a distribution from Saint Vincent northward to Puerto Rico, although this specific case has been attributed to the retention of ancestral genetic variation (i.e., incomplete lineage sorting) among insular populations (Carstens et al. 2004). Other species that are endemic to the West Indies, such as the Lesser Antillean tree bat (*Ardops nichollsi*), buffy flower bat (*Erophylla sezekorni*), and Greater Antillean long-tongued bat (*Monophyllus redmani*), do have structured populations within islands, which suggests limited or no movement among islands by these animals. These differences in population structure can ultimately help determine which species are more or less likely than others to disperse among islands in modern times and how communities are maintained (Carstens et al. 2004; Muscarella et al. 2011; Soto-Centeno 2013).

Deterministic factors, including island area, distance from a source, and habitat diversity, act simultaneously and presumably contribute to forming and maintaining bat communities in the West Indies (Willig et al. 2009; Hoffman et al. 2019). Species richness across the West Indies is primarily influenced by island area, with larger islands sustaining higher species richness than smaller ones (Fig. 2.2). However, among island groups (i.e., the Bahamas, Greater Antilles, and Lesser Antilles), regional factors such as habitat diversity or topography also influence local bat communities. Latitude prominently affects species richness in continental areas, but in the West Indies, the association of latitude with species richness surprisingly occurs only in the Greater Antilles (Willig et al. 2009).

Biologists often use elevation as a surrogate for describing habitat diversity on islands, because regions with high topographic

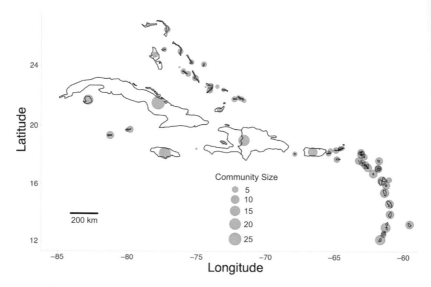

Figure 2.2. Size of bat communities in the West Indies. Circles indicate magnitude of species richness on each island (J. A. Soto-Centeno, unpublished data). Prepared by J. Angel Soto-Centeno and Camilo A. Calderón-Acevedo.

variation experience greater disparity in temperature, rainfall, wind speed, and soil structure, which results in more heterogeneous habitats (Ricklefs and Lovette 1999; Hoffman et al. 2019; see Chapter 1). For example, the Bahamas are characteristically flat, with a maximum elevation of only 63 m. Available habitats across this archipelago are few and primarily composed of groves of Bahamian pine (*Pinus caribaea*), mangroves, and isolated stands of coppice vegetation. Such homogeneous habitat sustains only a low diversity of bats, with six species, at most, occurring on any single island. In contrast, Saint Vincent, in the Lesser Antilles, is similar to many Bahamian islands in terms of area and distance from the mainland. However, Saint Vincent has a maximum elevation of 1245 m, twenty times higher than that of the Bahamas, and supports twice as many species as the most speciose of the Bahamian isles (i.e., Great Abaco and Long Island).

Caves in the West Indies form a significant and reliable habitat for bats, providing shelter, protection, and safe places for reproduction (Rodríguez-Durán 2009; see Chapter 3). In the West Indies, some bats roost exclusively in caves or only in trees, whereas other species are generalists that shelter in human-made structures, trees, or facultatively in caves. Over 40% of all West Indian bats roost underground on most islands; therefore, the presence of caves may also be a factor influencing the geographic occurrence and community composition of bats in this archipelago (see Chapter 3). Close examination of regional bat communities across the West Indies reveals that the proportion of obligate cave-dwelling species is positively associated with maximum elevation on an island (Fig. 2.3).

For example, rising and falling sea levels shaped large expanses of eolianite karst in the Bahamas, eroding the rock and aiding in the formation of numerous small caves—but not the large, complex caves commonly found on islands with greater topographic relief and elevation (Soto-Centeno et al. 2015). Bat communities in the Bahamas and other low-elevation West Indian islands are primarily composed of generalist species. For example, even species that roost only in subterranean sites on the Greater Antilles, such as the buffy flower bat, commonly shelter in buildings on the Bahamas (Speer et al. 2015); this shift in roosting preference presumably relates to the typically smaller and less intricate caves available on the Bahamas. Unlike obligate cave dwellers, the proportion of generalist species decreases as elevation increases (see Fig. 2.3). This negative association suggests that islands with greater elevation provide a wider variety of habitats, including roosts, which support different community associations.

The Future: Understanding West Indian Bats through Conservation Biogeography

Bat communities have influenced the development of the cornerstone theory of island biogeography, and the study of West Indian bats has contributed to an understanding of broad patterns of insular biodiversity and community assembly, from biogeographic,

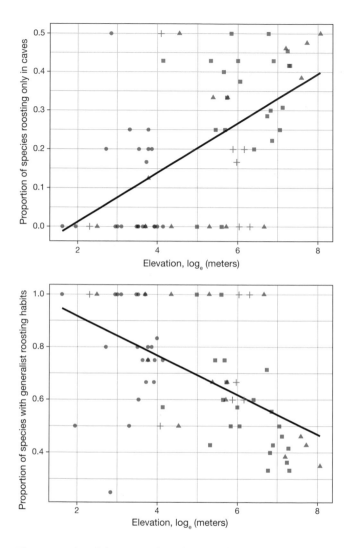

Figure 2.3. Plot of the proportion of bat species on an island that use only caves as roosts (*top*) or are roost generalists (*bottom*) as a function of maximum elevation in meters (log$_e$ transformed) on that island (J. A. Soto-Centeno, unpublished data). Circles = Bahama Islands; triangles = Greater Antilles; squares = Lesser Antilles; and crosses = Virgin Islands. Smaller islands tightly associated with the Greater Antilles, such as Culebra, Isla de la Juventud, Mona, and Vieques, are included within the Greater Antilles and also represented as triangles. Prepared by J. Angel Soto-Centeno and Camilo A. Calderón-Acevedo.

ecological, and evolutionary perspectives. Notwithstanding, the combination of different types of data (e.g., ecological, fossil, and genetic) can help uncover additional biogeographic patterns in West Indian bats. Future research focused on conservation biogeography, leveraging the principles and theoretical aspects of biogeography and disparate types of data applied to current problems of biodiversity conservation, could provide the integrative framework necessary to address pending questions about West Indian bats.

Humans have proven to be very good at extinguishing biodiversity, either directly or indirectly, and a primary concern for the conservation of biodiversity today pertains to the fast pace of habitat fragmentation and loss. Four of the top nine conservation threats to West Indian bats are linked to habitat modification (see Chapter 5). Some endemic species, such as the red fig-eating bat (*Stenoderma rufum*), respond negatively to environmental fragmentation, although the habitat requirements for most West Indian species remain unknown. Future studies should emphasize detailed assessments of the habitat needed by each species and the quantification of that which remains. Efforts to reconstruct the required conditions for each species, through the modeling of ecological niches or connectivity of habitat patches, can also produce information relevant to ameliorating the current threats that these bats face (e.g., Soto-Centeno et al. 2015; Calderón-Acevedo et al. 2021).

The interaction of multiple climatic, ecological, geographic, and geological factors has differentially shaped the composition of communities of bats across the West Indies. Although bioinventories conducted during the past century have produced important lists of species, thus improving our understanding of each insular community, processes related to the formation and maintenance of present-day communities are largely unexplored. Hispaniola, Jamaica, Puerto Rico, and many islands in the Lesser Antilles are important centers of habitat and species diversity (see Chapter 5), yet comparative paleontological assessments to complement modern bioinventories and document species turnover are lacking. Such assessments, when linked with radiocarbon dating of fossils, are important (e.g., Steadman et al. 2015; Stoetzel et al. 2016; Orihuela et al. 2020b) not only to provide context to species turnover but also to design a backdrop against which species' resiliency to change can be evaluated.

Knowledge of species- and population-level relationships is essential for understanding biologically relevant units for the development of accurate inventories and conservation plans. Although broad patterns of how different species relate to each other and how some populations are genetically structured exist for a few bats in the West Indies, considerable gaps remain in our understanding. Specifically, generating estimates of the time of colonization for widespread species across the West Indies, many of which are core members of the bat community, could aid in evaluating how and when communities were assembled. Genetic analyses of the divergence times of species, combined with thorough population-level estimates of genetic structure, would facilitate the reassessment of species limits and potentially uncover hidden diversity in need of conservation (e.g., Larsen et al. 2011). The application of modern computational methods to assemble and evaluate data from entire genomes is still rare in studies of West Indian bats. Nevertheless, the high level of resolution provided by these methods should help resolve outstanding questions about the timing of colonization and population structure with unprecedented reliability.

Among all extant mammals in the West Indies today, bats stand out as the most speciose, ecologically diverse (see Chapter 3), and often the only remaining native mammals. Although many bats are highly threatened (see Chapter 5), there is much to uncover about their diversity and evolution in this archipelago. The integration of modern approaches, combined with traditional inventories and ecological assessments of bat communities could reveal the information necessary to reevaluate current and future threats to extant species and to develop conservation plans to prevent the loss of populations or species in the West Indies.

CHAPTER 3

Ecology and Natural History of Bats in the West Indies

Armando Rodríguez-Durán

Bats are typically the only native mammals present on most islands of the West Indies. Each island supports a group, or an assemblage, of bat species, but the composition of that assemblage often differs among islands and over time (see Chapter 2). Over many millennia, the islands and their bats have been buffeted by natural disasters, such as hurricanes and volcanoes, and more recently, human activities have severely affected the structure and composition of ecosystems in the region (see Chapter 1). This chapter highlights differences between bats occupying the mainland and the islands and introduces two aspects of the biology of these animals that critically affect survival—their roosting ecology or dependence on shelter, and their aeroecology or use of the air above the earth (Kunz et al. 2008). Knowledge of these areas is fundamental to establishing successful programs for the future management and conservation of these intriguing creatures (see Chapter 5).

Differences between the West Indies and the Mainland

Although the field of island biology has had an enormous impact on our understanding of ecology and evolution, the study of bat ecology has focused mostly on continental assemblages. The West Indies, however, are a hot spot of bat biodiversity (Willig et al.

2009), and assemblages in the archipelago are not random subsets of mainland faunas (Fleming 1982; Rodríguez-Durán and Kunz 2001). For instance, the sac-winged bats (Emballonuridae), one of the oldest and most widely distributed families of bats, is absent from the West Indies, and forty-five of the insular species, 74% of the total, are endemic to one or more of the islands.

The preponderance of hurricanes in the region, combined with the abundance of caves or rocky shelters, influences broad differences in roosting behavior between bats of the continents and those of the Caribbean islands. To illustrate, most species in Mexico that shelter in caves are facultative cave-dwelling animals, sometimes roosting underground but often not (Arita 1993; Ávila-Flores and Medellín 2004), whereas most endemic taxa that use subterranean sites in the West Indies are obligatory occupants of caves (Rodríguez-Durán 2009). In addition, a greater proportion of all species in the West Indies, compared with the continents, are obligate cave inhabitants, and even facultative cave dwellers on the islands seem to roost underground whenever that is an option (Rodríguez-Durán 2009).

The foraging behavior of plant-eating bats differs, with insular species having more generalized diets than their mainland counterparts and a less uniform response to moonlight during their nightly foraging bouts. Although lunar phobia appears widespread in the mainland Neotropics, it is not common on most islands (e.g., Rodríguez-Durán and Feliciano-Robles 2016; but see Mancina 2008). Furthermore, dietary shifts occur in some insect-eating species that inhabit both the continents and the West Indies, such as the Brazilian free-tailed bat (*Tadarida brasiliensis*). In North America, this mammal preys heavily on moths and beetles, but on Cuba and Puerto Rico, it consumes large amounts of flies and has become the ecological equivalent of a species in the genus *Myotis*, a taxon that is missing from the Greater Antilles (Whitaker and Rodríguez-Durán 1999). Finally, energetics differs between mainland and island populations, with metabolic rates lower in insular species compared with continental animals (McNab 2009). Cave dwelling may facilitate this physiological adaptation in the West Indies, because the thermal stability and heat-storage capacity of subterranean sites allow bats to maintain a stable body temperature without a high energetic cost (Rodríguez-Durán 1995). Thus, the detailed study of the different habits of bats on these islands contributes toward our understanding of ecology and evolution in general.

Roosting Ecology

Bats spend over half their lives within roosts, and consequently, the presence of suitable resting sites represents a major factor in determining the distribution of various species. For instance, the northern edge of the range of the California leaf-nosed bat (*Macrotus californicus*), the most northerly species in the family Phyllostomidae, appears limited by the availability of underground sites that are geothermally heated (Bell et al. 1986). The importance of roosts in the biology of bats is underscored by the large number of reviews dealing with the subject over the past 40 years (e.g., Kunz 1982; Furey and Racey 2016; Rodríguez-Durán 2020). Most if not all social and reproductive activity, as well as interaction with parasites, occurs in the roost, and these shelters provide survival benefits, such as protection from predators and the elements, and warm sites for nurseries. Many bats have evolved structural traits that make them better adapted to use certain types of roosts, which attests to the intricate relationship between these animals and their preferred shelters. For instance, Spix's disc-winged bat (*Thyroptera tricolor*) from Central America evolved suction cups on its wrists and ankles to facilitate hanging inside the narrow funnels formed by the smooth unfurled leaves of plants such as *Heliconia* (Riskin and Fenton 2001). As another example, multiple species that hide in trees display dots or stripes on their face or body, which disrupt the outline of the animals and make them less conspicuous to day-hunting predators (Santana et al. 2011).

Biologists have generally neglected detailed characterization of roosts in the tropics, and this lack of work partly relates to the misconception that the stress of low ambient temperature is limited to temperate regions. Air temperature in the tropics, however, often drops to levels at which temperate zone bats normally resort to torpor, and many tropical species do not maintain constant body temperatures, that is, they are not continually homeothermic (Geiser and Stawski 2011). In addition, bats in the tropics typically seek roosts that vary little in temperature, suggesting that there is potential stress without such protection, and some bats even modify structures to increase their thermal stability (Rodríguez-Herrera et al. 2016).

For shelter, bats mainly use three types of roosts—phytostructures, lithostructures, and anthropic structures (Rodríguez-Durán 2020). Phytostructures include living or dead foliage, cavities in trees, spaces under loose bark, and even the arboreal nests of termites. Lithostructures are mostly caves but also undercut earthen banks and rock crevices, among others. Finally, bats use a myriad of human-made or anthropic structures that, from the viewpoint of the bats, probably differ little from roosts in the first two categories. For instance, there may be no distinction between sheltering in a small limestone cave or in an abandoned concrete bunker from World War II, or inside the walls of a wooden house versus within a hollow tree. Other anthropic roosts include underground spaces in cavelike mines and culverts and crevices in clifflike concrete bridges and stone buildings.

Several species in the region use phytostructures. For example, vesper bats in the genus *Lasiurus*, such as the minor red bat (*L. minor*) or Cuban yellow bat (*L. insularis*), consistently roost among leaves, as do endemic fig-eating bats in the genera *Ariteus*, *Phyllops*, and *Stenoderma*. In addition, the great fruit-eating bat (*Artibeus lituratus*) and a few other species hang within foliage occasionally. Although more than one thousand little goblin bats (*Mormopterus minutus*) may hide among the fronds of a single palm tree in Cuba, most species that rest among leaves are typically solitary or form small social groups of less than twenty individuals. Roosting in small numbers in cryptic sites makes it difficult for mammalogists to find and investigate the lives of these animals. However, radiotracking of two foliage-dwelling species—the red fig-eating bat (*Stenoderma rufum*) on Puerto Rico and the Jamaican fig-eating bat (*Ariteus flavescens*) on Jamaica—show that these solitary animals change resting sites every few days but remain loyal to a small patch of forest, with no evidence of territoriality (Gannon 1991; Hayward 2013).

At least seventeen continental species spend the day under the leaves of various philodendrons, palms, or heliconias that the bats modify into "tents," using their teeth (Rodríguez-Herrera et al. 2007). Species that occasionally find refuge in tents on the mainland, such as the Jamaican fruit-eating bat and the great fruit-eating bat, are present in the archipelago. However, ecologists have never discovered any tents on the islands, suggesting that these particular tent-roosting bats are not the species responsible for the actual construction of the tents that they occupy on the mainland.

Leaf roosts are not as substantial or as long-lived as caves or buildings, but they do have potential advantages. Leaves are plentiful and provide some protection from wind, rain, solar radiation, and the prying eyes of predators. The abundance of leaves in the landscape means that foliage-roosting bats are able to commute shorter distances to their feeding grounds, often just hundreds of meters or less, whereas cave-dwelling species, especially those roosting in large numbers, must frequently travel many kilometers to locate food. Moreover, ectoparasites often do not spend all their time on the body of the host, and frequent switching from one ephemeral leaf to another may aid in reducing the number of parasites harbored by each bat.

Although multiple types of bats in Europe and North America specialize in sheltering within tree cavities or under bark (Barclay and Kurta 2007), no species in the West Indies appears completely reliant on this type of phytostructure. The fact that more species depend on leaves than on hollow trees on the islands, either totally or occasionally, may seem curious, given that tree trunks are more substantial and provide better insulation and protection from rain and wind than do leaves (Kurta 1985; Clement and Castleberry 2013). Both phylogeny and hurricanes, however, could be at play here too; hollow trees are prone to destruction by windstorms, thus reducing the overall availability of this type of roost (Wiley and Wunderle 1993), and cave-dwelling families, such as the mormoopids, have a strong presence on the islands. Nevertheless, a few species use cavities in trees when caves are unavailable. This seems to be true for the greater bulldog bat (*Noctilio leporinus*) and possibly for the Jamaican fruit-eating bat, both of which can form large colonies composed of multiple social groups in a cave or just small colonies consisting of a single social unit in a tree. Based on the habits of their relatives in South America, some poorly studied species of the Lesser Antilles, such as Paulson's and Angel's yellow-shouldered bats (*Sturnira paulsoni* and *S. angeli*, respectively), probably also use tree cavities, at least on occasion.

Few species of bats in the West Indies select anthropic structures on a regular basis. Examples are big brown bats (*Eptesicus fuscus*) and Jamaican fruit-eating bats, which commonly dayroost in culverts, and a few other species that frequently occupy deserted military bunkers. In the West Indies, though, wooden and concrete houses or similar buildings are probably the most common human-made roost; building dwellers include big brown bats, Cuban evening bats (*Nycticeius cubanus*), and, to some degree, all species of free-tailed bats (C. Mancina, pers. comm., 2021). However, Pallas's mastiff bat (*Molossus molossus*) is by far the most common and wide-ranging species using such structures, followed at a distant second and third by the Brazilian free-tailed bat and Jamaican fruit-eating bat, respectively. Both the Brazilian free-tailed bat and Pallas's mastiff bat, as well as other species of mastiff, are generalized in their use of buildings, resting under roofing material, between outer and inner wooden walls, within stacked cinder blocks that frame a house, or tucked into the cracks and crevices of stone walls. Spaces under corrugated metal roofs exposed to the tropical sun attain high temperatures, whereas those within cinder-block walls are much cooler. Pallas's mastiff bat shows clear evidence of stress when exposed to an air temperature of 36 °C, but preliminary observations from Puerto Rico indicate that temperatures where these animals actually roost do not exceed 30 °C, suggesting that the bats thermoregulate behaviorally, moving from site to site within an individual building to select desired air temperatures.

Most West Indian bats, though, take refuge in natural caves, which often contain hundreds, thousands, or even tens of thousands of individuals (Rodríguez-Durán 2022). These underground shelters provide substantial protection against the rain and wind of tropical storms, which is likely one reason why so many species are cave dwellers on the islands. Although the fossil record shows that a few species have inhabited the same caverns over the past 27,000 years (Pelletier et al. 2017), the relative abundance of different bats in specific caves or even on whole islands can change significantly after a hurricane. Each of these storms has a potentially immense impact on small, insular populations but a lesser effect on larger, more widespread species on the mainland (Rodríguez-Durán 2009). Furthermore, individual species may react differently to tropical storms; some populations of some cave bats may be kept in check, and others may be allowed to persist because of periodic storms that impede one species from displacing the other. For instance, both the brown flower bat (*Erophylla bombifrons*) and the Puerto Rican mustached bat (*Pteronotus portoricensis*) occupy the same sections of the same caves. In 1998, at Culebrones Cave in Puerto Rico, Hurricane Georges reduced the population of brown flower bats, which may have facilitated occupation of the site by Puerto Rican mustached bats. In the 20 years after Georges, the mustached bat, the new species, continued to thrive, but the population of the brown flower bat, once the most abundant species in the cave, had not recovered to pre-Georges levels before it was downsized again by Hurricane Maria in 2017.

The internal environment of a cave varies widely, depending on the latitude, number of external openings, length and volume of passages, and elevation of internal chambers relative to the entrance. In general, air temperature within a deep underground site assumes the average annual surface temperature, which is mostly determined by latitude. Underground sites in the northern United States might exhibit a year-round temperature of only 4 to 6 °C (Kurta and Smith 2014), but typical caves in the West Indies experience "default" temperatures of 19 to 25 °C. However, complex subterranean systems on the islands may display an even wider gradient, ranging from 19 to 40 °C, and these latter caverns, the so-called hot caves, host many types of bats, especially endemic species (Ladle et al. 2012).

Although various physical processes can produce the warm temperatures of a hot cave, those in the West Indies result from the cave-dwelling organisms themselves. The source of the heat is the bodies of thousands of bats living inside the cave, and, to a lesser extent, the invertebrates and microbes involved in decomposing the guano that constantly rains to the floor. However, intrinsic to the concept of a hot cave is that this warm air does not readily dissipate and instead accumulates underground because of the structure of the cavern. This usually occurs when the entrance to the chamber where the bats live is small and situated at the base of the room; air inside the roost is warmed by the organisms, rises above the low entranceway, and becomes trapped, filling the chamber and producing the warm environment.

Biologists divide cave chambers into three broad categories, based on temperature and the types of bats that are present—the frigidarium, tepidarium, and caldarium (Fig. 3.1; Rodríguez-Durán 2009; Ladle et al. 2012). Bats resting within any of these areas may resort to various behavioral mechanisms, such as clustering or hiding in solution cavities, for further partitioning of the roosting space, depending on the specific microclimate that the animals require. The frigidarium can be a simple well-ventilated cave or a specific room within complex hot caves that are close to the outside entrance and where air temperature does not surpass 25 °C. Relative humidity typically exceeds 75% in the frigidarium. Several types of leaf-nosed, funnel-eared, free-tailed, and vesper bats, as well as the greater bulldog bat, frequent these cool caves. Obligate hot-cave dwellers, such as the Antillean ghost-faced bat (*Mormoops blainvillei*) and the greater Antillean long-tongued bat (*Monophyllus redmani*), usually roost in the hottest portion of the site but occasionally hang in this part of the cavern, although they too resort to clustering or shallow torpor because of the cool environment (Rodríguez-Durán 1995). Some species that consistently use the frigidarium, such as the Jamaican fruit-eating bat and the greater bulldog bat, are behaviorally flexible and also shelter in trees or anthropic structures.

The tepidarium provides an intermediate zone of temperature between the frigidarium and the cave's hottest room. Species capable of using either warm or cool caves may inhabit this part of a hot cave, which remains stable at 26 to 27 °C as long as the popula-

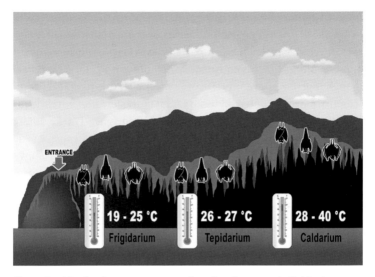

Figure 3.1. Idealized temperature gradient in a hot cave. Individual caves may contain one or all of these sections, depending on length and complexity. Courtesy of InterAmerican University of Puerto Rico, Bayamón Campus.

tion of bats remains steady. Thus, after a strong hurricane that decimates the local bats, the ambient temperature in the tepidarium actually decreases, because fewer heat-producing bodies remain. Species that prefer resting in the tepidarium include various leaf-nosed and mustached bats, such as the brown flower bat, Puerto Rican mustached bat, and Parnell's mustached bat (*Pteronotus parnellii*). Although some individuals of these species may roost in the frigidarium, the females give birth and raise their young only in the warmer confines of the tepidarium; nevertheless, these bats do not tolerate the higher temperatures past the heat trap (Rodríguez-Durán and Soto-Centeno 2003).

Finally, the caldarium is the hottest part of a hot cave. Air temperature generally ranges from 28 to about 32 °C, sometimes reaches 35 °C, and, in exceptional locations, hits an oppressive 40 °C. Relative humidity hovers between 90 and 100%; consequently, water vapor may condense in the saturated air, forming a cloud that limits the reach of a headlamp or fogs the cooler surfaces of eyeglasses or photographic lenses brought into the chamber. High levels of carbon dioxide exist in the caldarium of some caves, often resulting in labored breathing by human explorers, and although high levels of ammonia or hydrogen sulfide may also occur, scientists have not detected these gasses in the few sites where measurements have actually been attempted (Rodríguez-Durán 2009).

Several types of leaf-nosed and mustached bats roost exclusively in the caldarium, including the Cuban flower bat (*Phyllonycteris poeyi*), Jamaican flower bat (*P. aphylla*), and sooty mustached bat (*Pteronotus quadridens*). A few species, such as the sooty mustached bat, cannot survive 24 hours outside these hot, humid conditions; when forced to spend the day at normal ambient temperatures, the bat's body temperature falls, and the animal is unable to raise it back again (Rodríguez-Durán 1995). As with bats adapted to the tepidarium, individuals from species typical of the caldarium sometimes spend the day in cooler conditions, but parturition and pup rearing happen only within the hottest rooms. A complex cave system containing all three sections can shelter up to fourteen species (C. Mancina, pers. comm., 2021) that are nonrandomly distributed along the temperature gradient from the frigidarium to the caldarium (Rodríguez-Durán 1998). The combined population of all bats throughout such a hot cave varies from a few thousand to perhaps over six hundred thousand individuals, with most resting in the tepidarium and caldarium (Rodríguez-Durán and Lewis 1987).

There are drawbacks to living in very large groups, such as potential competition for local resources, as well as higher energetic costs and greater exposure to predators while commuting to foraging areas (Kunz 1982). In addition, large numbers of cavern-dwelling bats, flying through a few restricted exits, provide a predictable source of food for reptilian, avian, and mammalian predators, and these huge congregations of bats also yield an immense substrate for external parasites. Although the types of parasites are well documented (e.g., Silva-Taboada 1979; Kurta et al. 2007), their ecological interactions with bats in the West Indies are largely unknown (Rodríguez-Durán 2020).

In contrast, ecologists have described interactions between bats and predators in more detail. Owls, especially various types of barn owl (*Tyto*), often rest just inside caves, and throughout the West Indies, these birds are the most frequent predators of bats, consuming individuals from more than twenty species (e.g., Wiley 2010; Stoetzel et al. 2016). Unlike owls that hunt after dark, American kestrels (*Falco sparverius*), red-tailed hawks (*Buteo platypterus*), and especially migratory merlins (*Falco columbarius*) prey on bats as they depart from hot caves in dense columns around sunset (Rodríguez-Durán and Lewis 1985; Lenoble et al. 2014b). Naturalists have watched up to nine merlins sitting on surrounding trees, waiting for bats to emerge, and in response to this hunting pressure, the mammals sometimes change their exiting behavior, flying through vegetation instead of open space. Local boas (*Chilabothrus*) also congregate nightly at some entrances to capture departing bats or to climb the internal walls of the cave and prey on roosting animals; over twenty snakes visit some hot caves in a single night (e.g., Rodríguez-Durán 1996; Angin 2014; Newman et al. 2016). Feral cats, similarly, hunt both at the entrance and inside caverns, but unlike boas, whose attempts at catching bats are often unsuccessful, cats appear more effective; these invasive carnivores occasionally den in the cave, where they produce multiple generations that subsist mainly on bats and cockroaches (Rodríguez-Durán et al. 2010).

Airspace

Bats, of course, are the only mammals that fly, and their use of the airspace requires a number of physiological and anatomical modifications (see Introduction). The energetic cost of flight per unit of time is much greater than that of running or swimming, and bats have evolved several adaptations related to the efficient intake and distribution of oxygen. Bats, for example, have larger hearts, a higher proportion of red blood cells (hematocrit), and a greater concentration of oxygen-binding hemoglobin than nonflying mammals of similar size (Maina 2000; Rodríguez-Durán and Padilla-Rodríguez 2008). Some species, such as the hovering Greater Antillean long-tongued bat and the swift open-space insectivore

Figure 3.2. Airspace used by bats. (A) Open space over the canopy, used by open-space hunters. (B) Canopy and understory with various degrees of clutter, exploited by aerial foragers in vegetation and by stationary foragers. (C) Forest edges, used by some aerial foragers in vegetation. (D) Bodies of water, often brackish, where the single overwater strategist, the greater bulldog bat (*Noctilio leporinus*), commonly feeds. Other species of bats, though, such as Pallas's mastiff bat (*Molossus molossus*), are visible as they visit small freshwater ponds for drinking. Courtesy of InterAmerican University of Puerto Rico, Bayamón Campus.

Pallas's mastiff bat, have hearts that are even larger than those of most bats, presumably because of the high energetic demand of their flight styles (Rodríguez-Durán and Padilla-Rodríguez 2008).

Different species of bats partition their airspace, although probably not in such a fine-grained manner as with roosting space. Nevertheless, the aerodynamic characteristics of bats, as indicated by aspect ratio, wing shape, and wing loading, as well as the structure of signals used for echolocation, are important variables that dictate where each species can fly, at what speed, and potentially how far (see Introduction). Based on these traits and field observations (Silva-Taboada 1979; Vaughan Jennings et al. 2004), bats in the West

Indies fall into four preliminary flight categories (Fig. 3.2). The first group, "open-space foragers," includes all free-tailed bats and vesper bats in the genus *Lasiurus* that typically hunt above forests or buildings, over fields, or otherwise far from obstacles. "Aerial foragers in vegetation," in contrast, are bats that search for food within varying degrees of vegetative clutter, such as among shrubs, at the forest edge, or in the understory or canopy; this category comprises all mustached and funnel-eared bats and vesper bats of the genus *Eptesicus*. The third group is "stationary foragers in vegetation" that hover or alight on vegetation while gathering food; examples are the nectarivorous Miller's long-tongued bat (*Glossophaga longirostris*) or the gleaning insectivore Waterhouse's leaf-nosed bat (*Macrotus waterhousii*), as well as most fruit-eating bats. The last group, "overwater foragers," includes only the greater bulldog bat, which primarily feeds on fish. Although a number of mainland species are adapted to feeding over lakes and ponds, preying on adult insects that have aquatic larval stages, such specialization is not evident in the islands, perhaps because few standing bodies of water existed before humans constructed dams. As with any such system of classification, overlap occurs among groupings, and some fine-tuning will necessarily occur, as biologists learn more about the behavior of West Indian bats. Furthermore, these categories refer to where each species is most likely to feed. Many bats disperse over long distances to find food, and modern acoustic monitoring is revealing that foraging and commuting may take place in very different habitats; for example, understory foragers at times share the space of open-air species, presumably during the nightly commute.

In meteorological terms (Westbrook and Eyster 2017), the scale of motion by flying bats in the West Indies ranges from the microscale (≤ 2 km) to the mesoscale (up to a few hundred kilometers). Long-distance dispersal over water is how the ancestors of West Indian bats originally reached the islands (see Chapter 2), although most of those founding events occurred at a time when sea level was lower and continental landmasses extended closer to the islands than today. Nonetheless, Silva-Taboada (1979) captured a Brazilian free-tailed bat in Cuba carrying ectoparasites typical of mainland populations, rather than Cuban parasites, and a population of the Cuban fig-eating bat (*Phyllops falcatus*) recently appeared on Cozumel Island, off the coast of Mexico, 235 km from Cuba (Rivas-Camo et al. 2020); both events indicate substantial overwater movement in recent times. Moreover, Curaçaoan long-nosed bats (*Leptonycteris curasoae*) travel up to 79 km between islands in the southern Caribbean (Simal et al. 2015), and perhaps some bats in the West Indies make similar movements. No evidence, however, suggests that bats in the West Indies undergo extensive annual migrations, although short seasonal movements occur. For example, bats roosting in hot caves segregate by sex at the start of the breeding period; depending on location, one sex sometimes simply shifts to different chambers within the same cave or travels to a separate roosting locale up to 15 km away (e.g., Vela Rodríguez et al. 2019). Finally, many species flutter through the understory, dispersing each night just a kilometer or two from their dayroosts, whereas others fly resolutely for 5 to 10 or more kilometers, high over valleys, eventually feeding in the forests of distant mountain ranges.

In the Caribbean, multiple factors impinge on the aerospace through which bats navigate (Muñoz et al. 2008; Westbrook and Eyster 2017), but the implications for these airborne mammals are often unstudied. At the mesoscale, downhill and uphill breezes in mountains, onshore and offshore winds along the coast, and air flow resulting from daily thunderstorms are a few of the better-known atmospheric complexities that must, in some way, affect the flight of bats. At the microscale, the subtleties of our complex air space are illustrated by this passage from Harper Lee's (1960, 41) novel *To Kill a Mockingbird*:

> "What's a Hot Steam?" asked Dill.
> "Haven't you ever walked along a lonesome road at night and passed by a hot place?" Jem asked Dill.

Humans commonly perceive this "hot steam" in the karstic landscapes of the Greater Antilles. Local folklore associates the odd sensation with lost wandering souls, just as in Lee's novel, but modern biologists now wonder how hot steam and similar natural phenomena influence the distribution of insects or the dispersal of odors from fruits and flowers, thus influencing how and where bats forage.

What Lies Ahead?

Until recently, the secretive lifestyle of bats made them a mostly unknown component of the islands' wildlife. However, as ecological and environmental consciousness grows throughout the planet, residents of the West Indies are becoming increasingly aware of the relevance of their often-unique flying mammals. The ecosystem services provided by bats, especially large colonies roosting in caves, include important savings to agriculture, benefits to human health, and the restoration of native flora into disturbed ecosystems (see Chapter 4). Nevertheless, human activities are causing new intrusions into the night skies that the bats once freely roamed along with a few owls and insectivorous birds (Caprimulgidae). For example, the open air is becoming cluttered with moving obstacles of unfathomable size (to a bat), as gigantic wind turbines spring up on multiple islands (see Chapter 5). In addition, human-generated light pollution is allowing pugnacious birds, such as gray kingbirds (*Tyranus dominicensis*) and pearly-eyed thrashers (*Margarops fuscatus*), to extend their activity later into the night, leading to potential competition with bats for insects (Rowse et al. 2016). So far, it seems that bats in the West Indies have adapted to changes that people have inflicted on the animals' roosting sites and airspace, and hopefully, the current assemblages will keep pace and continue to coexist with humans into the Anthropocene (see Chapter 5).

~•~

CHAPTER 4

Bat-Human Interactions

Armando Rodríguez-Durán and Allen Kurta

Throughout space and time, the relationship between bats and humans has varied substantially, both within and among cultures (Low et al. 2021; Rocha et al. 2021). Early Christians often gave Satan batlike qualities, whereas Chinese societies typically viewed bats as symbols of prosperity. The Old Testament (Leviticus 11:13–19) declared that bats were "one of the birds . . . to regard as unclean and not to eat." In contrast, these mammals were a staple in the diet of many peoples in parts of Southeast Asia, such as the Solomon Islands, where the animal's canine teeth also served as a local currency (Lavery and Fasi 2019). Moreover, bats contributed to the traditional medicines of many regions; in India and Nepal, for example, bat flesh was a regular treatment for asthma (Riccucci 2012).

Bats and Indigenous Peoples of the West Indies

The people who occupied most of the Greater Antilles and the northern Lesser Antilles at the time of the European settlement, typically referred to as the Taíno, were the product of local cultural evolution and admixture among successive waves of settlers from mainland South America. For the Taíno and their forerunners, the Ostionoids, bats served an important magical-religious role, as evidenced by an abundance of stylized representations of these animals that appeared in pictographs and on pottery and stone carvings (Fig. 4.1; Rodríguez-Durán 2002). To those early inhabitants, bats were symbols of Maquetaurie Guayaba, the Lord of

Figure 4.1. Bat motifs are ubiquitous in pre-Columbian pottery vessels throughout the West Indies. This vessel came from Tecla, in Guayanilla, Puerto Rico, and is deposited at the Center for Archaeological Research, University of Puerto Rico. Photo by R. L. Joglar, Proyecto Coquí.

the Underworld and the first soul to inhabit the pantheon of the dead, and of the *opía*, which were spirits of the dead who followed (Stevens-Arroyo 2006; Garcia-Arevalo 2019). Many images combined structural elements of both bats and humans, which presumably stressed the close relationship between bats and the spirits. Just like bats, the opía spent the day sleeping underground, becoming active only at night, when they emerged from their caves to eat the fruit of the guava (*guayaba*) tree (*Psidium guajava*).

In addition to stylized images, archaeologists have discovered bat bones, mostly from the large-bodied Antillean fruit-eating bat (*Brachyphylla cavernarum*), in middens at various pre-Columbian sites in the region (Orihuela and Tejedor 2012; Lenoble 2019; Rodríguez-Ramos et al. 2019). These discoveries suggest that indigenous people ate bats, but it is not clear whether the mammals were consumed for sustenance, either systematically or opportunistically, or for ceremonial purposes (Oliver and Narganes 2003; Lenoble 2019; Rodríguez-Ramos et al. 2019). Evidence for consumption comes mainly from the islands of the Lesser Antilles, most of which were occupied by a different people, the Island Caribs, and from pre-Taíno sites on Puerto Rico, suggesting that bat eating may not have been part of Taíno culture. Alternatively, the dearth of bat remains from Taíno middens may reflect access to meatier mammalian prey in the Greater Antilles. For example, although Antillean fruit-eating bats are as large as 54 grams, modern hutias weigh up to 6 kg, and humans still hunt these burly rodents even today (Silva-Taboada et al. 2007).

Europeans and Bats of the West Indies

Despite common misperceptions, the European portrayal of bats is not limited to characters such as Dracula but over the centuries has spanned the cultural milieu from art to music to poetry (Laird 2018). The first modern novel of the western world, *Don Quixote*, by Miguel de Cervantes Saavedra, in 1615, contains a humorous passage about a multitude of crows and "night-birds like the bats" temporarily preventing the protagonist from entering a cave. Francisco de Goya, one of the most important European artists at the dawn of the nineteenth century, prominently depicts bats as symbols of the negative aspects of human nature, such as evil or ignorance, in his engraving *The Sleep of Reason Produces Monsters*. Nevertheless, Federico García Lorca, in the early twentieth century, gives bats a romantic place in his poems, in which these animals often announce the night, and he describes them as "true lovers of the stars." The classic operetta *Die Fledermaus* (*The Bat*), with music by Johan Strauss II, includes an inebriated man dressed as a bat at a costume ball; the work premiered in Vienna in 1874, but the concept has been reworked into multiple modern films and even adapted to cartoons featuring Tom (a cat) and Jerry (a mouse). In addition to cultural expression, bats influenced engineering, as Leonardo da Vinci in the fifteenth century and Clément Ader in the eighteenth century, designed flying machines inspired by bat wings.

Life in the West Indies, though, was more pragmatic than in the cultural and scientific centers of Europe, and early colonists paid little attention to flying mammals on the islands (Rodríguez-Durán and Santiago-Valentín 2014). The first description of the flora and fauna of the West Indies, written by someone with a scientific background, was a letter composed in 1494 for the municipal council of Seville by the Spanish physician Diego Álvarez Chanca, who accompanied Columbus on his second voyage. The dispatch, though, made no reference to bats and dealt only with findings deemed economically important at the time. In succeeding decades, several Christian missionaries briefly chronicled the biota of various islands, but it was not until the eighteenth century that extensive descriptions of the region's fauna began to appear. For example, Friar Iñigo Abbad y Lasierra (1788) wrote intensively about Puerto Rico, whereas André Pierre Ledrú (1810), a member of the expedition led by the French explorer Nicolas Baudin, contributed additional information on the plants and animals of Puerto Rico, as well as the Virgin Islands. Similarly, Patrick Browne (1789) and later Henry Gosse (1851) detailed the natural history of Jamaica, and Charles Plumier helped chronicle life in the Lesser Antilles (Pietsch and Marx 2021). These early accounts were broad ranging but often mentioned bats. Plumier, for instance, made the first scientific description of the greater bulldog bat (*Noctilio leporinus*) from animals on Saint Vincent (Pietsch and Marx 2021), and Gosse (1851) described behaviors of the Jamaican red bat (*Lasiurus degelidus*) that, more than 175 years later, remain as some of the few published observations on the species.

Nevertheless, it was Johann Christoph Gundlach, during the nineteenth century, who became the first serious student of bats in the region (Rodríguez-Durán and Santiago-Valentín 2014). Gundlach was born in Marburg in what later became Germany but eventually changed his name to the Spanish equivalent, Juan Cristóbal, and became a citizen of Cuba. He spent more than 50 years in the region, observing bats and numerous other animals on Puerto Rico and especially Cuba. His work included the first scientific descriptions of Pfeiffer's red bat (*Lasiurus pfeifferi*), the buffy flower bat (*Erophylla sezekorni*), and the Cuban evening bat (*Nycticeius cubanus*), among others (Silva-Taboada 1979).

The Economics of Guano

In the late 1800s and early 1900s, well after the European invasion, bats in the West Indies unexpectedly piqued the economic interest of entrepreneurs, as guano became widely used as fertilizer and as a source of nitrate for making gunpowder (Campbell Wolfmeyer 2010; Gallant et al. 2019). Miners worked in bat caves throughout the islands, from Jamaica to the Bahamas to the Lesser Antilles, and some operations, such as the one on Mona Island, harvested as much as 100 tons per day (Frank 1998). On Puerto Rico alone, government scientists evaluated the potential yield and nutrient composition of guano from 110 different subterranean sites to help guide extraction (Gile and Carrero 1918). Most cave deposits on the islands, however, were small. Consequently, the material at many locations was quickly exhausted, and with the introduction of synthetic fertilizers after World War I, large-scale guano mining in the Caribbean slowly disappeared.

Although the historical literature thoroughly documents the human effort in and the profits from guano mining, these articles and books never mention the associated toll on the bats. Sinking new shafts into bat caves or enlarging openings inevitably changes the temperature, humidity, and airflow within the roosts, often making them unsuitable for continued habitation. Bats also are sensitive to any physical disturbance within their roosting site, especially during the reproductive season, and even seemingly minor activities, such as recreational caving or casual tourism, can cause mothers to abandon their pups. A huge intrusion, like mining on a daily basis, likely forced any bats residing in these caves to flee, with unknown effects on the animals' survival or reproduction. Guano extraction still occurs, albeit on a smaller, local level, but even this activity can be detrimental. For example, only one nursery cave is currently known for the endangered Jamaican flower bat

(*Phyllonycteris aphylla*), and repeated visits by guano collectors is one factor threatening the continued existence of the species.

Bats and Human Health

In the twenty-first century, many bat-human interactions revolve around occupancy of shared living space and potential health issues. Worldwide, wild and domestic animals act as a reservoir of many bacteria or viruses that can pass to people, through bites, parasites, or other contact, and ultimately cause human disease (Recht et al. 2020). These include, among many others, bubonic plague from ground squirrels and rats, flu from swine and birds, tuberculosis from cows and deer, leprosy from armadillos and primates, and toxoplasmosis and cat-scratch fever from the family pet. Ever since Ebola virus was associated with bats and the 2011 movie *Contagion* made flying mammals into villains, a frenzied search for bat viruses has occurred. Bats, however, are no more likely to introduce a new disease to humans than any other group of mammals or birds (Mollentze and Streicker 2020).

Nevertheless, bats are sometimes associated with the virus that causes rabies and the fungus that results in histoplasmosis. Rabies is caused by a virus that proliferates in the nervous system of mammals and is usually fatal (Fooks et al. 2017). Unlike the mainland Americas, where rabies occurs in every country, the health departments on most islands in the West Indies declare that their jurisdictions are free of the disease, with only Cuba, the Dominican Republic, Haiti, and Puerto Rico reporting cases (Seetahal et al. 2018). On Hispaniola, the primary reservoir for the virus is the domestic dog, whereas in Cuba and Puerto Rico, it is the Indian mongoose (*Herpestes javanicus*). Only Cuba, Haiti, and the Dominican Republic have confirmed any cases of rabies in bats specifically. Antibodies against the virus also occur in Antillean fruit-eating bats from Puerto Rico, indicating exposure to the disease (Hirsbrunner et al. 2020), although none of the bats from the island that health officials have tested over many years actually harbored the virus. Based on the potential of different species for interisland movement and the current location of countries with reported rabies, epidemiologists suggest that the Cayman Islands, Jamaica, Dominica, and Saint Vincent and the Grenadines have a high potential for developing bat rabies and should be particularly vigilant (Morgan et al. 2020). Wildlife biologists are typically vaccinated against rabies, but to be safe, anyone in the West Indies who handles wild mammals, including bats, should wear leather gloves and seek medical attention if bitten; the disease is entirely preventable when a person receives prompt treatment.

Residents of the West Indies should also be aware of histoplasmosis, caused by the fungus *Histoplasma capsulatum*. This organism grows and reproduces (produces spores) in rich organic soil, bird droppings, and bat guano, and the disease is most often associated with pigeon roosts, seabird colonies, chicken coops, and bat caves (Nieves-Rivera et al. 2009; Benedict and Mody 2016). Humans entering contaminated areas may inhale spores, leading to a lung infection that is usually asymptomatic or mild but occasionally deadly (Kauffman 2007). The severe form of the disease is uncommon in the Caribbean, where mortality is most often associated with immunosuppressed individuals (Cano-Torres et al. 2019). Nevertheless, anyone visiting a site known to harbor the fungus should consider wearing an appropriate respirator, and anybody experiencing respiratory symptoms after entering a place with an accumulation of bird or bat guano should seek medical advice.

Bats and Buildings

Throughout the world, conflicts often occur between humans and animals that share the same living space, although these tensions generally diminish as people become more educated about and respectful toward wildlife. In the West Indies, a common conflict occurs when bats seek shelter in buildings (see Chapter 3), and the resident humans understandably become annoyed by a faint odor, the presence of droppings, scratching noises in the wall, or the occasional midnight visitor flying around the bedroom. Decades ago, the solution would be an attempt to exterminate the animals, but such drastic measures rarely provided a permanent solution and needlessly killed these beneficial mammals. Today, biologists know that the simplest, most long-lasting, and often cheapest solution is to exclude bats from the building and then seal the openings that the bats are using to access the structure.

An exclusion typically involves a "one-way valve." These devices are simple contraptions made from nylon screening or plastic pipe, placed over the entrance hole that the bats are using, and temporarily secured with tape, tacks, or caulk. Such devices allow the animals to leave at dusk, but not return, and are adaptable to many of the architectural styles found in the West Indies (e.g., Blumenthal 2011). Do-it-yourself videos and diagrams are freely available on the Internet, by searching with the phrase "bat exclusion device," and in various publications (e.g., Gomes and Reid 2015). Exclusions are highly effective but should not be performed during the reproductive season, because mother bats rarely carry their flightless offspring on foraging flights and preventing adults from returning to the roost inevitably causes the death of dependent youngsters trapped inside. On some islands, it is actually illegal to evict bats during the breeding season; in the Cayman Islands, for instance, exclusions are prohibited between 1 June and 15 November. To determine the optimal dates and the best way to perform an exclusion, seek the advice of your local government agency in charge of wildlife or the nearest bat conservation program (Table 4.1).

The Benefits of Bats

As continued research dispels myths and modern interest in bats grows, human curiosity has turned these mammals into opportunities for citizen scientists and even tourists (Geiling 2019). Although not to the extent of Carlsbad Caverns or the Congress Avenue Bridge in Austin, Texas, managers of Río Camuy and Ventana caves in Puerto Rico, Valle de Viñales in Cuba, and Cueva los Patos in the Dominican Republic include learning about bats and how to protect them as part of their tours. At Cueva los Patos, bats became a significant attraction after the cave was declared a "Site of Importance for Bat Conservation" by the Latin American and Caribbean Network for the Conservation of Bats (known by its Spanish acronym, RELCOM).

Table 4.1 Organizations involved with conservation of bats in the West Indies and the broader Caribbean region

Region	Organization
ABC Islands	Bat Conservation Program, Islands of Aruba, Bonaire and Curaçao
Cayman Islands	The National Trust for the Cayman Islands
Cuba	Programa de Conservación de Murciélagos de Cuba
Dominican Republic	Programa de Conservación de Murciélagos de República Dominicana
Florida	Florida Bat Conservancy
Guadeloupe	Groupe Chiroptères de Guadeloupe
Jamaica	Windsor Research Centre
Latin America and the Caribbean	Red Latinoamericana y del Caribe para la Conservación de los Murciélagos (RELCOM)
Puerto Rico	Para la Naturaleza
	Programa de Conservación de Murciélagos de Puerto Rico
Saint Eustatius	St. Eustatius National Parks Foundation
Trinidad	Trinibats
Virgin Islands	St. Croix Environmental Association
Worldwide	Bat Conservation International

Several institutions in the West Indies now recruit volunteers to assist in citizen-science projects, thus providing a wonderful opportunity to learn about these mammals and simultaneously contribute data to their long-term preservation (see Table 4.1). A "Program for the Conservation of Bats" (Programa de Conservación de Murciélagos) is a local organization that often relies on volunteers to conduct field studies; within the West Indies, these programs are currently operating in the Dominican Republic,

Website
https://www.facebook.com/Bat-Conservation-Program-Islands-of -Aruba-Bonaire-and-Cura%C3%A7ao-1541961716038107
https://nationaltrust.org.ky
https://relcomlatinoamerica.net/miembros.html
https://relcomlatinoamerica.net/miembros.html
https://floridabats.org
https://www.facebook.com/karubats
https://www.facebook.com/ Windsor-Research-Centre-1393077804272163/
https://relcomlatinoamerica.net
https://paralanaturaleza.org
https://facebook.com/PCMPuertoRico
https://www.statiapark.org/
https://trinibats.com
https://www.stxenvironmental.org/
https://batcon.org

Cuba, and Puerto Rico. Also in Puerto Rico, the environmental group Para la Naturaleza (For Nature) has hosted thousands of citizen scientists helping with bat-related projects (e.g., Rodríguez-Durán and Otero 2011). Other organizations that offer volunteer opportunities include the Groupe Chiroptères de Guadeloupe (Bat Group of Guadeloupe) in the Lesser Antilles and the Windsor Research Centre in Jamaica.

This surge in public interest in bats has coincided with recent scientific discoveries that document many of the more subtle but substantial benefits to humans that these winged creatures provide (Kunz et al. 2011; Ghanem and Voigt 2012; Ramírez-Fráncel et al. 2022). Throughout the archipelagos, nectar-feeding bats are indispensable to our ecosystem in that they pollinate the flowers of dozens of species of plants, including agaves, cacti, vines, and trees, such as the iconic silk cotton (*Ceiba pentandra*) and the majestic royal palm (*Roystonea*) (Silva-Taboada 1979; Gannon et al. 2005). Furthermore, a third of all bats in the West Indies consume fruits, and many of the seeds are later defecated intact and far from the parent tree or shrub. By dispersing these seeds, bats play an incredibly important role in maintaining the health and diversity of tropical forests and in helping revegetate areas ravaged by volcanos or hurricanes or recovering from use as farmland or pasture (Lobova et al. 2009; Abelleira Martínez et al. 2015; Villalobos-Chaves and Rodríguez-Herrera 2021). The maga tree (*Thespesia grandiflora*), the flowers of which are the national symbol of Puerto Rico, and the ubiquitous pepper (*Piper*) shrub are just two common plants that utilize bats for seed dispersal (e.g., Soto-Centeno and Kurta 2006).

Most bats in the West Indies are insectivorous, but they too provide important services to humans, in both urban and rural areas. The high energetic demands of these small flying mammals require that they consume large numbers of insects. For instance, mothers with large dependent young can eat more than the adult's body weight in insects during a single night (Kurta et al. 1989), and bats from just one cave in western Puerto Rico devour over 20 tons of insects per month (Rodríguez-Durán and Lewis 1987). Molecular biologists, like modern crime-scene investigators, now extract insect DNA from bat feces, and based on that DNA, identify the species of insect that was eaten. By using such techniques, scientists know that bats in the Caribbean consume insects that spread the nematode causing canine heartworm and the viruses resulting in West Nile disease and eastern equine encephalitis. Moreover, bats prey on defoliators of silk cotton, as well as agricultural pests that threaten corn, sugarcane, and rice (Rolfe et al. 2014; Aguiar et al. 2021). Furthermore, other studies demonstrate that excluding bats from coffee and cacao plantations, as well as cornfields, actually results in an increased number of insects, which potentially leads to added crop damage, lower yields for the grower, and ultimately higher costs for the consumer (Boyles et al. 2011; Maine and Boyles 2015; Castillo-Figueroa 2020).

Although bats in the West Indies are no longer symbols of gods or spirits, the benefits provided by these winged mammals appear essential to human health and the economy, as well as the functioning of our shared ecosystem. Multiple species in the region, however, are declining or even in danger of extinction, largely through human-mediated factors, such as climate change and deforestation (see Chapter 5). Continued research into the services provided by these animals and concerted actions by governmental wildlife agencies, conservation groups (see Table 4.1), and individual citizens are needed to ensure that West Indian bats and their beneficial actions continue into the future.

Global Change and the Conservation of Caribbean Bat Communities

J. Angel Soto-Centeno and Camilo A. Calderón-Acevedo

B ats are widely distributed, existing on all continents. Their ability to disperse is unparalleled among terrestrial mammals, and many species also live on islands far from the mainland, such as the Galápagos in the Pacific Ocean, the Azores in the Atlantic Ocean, and the Seychelles in the Indian Ocean (McCracken et al. 1997; Uyehara and Wiles 2009; O'Brien 2011). On many islands, bats are the only native terrestrial mammal, and in the West Indies, nearly three-quarters of all bats are endemic, either to single islands or to the region, making this archipelago a hot spot of diversity. The highest levels of endemism in the West Indies occur on the Greater Antilles (twenty-six species) followed by the Lesser Antilles (thirteen species).

Bats are integral to the functioning of their island ecosystems. Some individuals form an important part of the food chain, contributing to the diet of multiple predatory animals, such as snakes, raptorial birds, frogs, and even spiders and centipedes. Some bats are insatiable insect predators, whereas others are efficient seed dispersers and pollinators of native and commercial plants (see Chapter 4). Many of the roles that bats fulfill are ecosystem services that provide a direct or indirect economic benefit to humans. Despite the importance of bats, many species on islands are vulnerable to the effects of global change, including human-driven and natural causes. Our goal is to provide an overview of the conservation status of bats in the West Indies and describe the imminent threats faced by these beneficial flying mammals.

Trends in Vulnerability of Species

Generally, bats in the West Indies have been more resilient to extinction than nonflying terrestrial mammals. Following the last glacial period, the West Indies lost 80% of their nonflying mammalian fauna, whereas only 18% of the bats from this archipelago vanished (Cooke et al. 2017). During the late Quaternary period, an additional 15% of West Indian bats were extirpated from one or more islands, although other populations of those species continue to exist on the mainland or on different islands (Soto-Centeno and Steadman 2015; Soto-Centeno et al. 2017). The specific causes for the loss of bats from various islands are difficult to determine, although thorough examination of fossils shows that many species were extirpated between 4000 and 500 years ago, a time of active human colonization throughout the Caribbean (Soto-Centeno and Steadman 2015; Orihuela et al. 2020b). This information about past events in the West Indies provides a baseline to assess the vulnerability of present-day species to extinction or extirpation and the potential factors that may contribute to their loss.

To examine the types of threats and the conservation status of modern West Indian bats, we used data from the International Union for the Conservation of Nature (IUCN; www.iucnredlist .org). The mission of the IUCN is to preserve nature and categorize the risk of extinction for all species, via the Red List of Threatened Species. The IUCN evaluates risk for each species, based on the interaction of various quantitative factors, including the number of individuals thriving in the wild, the rate of population decline, and the size of the organism's current geographic range, and then assigns each species to a category, reflecting that degree of peril (IUCN 2019; Table 5.1). For example, depending on rate of decline, a species could be labeled Vulnerable, if its total population is less than 10,000 mature animals; Endangered, if less than

Table 5.1 Conservation categories and associated criteria developed by the International Union for the Conservation of Nature (2019)

IUCN category[a]	Criteria
Not Evaluated	Has not been assessed.
Data Deficient	Inadequate information available to assess the species' distribution or population status, and, hence, its risk of extinction.
Least Concern	Has been evaluated and does not qualify for Critically Endangered, Endangered, Vulnerable, or Near Threatened. Widespread and abundant taxa frequently fall into this group.
Near Threatened	Has been evaluated but does not qualify for Critically Endangered, Endangered, or Vulnerable now but is likely to qualify in the near future.
Vulnerable	Meets any of the criteria for Vulnerable; facing a high risk of extinction in the wild.
Endangered	Meets any of the criteria for Endangered; facing a very high risk of extinction in the wild.
Critically Endangered	Meets any of the criteria for Critically Endangered; facing an extremely high risk of extinction in the wild.
Extinct in the wild	Only surviving in captivity or as naturalized populations outside the historical range.
Extinct	No reasonable doubt that the last individual has died.

[a]Placement into one of the three categories that indicate a species is threatened with extinction (i.e., Vulnerable, Endangered, or Critically Endangered) requires a quantitative assessment of the geographic distribution, population size, rate of population decline, and other factors.

2500; and Critically Endangered, if its total population is less than 250. Qualitatively, these three categories reflect a high, very high, and extremely high risk of extinction, respectively. Our approach in using data of the IUCN to explore patterns of threats to bats in the West Indies is like that of Frick et al. (2020); however, we provide a reevaluated and updated species list focused on the West Indies, a region not well emphasized in that earlier global assessment.

Of the sixty-one extant species of bats in the West Indies, the IUCN considers 52% as species of Least Concern, meaning that they have a low risk of extinction. Two categories, Near Threatened and Vulnerable, each include 13% of the species, whereas the Endangered and Critically Endangered categories each have just over 3% of the species. The proportion of West Indian bats in each IUCN category roughly mirrors that of the global assessment of bats— 58% Least Concern, 7% Near Threatened, 8% Vulnerable, 6.3% Endangered, and 1.7% Critically Endangered (Frick et al. 2020). Only 3% of West Indian bats are Data Deficient, which seems remarkably low compared with the global assessment (15%), although an additional 12% of bats from the islands fall into the category of Not Evaluated. All species in these latter two groups are endemic to single islands; many of them, such as the Saint Vincent big-eared bat (*Micronycteris buriri*), have been described only recently, whereas others, such as Koopman's pallid bat (*Antrozous koopmani*), are some of the rarest mammals in the Neotropics, with only a handful of sightings over the last century.

Areas of the West Indies containing a high density of species that are threatened with extinction (i.e., classified as Critically Endangered, Endangered, or Vulnerable) are distributed across the archipelago (Fig. 5.1). Although data on species richness can be biased toward islands that are well studied, important patterns are still evident after examining these concentrations of threatened species. The densest hot spots include Jamaica, southeast Haiti, Dominica, and, to a lesser extent, Barbados. These hot spots are fueled primarily by species that are rare and live only on one island (single-island endemics) or on multiple islands of the West Indies (regional endemics) but nowhere else in the world.

Jamaica stands out as having some of the most imperiled and

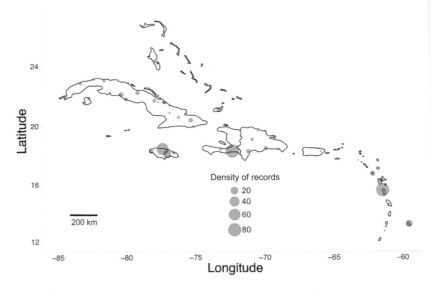

Figure 5.1. Map indicating density of threatened species of bats in the West Indies, based on records of bats currently classified as Critically Endangered, Endangered, or Vulnerable, by the International Union for the Conservation of Nature (IUCN.org, accessed May 2021). Prepared by J. Angel Soto-Centeno and Camilo A. Calderón-Acevedo.

least-studied species that are considered threatened. Four bats on the island fall within the three highest threat categories—the lesser funnel-eared bat (*Chilonatalus micropus*: Vulnerable), Jamaican funnel-eared bat (*Natalus jamaicensis*: Critically Endangered), Jamaican flower bat (*Phyllonycteris aphylla*: Critically Endangered), and Jamaican red bat (*Lasiurus degelidus*: Vulnerable). All threatened species in Jamaica are uncommon, single-island endemics, except the lesser funnel-eared bat, which also dwells on two other islands that are tiny and isolated. The Jamaican funnel-eared bat lives in only a single cavern, and an estimate from 1970 suggests that the population contains fewer than one hundred individuals (Goodwin 1970; Tejedor et al. 2005b). Similarly, the Jamaican flower bat inhabits only two caves; although one report estimates only 250 bats remaining (Koenig and Dávalos 2015), surveys in

2019 suggest that the population is somewhat larger, certainly over five hundred animals (J. A. Soto-Centeno et al., unpubl. observ.).

The West Indies is an important center for the conservation of global biodiversity of all kinds, because of high rates of endemism on the islands. Many endemic species in this archipelago inhabit designated Key Biodiversity Areas, or KBAs, which are defined as sites that make a significant contribution to the worldwide persistence of biodiversity (IUCN 2016). Current KBAs contain over 50% of the region's threatened species, and Jamaica, the second-smallest of the Greater Antilles, has thirty-eight KBAs, encompassing roughly 445,000 ha and sheltering over ninety species, ranging from plants to vertebrates, that are considered Vulnerable, Endangered, or Critically Endangered; nevertheless, about 45% of the Jamaican KBAs still need protection (Anadón-Irizarry et al. 2012). The most imperiled areas within the country also serve as home to threatened bats. These include Jamaican red bats and Jamaican flower bats in St. Ann and Trelawny Parishes, along the northern coast; Jamaican red bats and lesser funnel-eared bats in Westmoreland Parish, in the western part of the island; and Jamaican funnel-eared bats and Jamaican flower bats in St. Catherine Parish, in central Jamaica. All of these critical regions lie within KBAs that remain unprotected.

Three other countries contain high densities of threatened bats. Located in the southeastern part of Haiti, the Massif de la Selle is one of the largest protected KBAs in Hispaniola and a hot spot of special importance for bats. This protected area contains diverse habitats, including rain, pine, and dry forests, and is inhabited by two Vulnerable species, the lesser funnel-eared bat and the minor red bat (*Lasiurus minor*). Wooded patches within Massif de la Selle interconnect with others in the southwestern Dominican Republic, near Lake Enriquillo and within the national parks of Jaragua and Sierra de Bahoruco. Together these international sites form a large set of protected areas important for maintaining the bat diversity of Hispaniola. In the Lesser Antilles, the small islands of Dominica and Barbados are also among the top hot spots of threatened species of bats. This density, however, is driven primarily by two regional endemics, the Dominican myotis (*Myotis dominicensis*) and the Barbadian myotis (*Myotis nyctor*), that have been reported from many localities on each island. Highlighting these species is important, though, given that only 50% of 11,000 ha and 0% of 6000 ha of KBAs receive protection on Dominica and Barbados, respectively (Anadón-Irizarry et al. 2012).

Some potential hot spots of threatened species are missing from Figure 5.1, notably, the Bahamian Archipelago and some Lesser Antillean islands. These absences should be viewed with caution, because our quantification of hot spots is based on data currently held by the IUCN and focuses only on animals presently classified as Critically Endangered, Endangered, or Vulnerable. For example, although the minor red bat is considered Vulnerable, the IUCN lists Bahamian populations as occurring only on the island of Great Inagua. However, this species is often misidentified as the eastern red bat (*L. borealis*), a species of Least Concern, and is actually known from multiple islands (Speer et al. 2015; J. A. Soto-Centeno, unpubl. observ.); this taxonomic discrepancy in the IUCN data results in the exclusion of most Bahamian islands as a hot spot in our analysis. Also, some West Indian bats lack thorough assessments. The Bahamian funnel-eared bat (*Chilonatalus tumidifrons*) is a regional endemic considered Near Threatened. This species occupies only nine caves, and it is not abundant, with colonies likely comprising fewer than five hundred individuals (Speer et al. 2015; J. A. Soto-Centeno, unpubl. observ.). Still, recent field-based assessments of these populations, which are necessary for proper classification, are lacking, especially since 2019, when Hurricane Dorian caused US $3.4 billion in damages to the northern Bahamas. Additionally, Lesser Antillean species such as the Saint Vincent big-eared bat (*Micronycteris buriri*), which was first described in 2011, are classified as Data Deficient. Therefore, such species could not be included in this analysis.

Primary Threats

Examination of major threats to West Indian bats provides additional insight into the conservation needs of these animals, beyond just geographic patterns of vulnerability to extinction (see Fig. 5.1). The IUCN (2012) groups specific threats into nested categories,

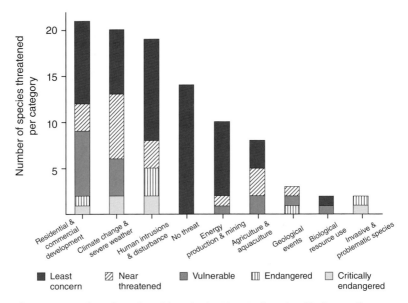

Figure 5.2. Number of species of bats in the West Indies, classified according to their present risk of extinction, that are facing various threats, as determined by the International Union for the Conservation of Nature (IUCN.org, accessed May 2021). Threats are not independent, and each species can be included in more than one category. Species categorized as Data Deficient are omitted. Prepared by J. Angel Soto-Centeno and Camilo A. Calderón-Acevedo.

which include both anthropogenic (e.g., agricultural or commercial development) and natural (e.g., geological events or severe weather) factors. Each species can fit into more than one category, because threats are not independent from one another. For example, a bat negatively affected by agricultural development can also be threatened by anthropogenic climate change. By examining the number of species within each category of threat (Fig. 5.2), we provide an overview of where conservationists could direct their efforts at a more local level.

Islands of the West Indies have often been described as natural laboratories, due to their unique geologic origins, geographic characteristics, sizes, levels of isolation, topographical reliefs, and climatic and colonization histories. Not surprisingly, bats that are endemic, either to single or multiple islands, and insular populations of widely distributed continental species face conditions and threats that mainland bats do not experience, such as regular disturbance from hurricanes (see Chapter 3). The threats to West Indian bats considered Critically Endangered, Endangered, or Vulnerable all relate to human-driven factors, such as residential and commercial development, climate change and severe weather, human intrusion and disturbance, and invasive and other problematic species (see Fig. 5.2). To explore how specific threats affect West Indian bats, we highlight the widely distributed Jamaican fruit-eating bat (*Artibeus jamaicensis*) and the red fig-eating bat (*Stenoderma rufum*), which, in contrast, lives only on Puerto Rico and some of the nearby Virgin Islands.

Bats, like other organisms in the West Indies, are affected not only by human-mediated factors, such as habitat fragmentation and conversion of natural habitats to developed land, but also by recurring natural disturbances, such as hurricanes and, to a lesser extent, volcanoes (Jones et al. 2001; Fleming and Murray 2009; Gannon and Willig 2009; Pedersen et al. 2009, 2012; Calderón-Acevedo et al. 2021). The red fig-eating bat is an uncommon and secretive foliage-roosting frugivore, although knowledge of its natural history has greatly improved in the last 30 years, as the number of capture localities has increased from two to forty and the number of individuals examined by biologists has surpassed 250. This animal undergoes drastic population declines following hurricanes, which destroy fruiting trees and shrubs and obliterate roosting sites, and local subpopulations of the bat may take 5 or more years to recover from powerful storms (Gannon and Willig 1994).

Despite multiple studies documenting a reduction in species richness and the size of bat populations following hurricanes, some biologists still consider these species safe from any long-term harm caused by such natural disturbances (e.g., Mancina et al. 2007b). The main argument for this position is that these well-established communities of bats evolved alongside frequent hurricanes and, thus, the animals should be resilient to weather-related extinction or extirpation. Although this argument is valid, it ignores another major threat to communities of West Indian bats—humans. Pressure exerted by human intrusion or disturbance and human-mediated habitat fragmentation and destruction has continually increased during the last 500 years (e.g., Orihuela et al. 2020b). Major urban centers now limit population connectivity for species such as the red fig-eating bat, and although large state forests exist throughout Puerto Rico, the remaining wooded patches that connect government-owned tracts have little protection and are vulnerable to development (Calderón-Acevedo et al. 2021). Human-driven deforestation and habitat fragmentation, which prevents storm-distressed animals from reaching less affected sites, combined with stronger and more frequent hurricanes brought on by climate change, increase the risk of local extirpation. Maintaining forest integrity by emphasizing the considerable conservation value of interconnecting forest patches is crucial for the preservation of the red fig-eating bat and likely other species on the islands, especially foliage-roosting species, such as the Cuban fig-eating bat (*Phyllops falcatus*) and the Lesser Antillean tree bat (*Ardops nichollsi*) (Calderón-Acevedo et al. 2021; Gonçalves et al. 2021).

Species with very narrow distributions are not the only animals susceptible to local pressures in insular systems. Bats with wide distributions that include the mainland and oceanic islands are often considered abundant and not in peril, but this judgement is typically based on generalized accounts of continental populations and ignores the smaller groups dwelling on islands. For example, the distribution of the Jamaican fruit-eating bat extends from southern Mexico to northern South America and into the West Indies, but the conservation needs of Caribbean populations receive little attention, because the species is globally abundant. Nonetheless, populations of the Jamaican fruit-eating bat on Grand Cayman were severely depressed following Hurricane Ivan in 2004 (Fleming and Murray 2009), and similarly, surveys following Hurricanes Irma and Maria on Culebra suggest the island-wide extirpation of the species (Rodríguez-Durán et al. 2020). This disregard of populations from islands may bias the classification of insular bats toward the lowest categories of extinction risk, even though these isolated demes (subpopulations) can contain important genetic or morphological variability and can be on independent evolutionary trajectories. Examining the overall responses of different species to various phenomena requires intensive work and continuous monitoring to provide a better understanding of their effects on isolated populations. Conservation initiatives leading to the reevaluation of local bat faunas may play a strong role in the effective management of widespread as well as endemic bats in the West Indies.

The threats affecting bats classified as Least Concern and Near Threatened are similar to those affecting species in the other categories (see Fig. 5.2). These factors include invasive and other problematic species, climate change and severe weather, residential and commercial development, geological events, agriculture, and energy production and mining. Within the latter category, a particularly recent threat for bats is the development of modern wind-energy facilities or wind farms, which often include dozens of turbines, with masts that are 150 m tall and rotors that sweep through a circle that is 100 to 200 m in diameter (Wilburn 2011). Collisions at wind farms are a major cause of death for bats in temperate North America and Europe, with hundreds of thousands of animals perishing annually (Hein and Schirmacher 2016).

Bat collisions with wind turbines are not chance events, and some species are killed disproportionately (Arnett et al. 2016; Hein and Schirmacher 2016). For example, even though 45% of the forty-six species of bats that occur in the United States and Canada have died at wind farms, about 78% of the fatalities involve only three species that perform continental-scale migrations—eastern red bats, northern hoary bats (*Lasiurus cinereus*), and silver-haired bats (*Lasionycteris noctivagans*). Similarly, in one region of Mexico, members of thirty-two of forty-two resident species were killed, although 52% of the deaths involved just two species—Davy's naked-backed bat (*Pteronotus davyi*) and Peter's ghost-faced bat (*Mormoops megalophylla*), both of which forage high in open space and congregate in large numbers in nearby caves.

The IUCN lists energy production as a potential threat for only nine (15%) West Indian species, but that assessment seems outdated. Given that wind energy is a fast-growing industry, detailed surveys at wind farms are needed for a better understanding of the diversity of species affected and the magnitude of the problem (Agudelo et al. 2021). Although modern wind-energy facilities span the Greater and Lesser Antilles, with developments in the British Virgin Islands, Dominican Republic, Guadeloupe, Jamaica, Martinique, and Nevis, biologists have examined the interaction of bats and wind power only on Puerto Rico, documenting the death of eleven of the thirteen species living on the island (Rodríguez-Durán and Feliciano-Robles 2015). Unlike North America, where most fatalities involve species in the genus *Lasiurus*, only a single minor red bat has been killed on Puerto Rico; although this is seemingly a low number, the event is significant, because this mammal has been captured only six other times on that island. An enigma worth further investigation is that many of the species involved in these collisions are fruit- and nectar-feeding animals that do not forage in open areas or fly at high altitudes, or so biologists believed. Why are these animals being killed so high above the ground and so far from the trees and shrubs at which they feed?

Population Trends

The high level of endemism and large proportion of threatened bats in the West Indies make this a priority region for conservation, yet little is known concerning population trends on each of the numerous islands. Currently, only one species in the West Indies, the big brown bat (*Eptesicus fuscus*), has an increasing population according to the IUCN (Fig. 5.3). This mammal, which globally is classified as Least Concern, occurs from Canada to northern South America and throughout the Bahama Islands and most of the Greater Antilles. Records of this species vary, though, ranging from dozens of localities on large islands, such as Cuba, to as little as one or two locations for smaller isles in the Bahamas and Cayman Islands. Although the big brown bat forms colonies that may exceed one hundred individuals on large islands (e.g., Silva-Taboada 1979; Gannon et al. 2005), only a handful of solitary individuals roost in some Bahamian caves (Speer et al. 2015), which attests to the low abundance of this species on certain islands. Furthermore, a survey in 2008 on Vieques, east of Puerto Rico, failed to find this

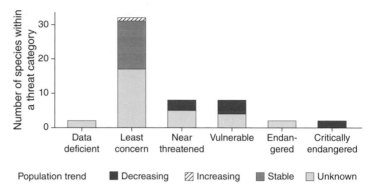

Figure 5.3. Population trends of fifty-four species of bats in the West Indies, grouped by their conservation status, as determined by the International Union for the Conservation of Nature (IUCN.org, accessed May 2021). Seven species classified as Not Evaluated are excluded. Prepared by J. Angel Soto-Centeno and Camilo A. Calderón-Acevedo.

species, despite its previous occurrence there (Alexander and Geluso 2013). Such interisland variation in abundance highlights the importance of developing localized assessments instead of relying on broad evaluations produced from information collected on the continent or even from neighboring islands.

West Indian bats with decreasing populations occur in the Near Threatened (5% of sixty-one species), Vulnerable (6%), and Critically Endangered (3%) categories (Fig. 5.3). All of these species are threatened by factors related to human activities and habitat loss, such as residential and commercial development, intrusions and disturbance, and agriculture. Only fourteen bats (23%) have stable populations, and all are classified as Least Concern; seven of these species are endemic to the broader West Indies, although none is restricted to a single island. However, whether populations are increasing or decreasing is unknown for most bats (60%) in the archipelago, which is similar to the global pattern (57%; Frick et al. 2020). In the West Indies, scientists have not determined the demographic status of over half the species classified as Near Threatened and Vulnerable, as well as the two species considered Endangered—the Guadeloupean big-eyed bat (*Chiroderma improvisum*) and the

Guadeloupean big brown bat (*Eptesicus guadeloupensis*). Further-more, the trend is unknown for the seven species not yet evaluated by the IUCN and the two species categorized as Data Deficient, which are the Saint Vincent big-eared bat and the Cuban lesser funnel-eared bat (*Chilonatalus macer*).

It is alarming that forty-six of sixty-one species of bats (75.4%) in the West Indies have populations that are decreasing or are of unknown status, particularly because most are endemic taxa facing the fastest rate of global change in human history. The distinctive levels of isolation and small size of many islands in the West In-dies, along with the high rate of human-driven habitat transforma-tion and increasing intensity and frequency of storms in the region, emphasize the need for swift conservation action at the local level. Moreover, many of the bats with undetermined population trends represent recently described species; these new discoveries further highlight the importance of thorough local assessments, because newly delimited species are often confined to single islands.

The Future of Bat Conservation in the West Indies

Developing evidence-based knowledge is critical for conservation of bats in the West Indies. This knowledge, as well as the future management strategy that stems from it, cannot be derived from "parachute-style" research, from which scientists base species as-sessments solely on a handful of short-term trips to the field. To be successful, conservation and research programs for bats on the islands generally require a jump start from locally driven ef-forts and local investment in biodiversity. The paucity of surveys led by resident teams on most islands and the infrequent visits by international scientists with the goal of developing local capacity, combined with the rareness of some species, may lead to mischar-acterization of population trends that can be corrected only by long-term monitoring.

Notably, some efforts toward overcoming these challenges are underway in different island nations, with groups focused on nar-rowing the existing gaps in data. Current efforts that promote the conservation of bats in the West Indies consist of multiple local networks that are driven primarily by volunteers. Established programs of varying capacity and focused on conservation, research, and the dissemination of information to the public have been developed in the Cayman Islands, Cuba, Dominican Republic, Guadeloupe, and Puerto Rico (see Chapter 4). The success of these local groups is illustrated by the establishment of twenty-nine "Areas" or "Sites of Importance for Bat Conservation," in Cuba (8), the Dominican Republic (5), and Puerto Rico (16) (Red Latinoamericana y del Caribe para la Conservación de los Murciélagos 2021). Efforts such as these instill in residents of each country a love for and national pride in local biodiversity, which is essential for the long-term success of any conservation program.

A Key for Identification of
Bats in the West Indies

When confronted with a new animal, field biologists often use a "dichotomous key" to help identify the organism. A key presents the reader with two contrasting sets of characters at a time. By choosing the traits that are most applicable to an unknown specimen, a person eventually proceeds, in a steplike process, to the correct identification of the animal.

Two keys are presented—one that distinguishes the six families of bats that live in the region, and a second that differentiates the species within each of those families. By using these tools, you can identify most bats from the West Indies using nothing more complex than a ruler and some simple observations. Knowing where the specimen originated is also important, because many species differ from those on neighboring islands in subtle ways, often involving molecular traits or skeletal characters that are not readily visible. Be sure to examine all traits that are listed, not just the first, and use the weight of the evidence to guide your choice. Once you make an identification, verify your conclusion using the photo and details that appear in the species account. Technical terms are described in the Introduction or listed in the glossary.

Keys to Adult Bats of the West Indies

KEY TO FAMILIES

1a. Small-to-prominent nose leaf present (Fig. 0.5); interfemoral membrane usually forming thin flap running along inside of the lower legs, typically passing from left to right side only near body and having overall appearance of a U- or V-shape, with the base of the letter pointing forward (see Fig. 0.2a–b); if midpoint of trailing edge of membrane extends as far as the ankles, then nose leaf is well formed.......**New World Leaf-nosed Bats: Family Phyllostomidae, I.**

1b. Not possessing above combination of characters................ 2.

2a (1b). Huge; forearm length > 75 mm; hindfoot length > 24 mm; upper lip split and ears long, pointed, giving the appearance of a hare ....**Bulldog Bats: Family Noctilionidae, II.**

2b. Smaller; forearm length < 70 mm; hindfoot length < 24 mm; not hare-like .... 3.

3a (2b). Posterior margin of interfemoral membrane perpendicular to long axis of body; tail shorter than membrane, protrudes from dorsal surface when animal at rest (see Fig. 0.2c)......................
........**Ghost-faced and Mustached Bats: Family Mormoopidae, III.**

3b. Posterior margin of interfemoral membrane forming a V; base of V points backward; tail as long or longer than membrane (see Fig. 0.2d–f) .... 4.

4a (3b). 50% or more of tail protrudes beyond posterior margin of interfemoral membrane (see Fig. 0.2d).. **Molossidae: Free-tailed Bats, IV.**

4b. Tail protrudes at most a few millimeters beyond posterior margin of interfemoral membrane (see Fig. 0.2e–f) .... 5.

5a (4b). Tail as long as head and body combined; ears appear funnel shaped; tragus twisted, sometimes helical (see Figs. 0.6d); obvious hairs on broad dorsoventrally flattened snout, giving appearance of a mustache; hind legs usually disproportionately thin and long; mostly hairless protuberance at junction of snout and forehead in adult males only ....**Funnel-eared Bats: Family Natalidae, V.**

5b. Not with above combination of characters.......................
................Vesper Bats: Family Vespertilionidae, VI.

KEY TO SPECIES WITHIN FAMILIES

I. New World Leaf-Nosed Bats: Family Phyllostomidae.

1a. Occurring in the Lesser Antilles, between Anguilla and Saint Vincent
.... 2.

1b. Occurring in the Greater Antilles, Bahamas, Cayman Islands, Turks
and Caicos, or Virgin Islands ....12.

2a (1a). White line down center of back, often faint...................
...... Guadeloupean big-eyed bat, *Chiroderma improvisum*, p. 248.

2b. No white line on back... 3.

3a (2b). Distinct, white patch of fur at shoulder ....
.... Lesser Antillean tree bat, *Ardops nichollsi*, p. 216.

3b. No white patch at shoulder................................... 4.

4a (3b). Forearm length greater than 50 mm .... 5.

4b. Forearm length less than 50 mm.............................. 8.

5a (4a). Distinctive, spear-shaped nose leaf present (see Fig. 0.5b–d);
snout not piglike... 6.

5b. Lancet of nose leaf a simple low ridge of skin; nostrils open forward
at end of piglike snout (see Fig. 0.5a)............................
........Antillean fruit-eating bat, *Brachyphylla cavernarum*, p. 238.

6a (5a). Found on Barbados and Martinique and farther north.........
.... Jamaican fruit-eating bat, *Artibeus jamaicensis*, p. 225.
(Hybrids between the Jamaican and Schwartz's
fruit-eating bats exist on Saint Lucia).

6b. Found on Saint Vincent or farther south ....7.

7a (6b). Facial stripes distinct; ventral hairs not tipped with white;
yellow tinge to ears and tragus; interfemoral membrane furred on
dorsal surface; slight hairy fringe on trailing edge of interfemoral
membrane near midline; forearm length ≥ 66 mm ....
.... Great fruit-eating bat, *Artibeus lituratus*, p. 230.

7b. Facial stripes indistinct; ventral hairs tipped with white; no yellow tinge to ears and tragus; interfemoral membrane not furred; lacking hairy fringe on edge of interfemoral membrane; forearm length ≤ 69 mm .... Schwartz's fruit-eating bat, *Artibeus schwartzi*, p. 234.
(Hybrids between the Jamaican and Schwartz's fruit-eating bats exist on Saint Lucia).

8a (4b). Interfemoral membrane greatly reduced, present only as thin, inconspicuous strip of hairy skin paralleling legs; yellowish tint to fur near shoulders, especially in adult males....................... 9.

8b. Interfemoral membrane obvious, broader, and not hairy; no yellowish tint to fur near shoulders ....10.

9a (8a). Found on Dominica, Guadeloupe, Martinique, and Montserrat. . ...Angel's yellow-shouldered bat, *Sturnira angeli*, p. 302.

9b. Found on Grenada, Saint Lucia, and Saint Vincent .... Paulson's yellow-shouldered bat, *Sturnira paulsoni*, p. 306.

10a (8b). Ears disproportionately large (≥ 50% as long as forearm), connected across top of head by noticeable ridge .... Saint Vincent big-eared bat, *Micronycteris buriri*, p. 273.

10b. Ears smaller (< 50% as long as forearm), not connected across the top of head ....11.

11a (10b). Stub of tail protrudes from the hind margin of the interfemoral membrane.. .... Insular single-leaf bat, *Monophyllus plethodon*, p. 276.

11b. Short tail totally encased in interfemoral membrane ....Miller's long-tongued bat, *Glossophaga longirostris*, p. 259.

12a (1b). Distinct white patch of fur at shoulder ....13.

12b. No white patch at shoulder...................................15.

13a (12a). Index (second) finger highly curved; membrane between second and third fingers translucent; found on Cuba and Cayman Islands..............Cuban fig-eating bat, *Phyllops falcatus*, p. 293.

13b. Not with above combination of characters.....................14.

14a (13b). Forearm length < 45 mm; yellowish tint to fur and ears; restricted to Jamaica . . . . .Jamaican fig-eating bat, *Ariteus flavescens*, p. 221.

14b. Forearm length > 45 mm; no yellowish tint to fur and ears; found on Puerto Rico and some Virgin Islands . . . .
. . . . .Red fig-eating bat, *Stenoderma rufum*, p. 297.

15a (12b). Ears disproportionately large (≥ 50% as long as forearm), joined across top of forehead; tragus taller than nose leaf; interfemoral membrane large, extending from ankle to ankle. . . . .
. . . . .Waterhouse's leaf-nosed bat, *Macrotus waterhousii*, p. 268.

15b. Not with above combination of characters. . . . . .16.

16a (15b). Forearm length ≥ 52 mm. . . . . .17.

16b. Forearm length ≤ 51 mm . . . . .19.

17a (16a). Large, spear-shaped nose leaf (see Fig. 0.5d); snout not piglike . . . . Jamaican fruit-eating bat, *Artibeus jamaicensis*, p. 225.

17b. Nose leaf a simple low ridge or disc of skin; snout piglike (see Fig. 0.5a) . . . . .18.

18a (17b). Forearm length ≤ 61 mm; found on Cayman Islands, Cuba, Hispaniola, and Turks and Caicos . . . .
. . . . .Cuban fruit-eating bat, *Brachyphylla nana*, p. 243.

18b. Forearm length ≥ 60 mm; found from Puerto Rico and Virgin Islands southward to Saint Vincent and Barbados . . . .
. . . . .Antillean fruit-eating bat, *Brachyphylla cavernarum*, p. 238.

19a (16b). Nose leaf prominent, spear-shaped on a narrow base (see Fig. 0.5c); forearm length ≤ 43 mm . . . . 20.

19b. Nose leaf a low triangle (see Fig. 0.5b) or simple ridge or disc of skin (see Fig. 0.5a); forearm length ≥ 42 mm. . . . . .21.

20a (19a). Stub of tail protrudes from the hind margin of the interfemoral membrane. . . . .
. . Greater Antillean long-tongued bat, *Monophyllus redmani*, p. 280.

20b.Short tail totally encased in interfemoral membrane . . . .
. . . . .Pallas's long-tongued bat, *Glossophaga soricina*, p. 264.

21a (19b). Interfemoral membrane extends to ankle; small calcar present
.... 22.

21b. Interfemoral membrane ends along tibia, above ankle; calcar absent
.... 23.

22a (21a). Found on Hispaniola and Puerto Rico ....
.... Brown flower bat, *Erophylla bombifrons*, p. 251.

22b. Found on Bahamas, Cayman Islands, Cuba, Jamaica, and Turks and
Caicos .... Buffy flower bat, *Erophylla sezekorni*, p. 255.

23a (21b). Found on Jamaica...
.... Jamaican flower bat, *Phyllonycteris aphylla*, p. 285.

23b. Found on Cuba and Hispaniola ....
.... Cuban flower bat, *Phyllonycteris poeyi,* p. 289.

II. Bulldog Bats: Family Noctilionidae...............................
.... Greater bulldog bat, *Noctilio leporinus*, p. 207.

III. Ghost-Faced and Mustached Bats: Family Mormoopidae.

1a. Occurring in the Lesser Antilles, between Anguilla and Saint Vincent
.... 2.

1b. Occurring elsewhere... 3.

2a (1a). Hairless wings meet at midline of back, giving naked appearance .... Davy's naked-backed bat, *Pteronotus davyi*, p. 144.

2b. Wings meet at side of body.......................................
.... Allen's mustached bat, *Pteronotus fuscus*, p. 148.

3a (1b). Flap under lower lip split, forming paired leaflike structures;
ears broad, rounded; eye appears inside funnel leading to ear opening; bizarre-looking ....
.... Antillean ghost-faced bat, *Mormoops blainvillei*, p. 139.

3b. Flap below lip not split, platelike; ears sharply or bluntly pointed...
.... 4.

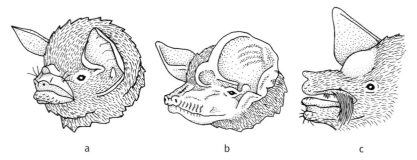

a   b   c

Figure 6.1. Useful facial characteristics. (a) Rostral protuberance present in many mustached bats (*Pteronotus*). (b) Grooved upper lip of many free-tailed bats (Molossidae). (c) "Four-lipped" profile of natalid bats in the genus *Chilonatalus*, as well as approximate position of the natalid gland in adult males of all funnel-eared bats (Natalidae).

4a   (3b). Forearm length > 46 mm; prominent fleshy bump (rostral tubercle) behind nostrils (Fig. 6.1a)................................... 5.

4b.   Forearm length < 46 mm; prominent fleshy bump (rostral tubercle) behind nostrils lacking ....7.

5a   (4a). Restricted to Puerto Rico...................................
..... Puerto Rican mustached bat, *Pteronotus portoricensis*, p. 160.

5b.   Found elsewhere... 6.

6a   (5b). Restricted to Hispaniola....................................
.... Hispaniolan mustached bat, *Pteronotus pusillus*, p. 163.

6b.   Found on Cuba and Jamaica ....
..............Parnell's mustached bat, *Pteronotus parnellii*, p. 155.

7a   (4b). Forearm length ≥ 39 mm; V-shaped notch between nostrils; small rectangular flap above each nostril; 7–10 small ridges on posterior ventral side of interfemoral membrane, perpendicular to trailing edge (Fig. 6.2); found on Cuba and Jamaica........................
...........Macleay's mustached bat, *Pteronotus macleayii*, p. 152.

7b.   Forearm length ≤ 39 mm; lacking characters mentioned above; widespread in Greater Antilles.......................................
..............Sooty mustached bat, *Pteronotus quadridens*, p. 167.

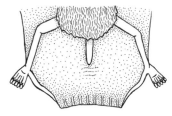

Figure 6.2. Ventral surface of the tail membrane in Macleay's mustached bat (*Pteronotus macleayii*), showing the ridges near the trailing edge.

IV. Free-Tailed Bats: Family Molossidae.

1a. Upper lip with deep vertical wrinkles or grooves (see Fig. 6.1b).... 2.

1b. Upper lip smooth, without deep vertical wrinkles or grooves ..... 4.

2a   (1a). Forearm length > 50 mm; restricted to Cuba .... Big free-tailed bat, *Nyctinomops macrotis*, p. 129.

2b. Forearm length < 50 mm; distribution varies .... 3.

3a   (2b). Base of ears continuous across top of head; restricted to Cuba .... Broad-tailed bat, *Nyctinomops laticaudatus*, p. 125.

3b. Ears separated by 1–2 mm on top of head; found throughout West Indies .... Brazilian free-tailed bat, *Tadarida brasiliensis*, p. 132.

4a   (1b). Forearm length > 50 mm; height of ear > 16 mm .... 5.

4b. Forearm length < 50 mm; height of ear < 16 mm .... 6.

5a   (4a). Dorsal hairs white or cream at base; tragus squared; restricted to Jamaica .... Fierce bonneted bat, *Eumops ferox*, p. 106.

5b. Dorsal hairs dark at base; tragus pointed; found on Cuba and Jamaica .... Black bonneted bat, *Eumops auripendulous*, p. 102.

6a   (4b). Forearm length < 34 mm; restricted to Cuba .... Little goblin bat, *Mormopterus minutus*, p. 121.

6b. Forearm > 34 mm; distribution varies............................7.

7a   (6b). Restricted to Hispaniola..................................... .... Hispaniolan mastiff bat, *Molossus verrilli*, p. 118.

7b. Found elsewhere... 8.

8a (7b). Found on Cayman Islands, Cuba, and Jamaica.................
.................Pug-nosed mastiff bat, *Molossus milleri*, p. 109.

8b. Broadly distributed on Puerto Rico, Virgin Islands, and throughout Lesser Antilles...... Pallas's mastiff bat, *Molossus molossus*, p. 113.

V. Funnel-Eared Bats: Family Natalidae.

1a. Occurring in Lesser Antilles ....
..... Lesser Antillean funnel-eared bat, *Natalus stramineus*, p. 196.

1b. Occurring in Greater Antilles and Bahamas .... 2.

2a (1b). Fringe of fine hairs on trailing edge of interfemoral membrane; forearm length ≥ 31mm; distribution varies .... 3.

2b. Fringe of fine hairs missing; forearm length ≤ 32 mm; restricted to Cuba..........Gervais's funnel-eared bat, *Nyctiellus lepidus*, p. 201.

3a (2a). Forearm length < 38 mm; bump on midline of snout behind nostrils; in profile, face appears to have four lips (see Fig. 6.1c) ...... 4.

3b. Forearm length > 38 mm; bump on midline of snout behind nostrils lacking; in profile, face normal, appearing to have two lips .... 6.

4a (3a). Restricted to Cuba...
.........Cuban lesser funnel-eared bat, *Chilonatalus macer*, p. 175.

4b. Found elsewhere... 5.

5a (4b). Found on Hispaniola and Jamaica ....
...........................Caribbean lesser funnel-eared bat,
Chilonatalus micropus, p. 178.

5b. Restricted to Bahamas ....
.... Bahamian lesser funnel-eared bat,
Chilonatalus tumidifrons, p. 182.

6a (3b). Restricted to Jamaica ....
.... Jamaican greater funnel-eared bat, *Natalus jamaicensis*, p. 185.

6b. Found elsewhere...7.

7a (6b). Restricted to Hispaniola..................................
.......Hispaniolan greater funnel-eared bat, *Natalus major*, p. 189.

7b. Restricted to Cuba ....
.... Cuban greater funnel-eared bat, *Natalus primus*, p. 192.

VI. Vesper Bats: Family Vespertilionidae.

1a. Occurring only in Lesser Antilles, from Guadeloupe southward ... 2.

1b. Occurring in Cayman Islands, Greater Antilles, Bahamas, or Turks and Caicos .... 5.

2a (1a). Forearm length > 40 mm; two pairs of upper incisors ....
.... Guadeloupean big brown bat, *Eptesicus guadeloupensis*, p. 320.

2b. Forearm length < 40 mm; three pairs of upper incisors .... 3.

3a (2b). Found on Dominica and Guadeloupe.........................
.... Dominican myotis, *Myotis dominicensis*, p. 345.

3b. Found elsewhere... 4.

4a (3b). Restricted to Martinique ....
.... Schwartz's myotis, *Myotis martiniquensis*, p. 349.

4b. Found on Barbados and Grenada ....
........................Barbadian myotis, *Myotis nyctor*, p. 353.

5a (1b). Interfemoral membrane hairless, except possibly near body ...
.... 6.

5b. At least 50% of dorsal surface of interfemoral membrane furred....
.... 9.

6a (5b). Ears disproportionately tall, > 23 mm in height; tragus tall, > 50% of ear height... Cuban pallid bat, *Antrozous koopmani*, p. 313.

6b. Ears not unusually tall, < 23 mm in height; tragus shorter, < 50% of ear height ....7.

7a (6b). Forearm length < 34 mm; only one pair of upper incisors.......
...................Cuban evening bat, *Nycticeius cubanus*, p. 357.

7b. Forearm length > 34 mm; two pairs of upper incisors............. 8.

8a (7b). Restricted to Jamaica ....
.... Jamaican brown bat, *Eptesicus lynni*, p. 324.

8b. Found on Bahamas, Cayman Islands, Cuba, Hispaniola, and Puerto Rico...................... Big brown bat, *Eptesicus fuscus*, p. 316.

9a (5b). Forearm length ≥ 57 mm; fur brownish-yellow; restricted to Cuba.................. Cuban yellow bat, *Lasiurus insularis*, p. 336.

9b. Forearm length ≤ 57 mm; fur color varies; distribution varies .....10.

10a (9b). Skin of face dark brown or black; light-colored ears rimmed with black or dark brown; body hairs mostly tan or brownish gray, possibly dark brownish red, often tipped with white or gray; restricted to Hispaniola........... Northern hoary bat, *Lasiurus cinereus*, p. 328.

10b. Skin of face not dark brown; ears not rimmed with dark color; body hairs usually yellowish-red, orange-red, or bright red; distribution varies..11.

11a (10b). Restricted to Jamaica .... Jamaican red bat, *Lasiurus degelidus*, p. 333.

11b. Found elsewhere...12.

12a (11b). Restricted to Cuba ....Cuban red bat, *Lasiurus pfeifferi*, p. 342.

12b. Found on Bahamas, Hispaniola, Puerto Rico, and Turks and Caicos .... Minor red bat, *Lasiurus minor*, p. 339.

FAMILY AND SPECIES ACCOUNTS

Free-Tailed Bats

Members of the Molossidae occupy tropical and warm temperate regions worldwide, and the family comprises 22 genera and 129 species. The number of species in this group continues to grow, and taxonomists have described or revalidated several since 2000. In the West Indies, there are five genera and nine species of molossids, and three of these, including the little goblin bat (*Mormopterus minutus*), are endemic to the region.

Bats in this family have a long tail, often about one-third of the animal's total length, and much of it extends beyond the posterior margin of the uropatagium (see Fig. 0.2); typically, 50% or more of the tail is free of this membrane in a free-tailed bat. These animals also have large ears that are wider than they are long, and the pinnae often point forward, rather than up. Like other bats in the West Indies, molossids have a tragus, but unlike other groups, most free-tailed bats also have a better-developed antitragus that is broad, rounded, and clearly visible, not far behind the eye (see Fig. 0.6). Their fur is typically dense and velvety, although the greater naked bat (*Cheiromeles torquatus*) of Malaysia is unusual in that it has very sparse body hairs and, from a distance, looks like it is wearing a well-used leather coat. Molossids vary in size from the blunt-eared bat (*Tomopeas ravus*) of Peru at a mere 2 to 4 g to the greater naked bat that weighs almost 200 g.

The structure of the skull in these animals is diverse. However, there are two general groups—species that have a well-developed sagittal crest along the dorsal midline of the skull, as well as a stout mandible with a well-built coronoid process, and those species that have a less robust build. These differences reflect the relative development of a major jaw-closing muscle, the temporalis, which attaches from the coronoid process on the lower jaw to the side of the skull and sagittal crest. The more stout construction, as shown in the drawing of the skull of Pallas's mastiff bat (*Molossus molossus*), is associated with a larger and stronger muscle and a bat that eats beetles and other hard insects, whereas the more delicate structures of other species suggest a greater reliance on softer prey, such as moths.

Long and narrow wings and powerful flight muscles allow some of these animals to commute at speeds up to 160 km/hr, forage at heights of 3000 m, and visit sites 50 km from home each night. They also are unusually agile on the ground or in a roost and capable

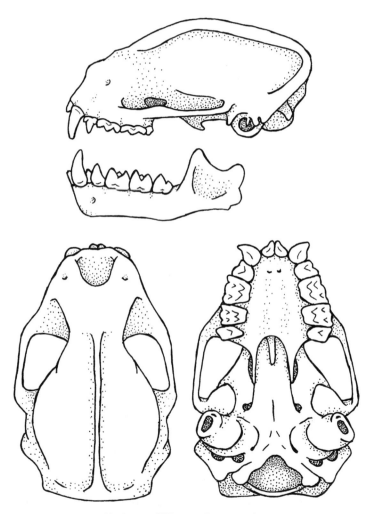

Pallas's mastiff bat, *Molossus molossus*

of quickly scurrying forward, backward, and even sideways. Free-tailed bats generally roost in hollow trees, rock crevices, and caves. Many species, though, have adopted buildings, bridges, and bat houses and have become well adapted to urban environments for both roosting and foraging. Molossids are aerial-feeding insectivorous bats that hunt in open spaces, sometimes close to the ground but often well above the forest canopy or surrounding buildings. Consequently, their echolocation calls are designed for long-range detection of prey. Search-phase pulses are typically prolonged (5–10 milliseconds), involve low frequencies (16–40 kHz), and are quasi-constant, meaning that the starting and ending frequencies vary by just 5 to 10 kHz.

Molossids normally give birth to one hairless pup at a time, and gestation might last from 60 to 120 days. In temperate zones, these mammals give birth once per year, but in tropical areas, many species annually have two and sometimes three reproductive events; the little free-tailed bat (*Chaerephon pumilus*) of equatorial Africa is remarkable in that it may produce five litters in a single year. Many of these bats have a scent gland, the gular gland, located near the junction of the throat and chest, that exudes an odiferous oily substance; when present, the gland is best developed in adult males, and the musky secretions are probably involved in mating or territorial behavior.

Selected References. Rodríguez-Durán et al. 1993; Jung et al. 2014; Taylor et al. 2019.

Eumops auripendulus

Black Bonneted Bat, Murciélago Cabecinegro, Eumops des Palmiers

Name. Frédéric Cuvier first used *mops* as the specific epithet for the Malayan free-tailed bat (*Mops mops*) in 1825. Many taxonomists have since speculated that the expression is a Malay term for "bat" and have incorporated the word into the names of various genera, including *Eumops*. However, a more likely explanation is that Cu-

vier, who was of Germanic origin, noted the resemblance of the Malayan free-tailed bat and a pug, a wrinkled-faced breed of dog; *mops* is the German word for a "pug." The first part of the generic name is the Greek root *eu*, denoting "true" or "real." In 1906, Gerrit S. Miller of the Smithsonian Institution coined the name *Eumops*, but he provided no explanation for using such a combination of terms. *Auripendulus* comes from the Latin *auris*, signifying "ear," and *pendulus*, indicating "suspended" or "hanging," and presumably refers to the forward-projecting ears that hang over the bat's face. Members of the genus *Eumops* are called bonneted bats, because their large ears project above the head in a manner reminiscent of a lady's hat or bonnet from the nineteenth century.

Distribution. In the West Indies, the black bonneted bat lives only on Jamaica, but the species is widely distributed on the mainland, from southern Mexico to northern Argentina and Uruguay.

Measurements. Total length: 80–85 mm; tail length: 43–54 mm; hindfoot length: 12–18 mm; ear height: 19–25 mm; forearm length: 55–68 mm; body mass: 23–35 g.

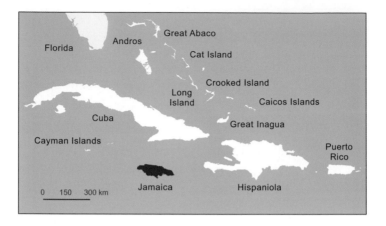

Description. Individual hairs are dark along their length, and over-all, the velvety fur appears black or dark brown. The ears are tall, very broad, and joined at the top of the head; the tragus is short, narrow, and pointed, although it is dwarfed by a prominent anti-tragus that is readily visible behind the eye. The upper lip appears smooth and lacks deep vertical wrinkles. Long hairs coat the feet and toes. An obvious throat (gular) gland occurs in adult males, but this scent organ is poorly developed in juveniles of both sexes and adult females. As in all members of the family Molossidae, half of the long tail continues freely past the back margin of the inter-femoral membrane. The black bonneted bat is larger than any other free-tailed bat on Jamaica, except the similar-sized fierce bonneted bat (*E. ferox*). However, that species is lighter in color overall, has dorsal hairs that are white or cream at the base, and possesses a tragus that is broad and square. The big free-tailed bat (*Nyctino-mops macrotis*), which lives on neighboring Cuba, also has simi-lar dimensions, but that animal sports an obviously grooved and wrinkled upper lip.

Natural History. The black bonneted bat is very uncommon on Ja-maica, with no published records since 2005 and site descriptions available for only two capture localities. Biologists collected two specimens from the densely packed dead fronds of a palm that also sheltered a colony of Antillean palm swifts (*Tachornis phoenicobia*); the tree was located in dry forest on the south coast of the island. Four other black bonneted bats were caught in mistnets set over an artificial pond near the north coast; the muddy pool, which supplied drinking water for local cattle, was bordered by an open grassy area and situated at the edge of a village. Both sites were less than 400 m in elevation.

On the mainland, black bonneted bats occupy an array of habitats, from coastal plains to the eastern slopes of the Andes, including savannah, dry deciduous forest, rain forest, and urban locales. These mammals roost in church belfries, inside house attics, and under corrugated metal roofs. In addition, some of the continental animals spend the day beneath the exfoliating bark of dead trees—a roosting site used by many species of bats in temperate North America.

No specific dietary information exists, but like the fierce bonneted bat and other free-tailed species, the black bonneted bat certainly is an aerial insectivore and likely forages high above buildings and trees. Echolocation calls consist of paired pulses that begin with a frequency-modulated sweep and end with a segment of quasi-constant frequency; the first call usually starts about 32 kHz and ends near 19 kHz, whereas the second pulse extends from 36 to 20 kHz. Knowledge of reproduction is also scarce, but scattered observations throughout the mainland suggest that females give birth to a single offspring twice per year and that juveniles develop a blackish pelage before molting to the brownish hairs typical of an adult. The gular gland of males probably enlarges during the mating season, as it does in other molossids, and the odiferous secretions presumably play a role in courtship or mating behavior.

Status and Conservation. Although seemingly rare on the Antilles, this bat is widespread and somewhat common on the mainland, and consequently, the species has a global ranking of Least Concern.

Selected References. Best et al. 2002; Genoways et al. 2005.

Eumops ferox

Fierce Bonneted Bat, Murciélago Mastín, Eumope Féroce

Name. The generic name is formed by the Greek root *eu*, meaning "true" or "real." *Mops*, in contrast, is a German word for a wrinkly-faced dog called a "pug" and was first used to describe the crinkled face of a Malayan free-tailed bat (*Mops mops*). However, the taxonomist who invented the name *Eumops* provided no explanation for combining *eu* and *mops*. The specific epithet *ferox* comes from the Latin adjective for "fierce" or "wild" and probably refers to the incessant screeches and powerful struggles that occur when one of these bats is held by a human hand.

Distribution. In the Antilles, this species inhabits only Cuba and Jamaica, but it also occurs on the mainland, from central Mexico to central Panama.

Measurements. Total length: 80–85 mm; tail length: 40–54 mm; hindfoot length: 10–15 mm; ear height: 17–23 mm; forearm length: 57–64 mm; body mass: 30–45 g. Males are slightly larger than females.

Description. Overall, the fierce bonneted bat appears dark brown with a hint of gray along the back and slightly lighter on the belly. Dorsal body hairs are short (5–7 mm long) and bicolored, with a brownish tip and pale to white base; these hairs continue onto the wings for about 12 mm. The upper lip lacks obvious wrinkles or furrows, and the ears are large, wider than high, and fused across the forehead. The tragus is broad, square, and smaller than the antitragus. Adult males have a well-developed throat (gular) gland whose secretions have a musky odor; this scent gland is rudimentary in adult females and subadults. A long tail, half of which projects beyond the uropatagium, identifies this animal as a member of the family Molossidae, and on Cuba, the fierce bonneted bat is larger than all other free-tailed bats, except the big free-tailed bat (*Nyctinomops macrotis*). However, that species has a wrinkled upper lip compared with the smooth lip of a fierce bonneted bat. The black bonneted bat (*E. auripendulus*) on Jamaica is also similar in size but looks darker in color overall, has dorsal hairs that are dark at the base, and possesses a short, pointed tragus.

Natural History. The fierce bonneted bat frequents subtropical and tropical, moist and dry forests, up to 500 m in elevation, although it seems most common in urban areas. On Jamaica, this animal roosts in buildings and caves, whereas on Cuba, it occupies buildings and trees, as well as fissures and hollows in wooden utility poles. Arboreal resting sites include inside an abandoned woodpecker hole in a royal palm (*Roystonea regia*), within hollow trunks of dagame (*Callycophyllum candidissimum*) and mastic tree (*Bursera simaruba*), and among the fronds of the bat palm, *jata de los murciélagos* (*Copernicia vespertilionum*). Fierce bonneted bats roost quietly during the day and tend not to flee from their tree when approached, although they may direct high-pitched vocalizations toward the intruder. At buildings, the preferred roosting site is under roofing material or within crevices and cracks in the walls. Each bat typically spends the day in a small group of mixed sex, containing from nine to thirty-two individuals. Other species of bats occasionally share the same roosting structure, whether it is a tree, building, or cave, although the fierce bonneted bat does not mingle with them.

These bats are crepuscular and have two main foraging periods; the first begins 5 to 30 minutes after sunset and the second one ends 15 to 30 minutes before sunrise. The middle of the night, though, is spent resting and socializing within the roost. These mammals fly rapidly, using their speed to catch insects in uncluttered airspace, often more than 50 m above the ground. Echolocation calls are clearly audible to nearby humans as a series of clicks. Search-phase sounds vary from only 23 to 15 kHz; call duration is long (14 milliseconds), and the interval between pulses is also lengthy (294 milliseconds), although highly variable. As with many insectivorous species, a sudden heavy rainfall forces these animals to put an abrupt end to the hunt and return to their roost. The diet consists of beetles, moths, crickets, and probably other insects as well.

In other free-tailed bats, sexually active males rub secretions from their gular gland on the walls of the roost as a way of advertising the male's attractiveness to females or as a warning to competing males, and presumably a similar behavior occurs in the fierce bonneted bat. Data on reproductive timing are fragmentary but suggest that females give birth twice each year. Litter size is one. Despite the high speed during flight, the long narrow wings of this species do not provide great maneuverability, and the fierce bonneted bat is occasionally prey for the American kestrel (*Falco sparverius*) or the American barn owl (*Tyto furcata*). Longevity is unknown.

Status and Conservation. The fierce bonneted bat has a moderately large geographic range and adapts well to human intrusion; consequently, the IUCN classifies this mammal as a species of Least Concern.

Selected References. Silva-Taboada 1979; Genoways et al. 2005; Mc-Donough et al. 2008; Mora and Torres 2008; De la Cruz Mora 2021.

Molossus milleri

Pug-Nosed Mastiff Bat, Murciélago Casero de Jamaica, Molosse de Jamaïque

Name. The name *Molossus* comes from the vague similarity between the face of these bats and Molossian shepherd dogs from ancient Greece. The name *milleri* is in honor of Gerrit S. Miller of the Smithsonian Institution, who studied the evolutionary relationships of New World bats during the early twentieth century.

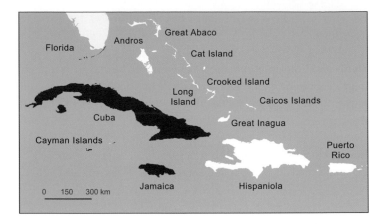

Distribution. The pug-nosed mastiff bat is endemic to Jamaica, Cuba, including Isla de la Juventud, and the Cayman Islands of Grand Cayman and Cayman Brac. The mastiff bats that occur throughout the Florida Keys also probably belong to this species, although they are not native to the American islands; these bats are descendants of animals that were either blown to the Keys from Cuba during a storm or introduced intentionally as a form of mosquito control in 1929.

Measurements. Total length: 97–113 mm; tail length: 32–42 mm; hindfoot length: 7–12 mm; ear height: 11–14 mm; forearm length: 37–40 mm; body mass: 13–19 g. Males are usually larger than females.

Description. Hairs on the back are short, from 3 to 5 mm in length, which gives the animal a velvety appearance. As in other mastiff bats in the West Indies, dorsal hairs are bicolored so that they look medium-to-dark brown at the tips and paler at the base. The belly appears slightly lighter than the head and back. Face, muzzle, ears, and membranes are also medium-to-dark brown. The upper lip is smooth, not wrinkled. The bases of the rounded ears contact each other on the forehead, and the distinct antitragus appears circular. Fifty percent of the tail continues past the uropatagium, and within this bat's range, it can be confused only with other free-tailed species—the Brazilian free-tailed bat (*Tadarida brasiliensis*), big free-tailed bat (*Nyctinomops macrotis*), and bonneted bats (*Eumops*). However, bonneted bats and the big free-tailed bat are two or three times larger than the pug-nosed mastiff, and although the Brazilian free-tailed bat is comparable in size, that animal has uni-colored hairs, a wrinkled and grooved upper lip, and ears separated by a gap of 1 to 2 mm on the forehead.

Natural History. This mammal is one of the most abundant species of bat on the Cayman Islands, Cuba, and Jamaica. The pug-nosed mastiff is plentiful in towns and cities but is also present in lowland wet forests, dry forests, shrublands, mangroves, and areas dominated by agriculture. On Jamaica, though, this bat apparently avoids the dry, southern coast, as well as high-elevation sites that are frequented by the Brazilian free-tailed bat.

The pug nosed mastiff relies primarily on human-made structures for shelter. On Cuba, for example, 81% of the dayroosts are in buildings, whereas 14% occur in cavities within trees, usually palms. This bat sometimes squeezes into rock crevices, although it very rarely shelters in caves. On the Cayman Islands, pug-nosed mastiff bats readily occupy a type of artificial roost that resembles a doghouse stuck on top of a utility pole; after Hurricane Ivan in 2004, 96% of the bat houses and their occupants remained intact, even though the storm destroyed many of the building roosts.

Males occasionally roost alone, but these mammals most often form colonies consisting of 30 to 500 and sometimes thousands of individuals. The composition of the assemblage inside each roost varies tremendously. Some colonies comprise only males, and others, only females; if both sexes are present, they often intermingle, but at some sites, the males and females remain apart. An old report from the 1860s even describes two occupied cavities in the same tree, one containing only females and the other, only males. Sexual segregation by bats often relates to reproductive events, although the reasons for such varied behavior in this particular mastiff bat are unknown. Other species, such as the Cuban lesser funnel-eared bat (*Chilonatalus macer*), occasionally reside in the same roosting structure.

Like other mastiff bats, this rapid-flying species forages for insects at heights of 20 to 40 m, well above the surrounding landscape, and never within dense forests or other cluttered environments. While searching for prey, the pug-nosed mastiff emits echolocation calls that consist of two or three pulses of quasi-constant frequency; the first pulse occurs near 30 kHz, and the following pulses are at successively higher frequencies of about 36 and 39 kHz. Beetles and true bugs dominate the diet, but moths, flying crickets, winged ants, and other groups contribute in smaller amounts. Pug-nosed mastiffs typically drink while in flight, by skimming the calm surface of a pond or stream with an open mouth.

The bats usually leave their roost between 24 minutes before and 18 minutes after sundown. Although a few individuals are active at any time of night, most return to the roost after only 40 minutes, and following a prolonged period of rest, they again leave near sunrise for a second feeding bout that is even shorter than the first. Despite the limited hunting time, these bats consume large amounts of insects; for instance, the stomach of one male from Cuba contained insects that weighed 25% of the animal's body mass. On cool nights, when the air temperature falls below 20 °C, the bats may curtail the first foraging session and totally forego the second bout; many individuals then go into shallow torpor to save energy, allowing their body temperature to fall from the normal resting level of 34 or 35 °C to as low as 17 °C. During prolonged cold spells, all members of a colony may remain torpid for days.

The pug-nosed mastiff gives birth to just one pup at a time. The reproductive cycle is best studied on Cuba, where small embryos, indicating early pregnancy, appear in late March, and large fetuses, suggesting imminent parturition, occur in late May and June. Adults apparently undergo a postpartum estrus, and small embryos again appear in mid-July, with these second litters born in late August or early September. Not all females, though, participate in this second round of mating and birth. Newborn appear naked, weigh about 4 g, and have a forearm that is 18 mm in length. Known predators in the Antilles include the Jamaican boa (*Chilabothrus subflavus*), bat falcon (*Falco rufigularis*), and American kestrel (*Falco sparverius*).

Status and Conservation. Most taxonomists included the pug-nosed mastiff as a subspecies of Pallas's mastiff bat (*M. molossus*), but the two groups were confirmed as distinct species, based on differences in their DNA, in 2018. The IUCN has not yet assessed the status of the newly described pug-nosed mastiff, although this animal appears common throughout its distribution.

Selected References. Silva-Taboada 1979; Genoways et al. 2005; Whitaker and Frank 2012; Loureiro et al. 2018, 2019b; Taylor et al. 2019.

Molossus molossus

Pallas's Mastiff Bat, Murciélago Casero, Molosse Commun

Name. The name *Molossus* comes from the similarity between the face of these bats and mastiffs, which are large-bodied heavy-boned

dogs with massive heads. The ancestors of modern mastiffs were bred as livestock guardians by an ancient Greek tribe called the Molossians. Using a specimen from Martinique, Peter Simon Pallas, a prolific Prussian naturalist, described this species in 1766, when he was just 25 years old; this bat also was the first taxon described for the genus *Molossus* and the family Molossidae.

Distribution. In the Caribbean region, Pallas's mastiff bat has a wider distribution than any other mammal except the Jamaican fruit-eating bat (*Artibeus jamaicensis*), Brazilian free-tailed bat (*Tadarida brasiliensis*), and greater bulldog bat (*Noctilio leporinus*). Pallas's mastiff inhabits Puerto Rico and the Virgin Islands and most major and many minor islands throughout the Lesser Antillean arc, including Anguilla, Antigua, Barbados, Barbuda, Dominica, Grenada, Guadeloupe, La Désirade, Marie-Galante, Martinique, Montserrat, Nevis, Saba, Saint Bartholomew, Saint Eustatius, Saint Kitts, Saint Martin, Saint Lucia, and Saint Vincent.

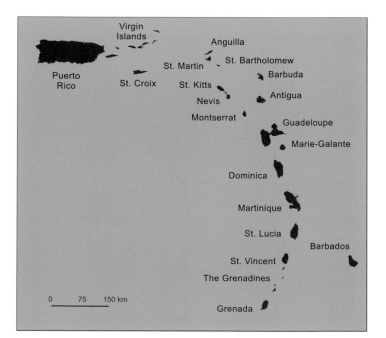

It also lives on most coastal islands, from Cozumel to Trinidad, and on the mainland, this mammal occurs from northern Mexico to northern Argentina.

Measurements. Total length: 84–124 mm; tail length: 30–46 mm; hindfoot length: 7–11 mm; ear height: 11–15 mm; forearm length: 36–43 mm; body mass: 9–17 g. Females are usually smaller than males.

Description. Dorsal hairs are bicolored—brown at the tips and whitish on the inner 25 to 50% of each shaft. Overall, the back appears medium brown and occasionally dark brown in some populations; the underside is lighter than the dorsum. The face, muzzle, ears, and membranes are also medium-to-dark brown. The prominent antitragus is nearly circular in outline. Like all members of the family Molossidae, the distal half of the tail extends beyond the uropatagium, and the only other species with a free tail that lives on the same West Indian islands is the Brazilian free-tailed bat (*Tadarida brasiliensis*). Pallas's mastiff, though, has bicolored hairs, a comparatively smooth upper lip, and external ears that meet at the top of the forehead, whereas the Brazilian free-tailed bat has unicolored hairs, a furrowed upper lip, and pinnae that remain separated by 1 to 2 mm.

Natural History. Pallas's mastiff is ubiquitous in the West Indies. It frequents a myriad of natural and seminatural habitats, including savannahs, tropical rain forests, evergreen forests, humid and dry semideciduous forests, pastures, banana groves, palm plantations, and botanical gardens. This molossid is one of the few species of bats that is common in human-dominated landscapes, ranging from small hamlets to bustling cities. It inhabits lowlands and mountains, up to about 1250 m of elevation.

Pallas's mastiff occasionally rests in hollow trees, under palm leaves, and inside bat houses, but it most often roosts in buildings. Structures adopted by this mammal can be old or new and deserted or actively occupied by humans. Anthropogenic roosts generally have a tile or corrugated-metal roof that allows a bat to crawl underneath, and once inside, the animal typically descends into the walls of the structure, where it is cooler. These bats sometimes tuck themselves into joints and cracks in concrete or stone structures but rarely occupy rock crevices or caves, although this bat does shelter inside cavelike military bunkers on Vieques.

Pallas's mastiff bats typically betray their presence to nearby humans by repeatedly vocalizing during the 15 to 30 minutes before evening departure begins. Individuals leave the roost before darkness falls, often before sunset, but foraging flight usually continues for merely 30 to 60 minutes before the bat returns home. Back in the roost, the animals noisily socialize with their colony mates for a few hours before finally becoming quiet. Just before dawn, though, each bat leaves for another, equally brief bout of feeding.

Like all free-tailed bats, Pallas's mastiff is an aerial insectivore. Long slender wings allow this mammal to fly swiftly, but they also limit maneuverability; hence, foraging flight most often occurs in open habitats, such as over pastures and parks or high above the urban environment. As in all insectivorous bats, this species uses echolocation to detect and capture its prey. When flying in the open, Pallas's mastiff emits calls consisting of two or three pulses of quasi-constant frequency in quick succession. The first pulse starts at about 36 kHz, whereas the second and third begin a little higher, at approximately 39 and 43 kHz, respectively. A Pallas's mastiff rarely forages alone and usually stays within hearing distance of other bats; consequently, one individual can eavesdrop on another and discover when and where other bats are successfully feeding. Beetles seem to be the primary food, at least on Puerto Rico, but flies, moths, bugs, termites, wasps, dragonflies, and crickets also are common fare.

Social organization varies considerably during the year and throughout the range of these bats. In some regions, groups of mixed sex containing up to four hundred animals occur during the mating season, although many adult males become solitary or join small bachelor colonies while the mothers are raising the young. After youngsters become independent, adult females may vacate the maternity roost and lead a solitary existence, form small single-sex colonies, or join a group of males. In other parts of the range, though, this species seems to form harems, each consisting of a single adult male roosting with a group of five to ten females that remains stable year-round.

Pallas's mastiff bats conceive once or twice each year, and ultimately produce a single hairless pup each time, usually during the rainy season. In late spring or early summer, some, but not all, females are simultaneously pregnant and lactating, indicating that many mothers quickly mate after the first birth of the season. On Puerto Rico, for example, a female copulates in March, parturition and a second mating happen in June or July, and another birth occurs in September. Although all females have a period of sexual inactivity between September and February, males have viable sperm present year-round. Nonetheless, reproductive patterns are dynamic, and the details likely vary among different locations throughout the large geographic range of this bat. Gestation in Pallas's mastiff is quite long for such a small mammal, up to 3.5 months, but lactation spans only about 6 weeks.

Median lifespan of females in Panama is 1.8 years, with an estimated maximum longevity of 5.6 years. Predators are diverse and include various hawks, giant centipedes (*Scolopendra*), the tokay gecko (*Gekko gecko*), the bat falcon (*Falco rufigularis*), and even the greater bulldog bat (*Noctilio leporinus*). Collisions with the rotors of modern wind turbines kill many bats worldwide, and on Puerto Rico, Pallas's mastiff accounts for 48% of the deaths.

Status and Conservation. Until 2018, Pallas's mastiff was the lone species of this genus identified for the Caribbean region. However, citing molecular evidence, biologists split the group into three species—the pug-nosed mastiff (*M. milleri*), Hispaniolan mastiff (*M. verrilli*), and Pallas's mastiff. The latter species is an abundant and widespread animal and is considered a taxon of Least Concern.

Selected References. Rodríguez-Durán and Feliciano-Robles 2015; Loureiro et al. 2018, 2019a; Catzeflis et al. 2019; Taylor et al. 2019.

Molossus verrilli

Hispaniolan Mastiff Bat, Murciélago Casero de La Española, Molosse de Hispaniola

Name. The structure of the head and face of these small bats is somewhat reminiscent of mastiff dogs, which are descended from large-headed shepherd dogs that were developed by a group of ancient Greeks called the Molossians. Alpheus Verrill, a naturalist and, later in life, a renowned science-fiction writer, collected the original specimen on 5 February 1907, near Samaná, in the Dominican Republic.

Distribution. This mammal is restricted to the Greater Antillean island of Hispaniola and adjacent Gonave.

Measurements. Total length: 85–126 mm; tail length: 30–46 mm; hindfoot length: 7–11 mm; ear height: 11–15 mm; forearm length: 37–41 mm; body mass: 13–20 g. Males are usually larger than females.

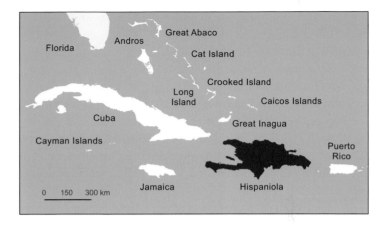

Description. The back of the animal appears medium-to-dark brown, and individual dorsal hairs are dark at the tips with a pale band covering the inner 25 to 50% of the shaft. The belly is slightly lighter than the head and back. The pinnae, wings, and uropatagium are also medium-to-dark brown. The external ears are rounded, and their bases contact each other on the forehead; the well-developed antitragus is somewhat circular in shape. The upper lip is smooth and not grooved or wrinkled. Half of the long tail extends beyond the tail membrane, readily identifying this animal as a free-tailed bat. The only other bats with free tails on Hispaniola are the Brazilian free-tailed bat (*Tadarida brasiliensis*) and the rare big free-tailed bat (*Nyctinomops macrotis*). However, the big free-tailed bat has a reddish-to-blackish appearance and is considerably larger, with a forearm length of at least 54 mm. Unlike the Hispaniolan mastiff, the Brazilian free-tailed bat has unicolored hairs, a semicircular antitragus, a wrinkled upper lip, and ears not in contact at their base.

Natural History. Compared with the lives of pug-nosed and Pallas's mastiffs (*M. milleri* and *M. molossus*, respectively), far less is known about the natural history of Hispaniolan mastiffs, although the three species likely are similar in overall behavior and ecology. The Hispaniolan mastiff occurs in various types of tropical forest, from lowlands near the sea to mountainous areas as high as 460 m in elevation. Nevertheless, humans most frequently encounter this mammal in urban areas, where it consistently roosts in buildings. These anthropogenic roosts undoubtedly offer a secure place of refuge, but the ubiquitous artificial lighting that accompanies buildings also attracts the insects on which these mastiffs might feed. The bats either hang alone, or they form small colonies consisting of twenty-three or fewer individuals. In addition to buildings, biologists occasionally find this animal resting inside caves, such as La Chepa and Trunicolás, and within hollow trees, especially palms.

As do other mastiffs, members of this species capture their insect prey while flying at high speed in open, uncluttered areas. Foraging typically occurs above nearby buildings or trees, although this nocturnal animal may shift its activity closer to the ground, depending on insect abundance. Foraging begins soon after sunset when there is still sufficient light to see, but activity ceases after only 1 hour when the bat returns to its roost; each individual, though, probably has a second brief foraging bout near sunrise, similar to mastiffs on other islands of the West Indies.

A typical echolocation call of the Hispaniolan mastiff consists of three pulses of quasi-constant frequency. The call starts with a pulse at about 30 kHz and is quickly followed by pulses at consecutively higher frequencies that are usually near 36 and 39 kHz. Many species of bats use sounds of higher frequencies (>40 kHz), but the frequencies used by the Hispaniolan mastiff allow this mammal to detect prey at a greater distance, thus giving the fast-moving mammal time to adjust course and intercept its prey. Furthermore, use of multiple frequencies probably allows these bats to detect prey of a wider range of sizes than if the animal used just a single frequency.

The Hispaniolan mastiff bat always produces just one pup at a time. Mammalogists have caught pregnant females in Haiti during May, July, and September, and apparently nonreproductive females during January, February, August, and December. Pallas's mastiffs on Puerto Rico give birth in March and September, and the same may be true for the Hispaniolan species.

Status and Conservation. Taxonomists previously classified the Hispaniolan mastiff as a subspecies of Pallas's mastiff bat, but analysis of nuclear and mitochondrial DNA confirmed that the two groups were separate species in 2018. The IUCN, however, has not yet assessed the status of the newly described Hispaniolan mastiff bat, although it seems common within its limited distribution.

Selected References. Timm and Genoways 2003; Núñez-Novas and Leon 2011; Loureiro et al. 2018, 2019b; Taylor et al. 2019.

Mormopterus minutus

Little Goblin Bat, Murciélago de las Jatas, Chauve-Souris de la Jata

Name. *Mormopterus* combines the Greek words for "goblin" (*mormo*) and "wing" (*pteron*); hence, this mammal is the "winged goblin." The specific name *minutus*, denoting "little," indicates that

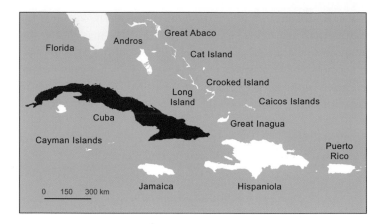

this species is the smallest member of the genus *Mormopterus* and one of the smallest representatives of the family Molossidae.

Distribution. The little goblin bat is one of only three species of *Mormopterus* in the New World and the only one within the Caribbean basin. This diminutive mammal is endemic to Cuba but is known from just thirteen localities in lowlands of the central and eastern part of the island.

Measurements. Total length: 68–82 mm; tail length: 22–34 mm; hindfoot length: 5–6 mm; ear height: 8–11 mm; forearm length: 28–33 mm; body mass: 4–8 g. In general, males are larger and heavier than females.

Description. Overall, the animal appears grayish-brown but somewhat darker on the back than the belly. Individual hairs are short, about 3 mm long, both dorsally and ventrally, and lighter at the base than the tip. The dense pelage extends onto the wings for 3 to 5 mm. The upper lip protrudes well beyond the mandible so that the distance from the angle of the mouth to the front midline is almost twice as long on the upper lip than on the lower lip. Lips are smooth, not wrinkled or grooved. The snout is long and pointed and ends with sharply outlined nostrils bordered by a horny ridge. The external ears appear triangular and remain separated by about 2 mm on the forehead. As in all free-tailed bats, the tail is long, and in this species, the vertebrae project from 10 to 20 mm beyond the free edge of the interfemoral membrane. A forearm less than 35 mm in length and a free tail are sufficient to separate this bat from all others on Cuba.

Natural History. The only known type of natural roost for the little goblin bat is a palm known locally as *jata de los murciélagos* or the bat palm (*Copernicia vespertilionum*). This tree is a natural hybrid, with a spotty distribution in central and eastern Cuba, where it is associated with small-leaved (microphyllous) semideciduous forests and savannahs, generally near the coast. Mature trees are about 14 m tall, with a diameter from 30 to 40 cm. The foliage of this palm is globose, compact, and restricted to perhaps the upper 20% of the trunk; the upper part of the foliage contains erect green plumes, whereas the lower half consists of a dense mat, many meters thick, of dry brown leaves that hang downward. Little goblin bats spend the day hidden among these dead fronds. Generally, colonies associated with palms range from two to five thousand bats, and over time, guano beneath an occupied tree can accumulate in a layer up to 60 cm deep.

This animal occasionally adopts a building as a diurnal shelter, although use of human-made structures may result from the scarcity of natural roosts, as developers increasingly remove native palms in coastal areas. The largest colony, estimated at fifteen thousand individuals, once inhabited the Cantero Palace in the city of Trinidad, along the southern coast of Cuba, although the bats disappeared after the structure was remodeled and became a museum. In buildings, little goblin bats often tuck themselves into tight spaces behind door and window frames, between walls and outer siding, or within roofing thatch. Whether in buildings or trees, these mammals roost with their backs and bellies touching the roosting material, and their sides are frequently in contact with one or more colony mates. Other species of free-tailed bat occasionally share a palm or building with the little goblin bat, including the broad-tailed bat (*Nyctinomops laticaudatus*), fierce bonneted bat (*Eumops ferox*), Brazilian free-tailed bat (*Tadarida brasiliensis*), and Pallas's mastiff bat (*Molossus molossus*).

The little goblin bat is an aerial insectivore, and its long and slender wings (i.e., high aspect ratio) suggest that this animal forages away from trees or human-made structures, as do most molossids. This bat eats various soft- and hard-bodied insects. Ants, especially carpenter ants (*Camponotus*) that often damage human structures, can be the most frequent item in the diet. Other prey include leafhoppers and, less commonly, small beetles, flies, and moths. Foraging activity occurs during short periods around dusk and dawn, when small insects are most abundant. The stomach of this small mammal can accommodate a chewed mass of insects equal to 20% of the bat's body weight.

When searching for prey, most insectivorous bats use a single type of echolocation call that varies little in frequency or timing within a single train of pulses, between different sequences of calls produced by the same animal, or among sounds made by different animals from the same species. However, the little goblin bat demonstrates a high level of plasticity in its echolocation calls. Individuals often switch from emitting a sequence of one type of call to a sequence of another kind in midflight, and 40% of echolocation sequences contain pulses that alternate in structure. When flying in open spaces, this mammal can produce calls that biologists describe as constant frequency, quasi-constant frequency, frequency modulated/quasi-constant frequency, or multiharmonic frequency modulated, among others. Presumably, this variation allows the little goblin bat to exploit a greater diversity of habitats or prey than it otherwise could.

Mothers bear a single young, once per year. Pregnant females are present from April to July, births begin in early June, and lactation continues into August. At palm roosts, juveniles typically rest close to the trunk rather than along the perimeter of the foliage. Known predators are the American barn owl (*Tyto furcata*) and Cuban pygmy owl (*Glaucidium siju*). The Cuban boa (*Chilabothrus angulifer*) is capable of climbing into the foliage of a bat palm and may be an additional threat.

Status and Conservation. The population of *jata de los murciélagos* is declining, because of selective felling, fires, and hurricanes, and the palm appears as threatened on the Red List of Cuban Flora. Consequently, the IUCN considers the little goblin bat as Vulnerable.

Selected References. Silva-Taboada 1979; García-Rivera and Mancina 2011; Mora et al. 2011.

Nyctinomops laticaudatus

Broad-Tailed Bat, Murciélago Oreja de Soplillo, Nyctinomope des Rochers

Name. *Nyctinomus* is an old name for a particular group of wrinkle-faced free-tailed bats that once included the Brazilian free-tailed bat (*Tadarida brasiliensis*), and *ops* is Greek for "appearance." Gerrit S. Miller, who invented the name *Nyctinomops* in 1902, emphasized that bats in his new genus bore a resemblance to those in *Nyctinomus*. The species name *laticaudatus* is Latin and indicates that this

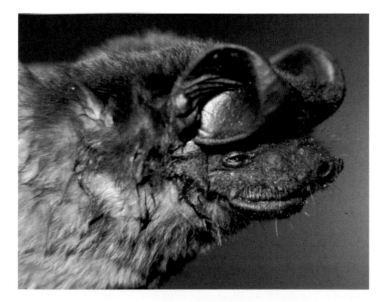

bat is "provided with" (*atus*) a "broad" (*latus*) "tail" (*cauda*). Some books refer to this species as the broad-eared bat.

Distribution. The broad-tailed bat is widely distributed across the Neotropics, from Mexico to Paraguay, but in the West Indies, it occurs only on Cuba.

Measurements. Total length: 90–116 mm; tail length: 30–45 mm; hindfoot length: 8–9; ear height: 15–17 mm; forearm length: 40–45 mm; body mass: 8–13 g. Total length includes specimens from the mainland that are slightly larger than those on Cuba.

Description. The broad-tailed bat is a medium-sized bat and the smallest of the four species in the genus *Nyctinomops*. The pelage is velvety in texture, dark chocolate-brown to blackish overall, and slightly darker on the back than the belly. Individual hairs are from 5 to 7 mm long with white bases. The upper lips are characteristically furrowed and overhang the lower lips, whereas the nostrils are circular and point somewhat laterally. The rounded ears are short, thick, and broader than high; the pinnae project forward, and their bases connect across the top of the forehead. Thick, recurved hairs coat the feet. The wings are naked, narrow, and pointed and may appear translucent. As in all free-tailed bats, the tail is long and extends freely beyond the hind margin of the uropatagium.

On Cuba, the only other species with a wrinkled upper lip are the Brazilian free-tailed bat and the big free-tailed bat (*Nyctino-mops macrotis*). Nevertheless, the big free-tailed bat is much larger; it can weigh twice as much as a broad-tailed bat and has a forearm that is more than 50 mm in length. The Brazilian free-tailed bat, in contrast, is generally smaller, although it overlaps in size with the broad-tailed bat; however, a gap of 1 to 2 mm separates the ears on the forehead of a Brazilian free-tailed bat, unlike the pinnae of a broad-tailed bat. In addition, the Brazilian free-tailed bat has three pairs of lower incisors, whereas the broad-tailed bat has only two.

Natural History. Most knowledge of this mammal's life history comes from the mainland, where the broad-tailed bat frequents various habitats, including tropical evergreen forests, subtropical dry forests, scrublands, and urban and peri-urban areas. Most captures occur in lowlands, although records exist from coastal sites to over 1500 m above sea level.

These bats typically form colonies consisting of 150 to perhaps 1000 individuals. They occasionally take refuge in caves but often rest inside crevices within steep cliffs or the stone walls of buildings, including Mayan ruins in the Yucatán Peninsula. On Cuba, biologists report one colony hiding between corrugated metal siding and the interior wooden walls of a dilapidated building. Another haven on that island is within a *jata*, which is the term for a palm that retains thick clusters of dead, dried, and drooping fronds. Broad-tailed bats roost among the dead leaves of a specific species of jata, the *jata de los murciélagos* or bat jata (*Copernicia vespertilionum*). This palm is also the favorite haunt of the more abundant little goblin bat (*Mormopterus minutus*), and the two species may simultaneously shelter in the same tree.

Like all free-tailed bats, the broad-tailed bat is an aerial hunter of insects, scooping its prey from midair with a wing or tail membrane. Echolocation recordings from the mainland indicate that these bats emit search-phase sounds, consisting of two or three pulses of quasi-constant frequency that differ slightly in starting frequency (27, 29, and 32 kHz), similar to mastiff bats. Use of low, almost constant frequencies and the bat's long, narrow wings indicate that the broad-tailed bat is a fast flier that hunts primarily in wide-open habitats—above woodlands, in savannahs, or over urban developments—where it can detect its prey at a distance. Unlike the related Pallas's mastiff, which concentrates most activity into a brief period near sunset, the broad-tailed bat forages throughout the night. The stomachs of eleven individuals captured in July on Cuba contained some moth scales but mostly the remains of beetles, including predaceous diving beetles (*Copelatus*).

Within the population, these bats form temporary associations consisting of ten to fifteen animals, both males and females, but the composition of each group seems fluid, which suggests a promiscuous mating system. Male broad-tailed bats apparently use courtship "songs" to attract females; the songs consist of a combination of chirps, trills, and buzzes, mostly in the ultrasonic range, although portions of these vocalizations are audible to humans. Litter size is one, and births probably occur in June on Cuba. Newborns are hairless, except the vibrissae and bristles on the snout and feet, and weigh about 3 g or 28% of the mother's mass; neonates are able to move their ears and open their eyes within hours of birth. Wind turbines kill broad-tailed bats on the mainland and probably do so in the Antilles as well, and the American barn owl (*Tyto furcata*) occasionally catches one of these mammals.

Status and Conservation. The IUCN classifies the broad-tailed bat as Least Concern throughout its extensive range on the mainland. However, naturalists have reported this animal from just a handful of localities in central and eastern Cuba, and its current status on the island is unknown. Foraging high in the air means that broad-tailed bats generally do not come near the ground-based mistnets that mammalogists use to capture these creatures, and perhaps an acoustic survey would better document the distribution and abundance of this animal in the West Indies.

Selected References. Ávila-Flores et al. 2002; MacSwiney González et al. 2008; Ortega et al. 2010; Bohn et al. 2016.

Nyctinomops macrotis

Big Free-Tailed Bat, Murciélago Viejo Grande, Grand Nyctinomope

Name. *Nyctinomus* is an old name for a group of free-tailed bats, and *ops* is Greek for "having the appearance of." The combination indicates that bats in the genus *Nyctinomops* superficially resemble those in *Nyctinomus*. *Macrotis* combines two Greek words to mean "large ears."

Distribution. The geographic range of the big free-tailed bat extends from the southwestern United States to Uruguay and northwestern Argentina. In the Greater Antilles, this mammal lives on Cuba, Hispaniola, and Jamaica.

Measurements. Total length: 120–139 mm; tail length: 40–63 mm; hindfoot length: 9–11; ear height: 24–32 mm; forearm length: 55–65 mm; body mass: 17–34 g. Cuban big free-tailed bats are sexually dimorphic, with males larger than females.

Description. The big free-tailed bat is the largest of the four species in the genus *Nyctinomops*. Body color varies from blackish to dark brown or reddish-brown, and is slightly darker above than below;

individual hairs are lighter at the base than at the tip. The huge ears point forward, and their inner edges unite to form a V-shaped ridge across the forehead. The rounded nostrils open laterally above a deeply wrinkled lip, which is separated from the eye by a distinct horizontal groove. As in all molossids, the outer half of the tail continues beyond the large interfemoral membrane. In the Antilles, this species could be confused with two other free-tailed bats that have grooved upper lips—the Brazilian free-tailed bat (*Tadarida brasiliensis*) and the broad-tailed bat (*Nyctinomops laticaudatus*); however, the big free-tailed bat is much larger than either of those species, which have forearms that are less than 50 mm in length.

Natural History. Biologists capture this species in tropical dry and moist forests, xeric scrublands, pine forests, savannahs, and urban zones. In North America, the big free-tailed bat most frequently spends the day inside crevices in steep rock faces, but virtually all diurnal retreats in the Greater Antilles are associated with caves or buildings. Roosts generally contain from 1 to 15 individuals, although colonies of 100 to 200 animals exist. When sheltering underground, this bat hangs alone or in groups of two to four individuals, either from a flat ceiling or in shallow depressions in the walls; it does not venture far underground but always rests close to the entrance in areas of semidarkness. When roosting in buildings, these animals hide in attics, within crevices in the walls, or behind moldings and cornices, often in association with the little goblin bat (*Mormopterus minutus*). There is a single record of a big free-tailed bat using a hollow tree on Cuba.

The big free-tailed bat is an aerial insectivore that feeds primarily on moths, as well as several other groups of insects, such as flying ants, stinkbugs, froghoppers, leafhoppers, crickets, and bush crickets. When searching for prey, this bat uses narrowband echolocation calls that typically begin at 29 kHz and fall to 16 kHz over 14 milliseconds; time between pulses is 20 milliseconds. Because the upper limit of human hearing is about 20 kHz, the low-frequency foraging sounds of the big free-tailed bat are clearly audible to people standing nearby. This mammal flies swiftly and generally hunts at heights of 50 m or more above the ground, where few if any obstacles block its outgoing sounds or the returning echoes. Using radio-tracking techniques, ecologists in Arizona determined that these bats flew at an average speed of 19 km/hr but were capable of moving as fast as 61 km/hr; some individuals hunted as far as 23 km away from their dayroost.

The big free-tailed bat apparently produces a single offspring once per year. On Cuba, mammalogists report pregnant females in May and June, with the earliest known birth on 13 June, and nursing mothers during June and July. Three bats captured in Grove Cave near Balaclava, Jamaica, were pregnant in June, whereas two adults from Jarabacoa and three individuals from La Vega, in the Dominican Republic, were lactating in July. In the United States and northern Mexico, this species is migratory, and most individuals that make the northward journey in spring are pregnant females that eventually gather in nursery colonies. However, there is little evidence of sexual segregation during the maternity season in the presumably nonmigratory bats of the Greater Antilles. Lifespan is unknown, and the only mortality factors recorded so far are collision with the rotors of a modern wind turbine and predation by the American barn owl (*Tyto furcata*).

Status and Conservation. The big free-tailed bat has a broad geographic range, and the IUCN considers its global status as Least

Concern. Nevertheless, resource managers know little of the natural history of this species in the Greater Antilles, and no data are available on population sizes or trends. Naturalists have captured this mammal at fewer than a dozen localities on Jamaica and Hispaniola combined, although the species seems more abundant on Cuba. Part of this apparent rarity, though, may be attributable to the great heights at which this animal typically flies, and an acoustic survey on each island may be more effective at determining the presence of these bats than mistnetting at ground level.

Selected References. Silva-Taboada 1979; Genoways et al. 2005; Corbett et al. 2008; Mora and Torres 2008.

Tadarida brasiliensis

Brazilian Free-Tailed Bat, Murciélago Viejo, Tadaride du Brésil

Name. Constantine Samuel Rafinesque first applied the word *Tadarida* to the European free-tailed bat (*Tadarida teniotis*) in 1814, while he lived on the island of Sicily. *Tadarida* is a general term in the Sicilian language meaning "bat." Isidore Geoffroy Saint-Hilaire, a French naturalist, created the specific epithet *brasiliensis* in 1824, after receiving specimens from the type locality of Paraná, Brazil.

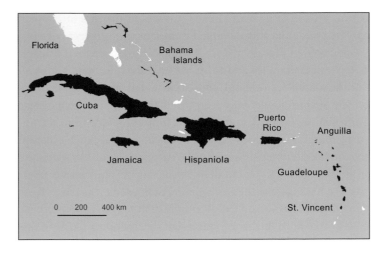

Distribution. The greater bulldog bat (*Noctilio leporinus*) and Brazilian free-tailed bat are the most widespread species of bat within the West Indies. The Brazilian free-tailed bat occurs throughout the Greater and Lesser Antilles, and most of the Bahamas, except Andros, Great Inagua, Long, New Providence, and San Salvador islands. On the mainland, this mammal ranges from the southern United States to central Chile and Argentina, although some biologists suspect that this broad-ranging taxon is actually a complex of cryptic species.

Measurements. Total length: 52–98 mm; tail length: 27–41 mm; hindfoot length: 7–11 mm; ear height: 8–15 mm; forearm length: 36–43 mm; body mass: 6–13 g. Males appear slightly larger than females on some islands, such as Cuba, but not on others. Animals from Puerto Rico down through the Lesser Antilles are a separate subspecies, and on average, they are smaller than bats from the Greater Antilles and Bahamas.

Description. The fur is short and dense and almost the same length over the entire body. Color is uniform across the back, varying among different individuals from dark brown to gray-brown and sometimes with an orange tinge; the belly is paler. Ears, wings, and tail membrane are dark brown. A tuft of stiff, dark hairs is pres-

ent on the top of a largely naked snout, and the toes are noticeably hairy. A gular (throat) gland is present in both sexes but most
pronounced in mature males. The upper lips are marred by multiple vertical wrinkles and protrude over the lower lips, thus giving the animal an "old-age" appearance and providing the basis
for the Spanish name, *murciélago viejo* (old bat). The wide pinnae
point slightly forward, and obvious bumps protrude from the anterior margins; the bases of the ears do not contact each other on
the forehead but remain separated by a small gap of 1 to 2 mm. As
in all free-tailed bats, about half of the elongated tail extends freely
past the back edge of the interfemoral membrane. The cheek teeth
of a Brazilian free-tailed bat, like those of other insectivorous bats,
have pronounced W-shaped ridges, but the last upper molar in this
species has an unusual Z-shaped crest.

The furrowed upper lip clearly separates this species from most
other free-tailed bats, such as the little goblin bat (*Mormopterus
minutus*) or any of the mastiff bats (*Molossus*). The broad-tailed
bat (*Nyctinomops laticaudatus*) and big free-tailed bat (*N. macrotis*) also have wrinkled lips; however, both species are much larger
than a Brazilian free-tailed bat and sport ears that are continuous
across the top of their head.

Natural History. The Brazilian free-tailed bat is an adaptable species found in a myriad of habitats, including dry forest, montane
forest, thorn scrub, subtropical moist forest, and urban areas across
the West Indies. This species has somewhat-flexible roosting requirements and spends the day in caves or human-made structures.
When sheltering underground, this small bat chooses fairly large
and open galleries, often in well-ventilated areas not very deep into a
cavern. In North America, the Brazilian free-tailed bat forms some
of the densest aggregations of mammals on earth, with colonies
sometimes surpassing one million individuals within the confines
of a single cave. However, throughout the Caribbean, populations
are typically much smaller than on that continent, ranging from
just a few dozen to a few thousand animals at any one location. Although these bats occasionally roost alone, they typically cluster
with others of their kind. In caves, small groups roost in cracks and

crevices or inside solution cavities, but hundreds of bats may form a teeming mass on the ceiling at some sites. When using buildings, Brazilian free-tailed bats tuck themselves behind window shutters, into deep fissures, under corrugated metal sheets or roofing tiles, and especially within the hollows of cinder-block walls.

The Brazilian free-tailed bat is the fastest flying animal on earth, achieving speeds as high as 160 km/hr for brief periods as it commutes to distant foraging grounds. Nightly activity begins between sunset and 30 minutes after sunset, and these bats typically have two main foraging bouts—after sundown and before dawn. Although the diet of Brazilian free-tailed bats in Texas and the Greater Antilles contains the same groups of insects, the different populations emphasize dissimilar types of prey. For example, the continental menu is dominated by moths and beetles, whereas Puerto Rican bats concentrate mostly on flies and flying ants, although the insular animals also consume some moths and leafhoppers but very few beetles. These dietary differences likely result from competition with mormoopid bats that are present on the islands but missing in North America. In tropical environments, insects are less abundant during the dry season, and Brazilian free-tailed bats store fat toward the end of the rainy season to tide them over the period of reduced abundance. Brazilian free-tailed bats on the mainland consume large numbers of pests that are injurious to crops, such as corn and cotton, and these mammals probably play a comparable role in the agricultural fields of the West Indies.

Foraging usually takes place far from any obstructions, well above the forest canopy or over open fields, but occasionally closer to the ground. Not surprisingly, the echolocation calls of these bats are suitable for detecting prey at long distances. When searching for flying insects on Guadeloupe, calls are almost constant in frequency, decreasing only slightly from 29 to 26 kHz; pulses last approximately 12 milliseconds, with an average of 268 milliseconds between successive calls.

The gular gland of adult males becomes hairless and hypertrophies just prior to the mating season and exudes an oily substance that the animals smear on the roosting substrate. While copulating, males often bite the neck of the female and emit a distinctive mating call that is audible to humans. During gestation and lactation, adult females form maternity colonies that are usually in sites different from those occupied by the males. The timing of reproductive events likely varies among islands and years, but in general, biologists report pregnant individuals from April through July and nursing females during July and August. Gestation lasts about 77 days, and most adults give birth to a single pup, probably once per year, in June or July. As in most bats, neonatal Brazilian free-tailed bats are huge relative to their mother; newborns in the West Indies represent as much as 30% of the mother's mass, and the forearm may be up to 52% as long as that of the adult. Mothers do not carry the naked pups on foraging trips but leave them behind in a large cluster or crèche within the roost. When the adults return, mothers locate their offspring using spatial memory and auditory cues. In addition, a female apparently marks her pups with secretions from the mother's gular gland, which helps discriminate her offspring from the dozens or hundreds of others in the colony.

In the West Indies, avian predators include the American kestrel (*Falco sparverius*), American barn owl (*Tyto furcata*), ashy-faced owl (*T. glaucops*), and Lesser Antillean barn owl (*T. insularis*). On Martinique, one individual fell prey to a tokay gecko (*Gekko gecko*). Collisions with wind turbines kill thousands of these bats in North and South America each year, and similar fatalities occur in the Antilles. Some colonies seem to disappear over short periods, and biologists speculate that this may be due to human disturbance or possibly the effects of pesticides. The Brazilian free-tailed bat lives up to 8 years in the wild and at least 12 years in captivity.

Status and Conservation. This mammal is abundant throughout its broad distribution and considered a species of Least Concern.

Selected References. Gustin and McCracken 1987; Whitaker and Rodríguez-Durán 1999; Keeley and Keeley 2004; McCracken et al. 2016; Speer et al. 2017; Morales et al. 2018.

Ghost-Faced and Mustached Bats

The Mormoopidae comprises eighteen species in two genera—the ghost-faced bats (*Mormoops*) and the mustached bats (*Pteronotus*). This family occurs only in the New World, from the southwestern United States to central Brazil and on many Caribbean islands. Six species are endemic to the West Indies. Mormoopids are all quite small, ranging in size from 4 g in the sooty mustached bat (*P. quadridens*) of the Greater Antilles to 27 g in Wagner's mustached bat (*P. rubiginosus*) of South America.

These bats share a number of distinctive features. All mormoopids have a protruding fold of tissue beneath the lower lip; in mustached bats, this flap is platelike and smaller than the lip, but in ghost-faced bats, it is notched in the middle, resulting in two leaf-like appendages that extend beyond the lip. Short bristle-like hairs occur on either side of the upper lip and form the characteristic "mustache." Unlike other bats in the Antilles, the tragus of a mormoopid has a distinct secondary fold, and the outer edge of the ear connects to the angle of the mouth by a ridge of tissue that runs beneath the eye. The uropatagium is large, and its rear margin forms a straight line, perpendicular to the axis of the body, rather than a V, as in other kinds of insectivorous bats, such as natalids and vespertilionids (see Fig. 0.2). The mormoopid tail is shorter than the membrane, and when the bat is at rest, the outer half protrudes from the middle of the uropatagium, not the trailing edge; however, when the animal spreads its legs for flight, all vertebrae disappear into a pocket within the membrane.

The rostrum is longer than wide. It is slightly upturned in most species and strongly so in ghost-faced bats. As in most insect-eating bats, the canines are long, and the cheek teeth have distinct W-shaped ridges, which provide a scissors-like shearing action to help slice through the external skeletons of their prey. The inner lower incisors are trilobed, whereas the inside upper incisors appear bulky and distinctly bifid.

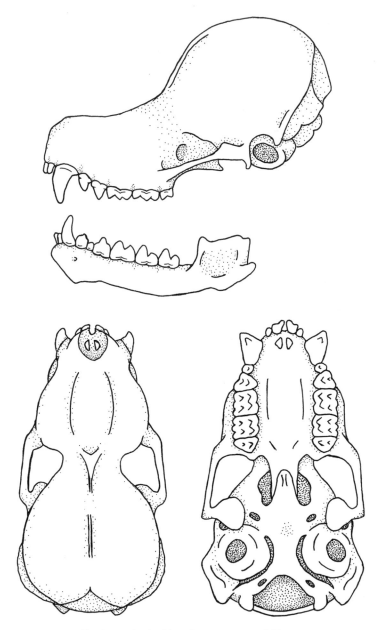

Sooty mustached bat, *Pteronotus quadridens*

Mormoopids are highly gregarious, usually forming colonies of several hundred to tens of thousands of individuals. These gatherings commonly include two or more kinds of mormoopids, as well as other species, particularly from the families Phyllostomidae and Natalidae. Most mormoopids in the West Indies are obligate cave dwellers and roost almost exclusively inside large hot caves, where the air is stagnant and temperatures range from 28 to 40 °C. Some species living on the mainland, including Allen's mustached bat (*P. fuscus*), appear more tolerant of cold and more flexible in their choice of roost, sometimes selecting small caves, human-made structures, or hollow trees.

All mormoopids are aerial predators. When hunting, these bats produce echolocation pulses that generally consist of a short component of constant frequency or quasi-constant frequency, followed by a steep frequency-modulated sweep, although a few species, such as Parnell's mustached bat (*P. parnellii*), emit calls with long segments (>15 milliseconds) of constant frequency. Most mormoopids have broad habitat associations and usually forage along woodland edges and in semiopen environments; however, a number of species, like Parnell's and the Puerto Rican (*P. portoricensis*) mustached bats, often visit more cluttered microhabitats in the interior of a forest.

These bats typically have one offspring per year. Young are born hairless and with their eyes closed. In the West Indies, births usually coincide with the onset of the rainy season, when insects become more abundant.

Selected References. Gannon et al. 2005; Pavan 2019.

Mormoops blainvillei

Antillean Ghost-Faced Bat, Murciélago Barbicacho, Mormoops de Blainville

Name. In ancient Greek, *mormo* can mean "ghost," "goblin," or "she-monster," and *ops* indicates "appearance" or "face." The name of the genus refers to the unusual facial anatomy of these bats,

which includes broad ears, multiple fleshy extensions, and numerous wartlike bumps. The specific epithet *blainvillei* honors Henri Marie Ducrotay de Blainville, a French zoologist and anatomist of the early nineteenth century who described several species of reptiles and also coined the term paleontology.

Distribution. This species is the lone ghost-faced bat in the West Indies, and it resides only in the Greater Antilles—on Cuba, Jamaica, Hispaniola, and Puerto Rico, as well as many of the smaller satellite islands, including Gonave and Mona. Fossils, however, occur on islands as far north as South Abaco, Little Exuma, and New Providence in the Bahamas and as far east as Anguilla, Antigua, Barbuda, and Marie-Galante in the Lesser Antilles, suggesting that the current range is much smaller than long ago.

Measurements. Total length: 78–87 mm; tail length: 21–31 mm; hindfoot length: 6–10 mm; ear height: 11–18 mm; forearm length: 45–49 mm; body mass: 8–11 g.

Description. The fur of the Antillean ghost-faced bat appears bright orange in most individuals, but some have a cinnamon or amber-brown pelage. Dorsal hairs are long, dense, and soft, whereas those on the ventral side are shorter and paler in color. Wing and tail membranes are dark, as is the outer edge of the external ear. The ears are short and broad, and their outer rims connect to the lower lip, creating the appearance of a funnel leading into the ear canal; the animal's beady black eyes actually appear to be inside this imaginary funnel. The upper lips and nostrils form a labionasal plate that is studded with fleshy bumps. The tubercle-covered lower lip extends outward from the jaw and lies above two large flaps of skin. Bristly hairs grow from the upper lip, similar to the mustached bats, although not as pronounced. Even though some humans consider the Antillean ghost-faced bat to be grotesque, the unique and intricate facial features of this gentle mammal are quite impressive when examined closely. The skull is also distinctive, with an upturned rostrum that forms nearly a 90° angle with the cranium.

Natural History. The Antillean ghost-faced bat occupies numerous habitats, from lush evergreen forests to dry deciduous landscapes. Although members of this species occasionally roost in small groups in cool environments, most individuals reside deep inside large hot caves, where the temperature generally exceeds 30 °C and moisture saturates the air. Although some caves support nearly fifty thousand Antillean ghost-faced bats, each animal hangs individually, usually from the ceiling, and away from other species that might share the cave. Individuals roosting in sites where the ambient temperature falls below 28 °C can become torpid, and their body temperature may dip at least as low as 25 °C.

The Antillean ghost-faced bat leaves its roost later than other species that shelter in the same cave. A few individuals depart as early as 22 minutes after sunset, but peak emergence is not until 2 or 3 hours later. These bats often return to and emerge from the cave multiple times throughout the night, with the final reentrance for most individuals occurring between 10 and 44 minutes before sunrise, when it is still dark. The flight membranes of the Antillean ghost-faced bat produce an odd sound that resembles the faint humming produced by a laser beam in a science-fiction movie; the sound is clearly audible to nearby humans as the bat passes overhead. These animals fly at speeds up to 41 km/hr, but average velocity is 32 km/hr.

By exiting its cave late in the evening, the Antillean ghost-faced bat matches its emergence with the peak activity of its primary prey—moths—which form about 67% of the diet by volume. Among the moths consumed, there are at least six species that cause considerable damage to human crops; these injurious insects include the yellow mocis moth (*Mocis disseverans*) and the striped grass looper (*M. latipes*), both of which attack sugarcane, and the fall armyworm (*Spodoptera frugipera*), which is the most important pest of corn on Puerto Rico. This mammal also feeds on beetles, bugs, and flying ants.

The Antillean ghost-faced bat flies faster and has more pointed wings than some other mormoopids, so it is less able to maneuver in dense vegetation; consequently, this animal primarily hunts along forest edges and around tree crowns. While searching for prey, this bat emits calls that sweep downward from 68 to 40 kHz; the frequency decreases less rapidly at the beginning of a pulse and more quickly at the end, making the call appear convex on a spectrogram (see Fig. 0.4). The duration of each call is only about 2.9 milliseconds. The flying bat scoops each insect out of the air using its large tail membrane or wings and then transfers the prey to the mouth for processing.

The Antillean ghost-faced bat usually has just one pup per year, although rare records of twins exist. Neonates are naked, except for bristly hairs around the snout and feet. Females in various stages of pregnancy are found from March to June but occur with the highest frequency in May; parturition begins in June, and some individuals are still lactating in September. Beginning in May, males and females often roost in different caves, which may reduce competition for food between adult males and the mothers and their offspring.

Ashy-faced owls (*Tyto glaucops*) and American barn owls (*T. furcata*) prey on these bats in the Dominican Republic, and house cats and the Puerto Rican boa (*Chilabothrus inornatus*) are consistent predators on Puerto Rico. The spinning blades of modern wind turbines cause some deaths, although the impact of this emerging threat on the population is unclear. Compared with sooty mustached bats (*Pteronotus quadridens*) that roost in the same caves, fewer Antillean ghost-faced bats are afflicted with external parasites, and those animals that are infested have a much smaller number of parasites. Chiggers that cling to the tail and wing membranes are the most frequent pests.

Status and Conservation. Although this bat depends on an uncommon resource (hot caves), it inhabits many different islands and is generally abundant. The IUCN, therefore, considers the Antillean ghost-faced bat as a species of Least Concern.

Selected References. Kurta et al. 2007; Mancina et al. 2012; Rolfe and Kurta 2012; Rolfe et al. 2014; Rodríguez-Durán and Feliciano-Robles 2015.

Pteronotus davyi

Davy's Naked-Backed Bat, Murciélago de Davy,
Ptéronote de Davy

Name. The generic name comes from the Greek words *pteron*, meaning "wing" and *notus*, referring to "rear" or "back." These words describe how the hairless wings attach to the back of this animal's body, not to the sides, thus imparting a naked look. Joseph Gray, who described the species in 1838, based on a specimen from Trinidad, chose *davyi* to honor Dr. John Davy, a renowned physiologist of the time.

Distribution. Davy's naked-backed bat inhabits the mainland from southern Nicaragua and Costa Rica southward to the coastal lowlands of Venezuela and Colombia. This species also occurs on a number of continental islands, including Trinidad and Curaçao, and ranges northward into the West Indies, where it lives on Dominica, Guadeloupe, Marie-Galante, Martinique, and Saint Lucia.

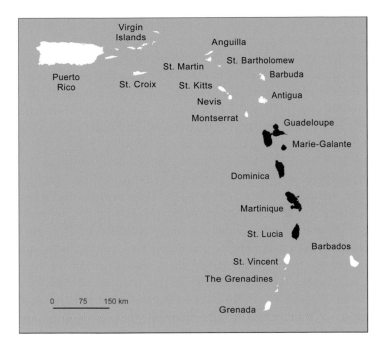

One specimen from 1894 suggests that this bat may have lived on Grenada at one time. Nevertheless, the lack of modern captures on Grenada and Saint Vincent is intriguing, given the presence of Davy's naked-backed bat to the south of those islands, on Trinidad, and to the north, starting with Saint Lucia.

Measurements. Total length: 73–84 mm; tail length: 15–24 mm; hindfoot length: 10–12 mm; ear height: 14–17 mm; forearm length: 43–48 mm; body mass: 7–9 g.

Description. Wings uniting in the middle of the back make this mammal unlike any other bat in the West Indies. The animal in the photo is actually a Thomas's naked-backed bat (*P. fulvus*) from Mexico, which taxonomists considered a subspecies of Davy's naked-backed bat until 2016; the two animals differ primarily in genetic characters. In Davy's naked-backed bat, the dorsal fur, which is present underneath the wing membranes, is grayish- to reddish-brown, and the belly is lighter. Individual dorsal hairs are unicolored and longer than the ventral hairs, which are bicolored and have grayish-white tips. Membranes, tragus, and ears are dark. The pinnae are broad at the base and pointed at the tips. Like other members of the genus, the enlarged upper lip incorporates the nostrils, thus forming a labionasal plate. The upper lip also sports bristle-like hairs, giving the appearance of a mustache, whereas the broad lower lip bears numerous small bumps.

Natural History. In South America, Davy's naked-backed bat typically roosts in hot caves, where it selects regions with the highest temperatures and poorest ventilation. Males and females commonly form mixed colonies and share their roosts with other mormoopids, phyllostomids, and sometimes natalids. Although mammalogists frequently detect the echolocation calls of these bats or capture them while they forage, few roosting sites have been found. The only known dayroost on Dominica is a cave that is, at most, 5 m wide and 2.5 m high, but despite the small size, this underground retreat apparently shelters several thousand of these bats. On Marie-Galante, biologists estimate several hundred animals living in Le Grand Trou à Diable, and on Martinique, up to ten thousand individuals occupy La Grotte de Bellefontaine. However, daytime shelters are unknown on Guadeloupe.

On Martinique, biologists have intensively studied the activity and habitat preferences of this mammal using acoustic monitoring and mistnetting. Davy's naked-backed bat appears ubiquitous on that island. The bat is associated with all forest types, including mangroves, riparian woodlands, and swamp forests, especially those with a low density of understory trees. Acoustic recordings often indicate the presence of this species in forest gaps and along edges, and it is more active if a stream or pond is nearby. Unlike Schwartz's myotis (*Myotis schwartzi*), which favors deep forests and shuns human-modified environments on that island, Davy's naked-backed bat and other island residents, such as Pallas's mastiff (*Molossus molossus*), are opportunistic and readily seek out insects attracted to artificial light sources, such as street lamps. This animal occurs at all elevations, from sea level up to 950 m on the windswept slopes of Piton Boucher. The results from mistnetting indicate that Davy's naked-backed bat makes up 2% of the bats caught on Martinique and that this species may be more active in the wet season. On Guadeloupe, acoustic monitoring reveals that these mammals are most abundant in humid forests, rain forests, and mangroves, near the northern coast of both Grande-Terre and Basse-Terre, whereas limited data from mistnetting on Saint Lucia and Dominica suggest that these bats follow pathways through secondary forests or along stream corridors and are tolerant of disturbed sites.

At Le Grand Trou à Diable, isolated individuals begin leaving the cave near sunset, and a cloud of bats starts exiting by 8 to 12 minutes past sundown. Activity at foraging sites on Martinique, though, peaks later in the evening, during the second and third hours after sunset; a second bout of activity likely occurs before sunrise. This species is strictly an insectivore and eats mostly moths and flies. In Venezuela, these mammals include many agricultural pests in their diet, and the same is probably true in the West Indies. Echolocation pulses typically last 4.6 milliseconds, with 13 to 85 milliseconds between calls; they begin with a long segment of constant frequency at about 70 kHz and end with a short sweep down to about 51 kHz.

Reproductive data are sparse. On Saint Lucia, mammalogists have caught pregnant females on 16–17 March and 6 April and nursing mothers on 1–4 August; on Dominica, the only date for a gravid individual is 13 March. Copulation likely happens in January and February, and subsequent births occur in May and June, at the start of the rainy season, as flying insects become more plentiful; lactation probably continues through July and into early August. Examination of embryos indicates that litter size is invariably one, as in other mormoopids.

American kestrels (*Falco sparverius*) prey on these bats as they emerge from their cave on Marie-Galante. The birds arrive near sunset, sit on a utility pole, and wait for the column of departing bats to cross low over an open road. Each hawk then dives into the swirling mass of mammals, plucks one from the air, and flies to a nearby perch, where the predator dismantles the hapless animal. A kestrel is successful in capturing a bat 16% of the time. Giant centipedes (*Scolopendra*) are occasional predators in northern Venezuela.

Status and Conservation. Davy's naked-backed bat is common on the tropical mainland, and the IUCN categorizes the species as Least Concern.

Selected References. Lenoble et al. 2014b; Pavan and Marroig 2016; Barataud et al. 2017; Pedersen et al. 2018a.

Pteronotus fuscus

Allen's Mustached Bat, Murciélago Bigotudo de Allen, Ptéronote d'Allen

Name. *Pteronotus* comes from two Greek words, *pteron* ("wing") and *notus* ("rear" or "back"), and refers to Davy's naked-backed bat (*P. davyi*), which is the original member of the genus. In that animal, the naked wings join at the middle of the back, instead of at the side of the body, as they do in most other bats, including Al-

len's mustached bat. The specific epithet *fuscus* is the Latin adjective for "dusky." Joel Allen, who first described this taxon in 1911, noted that it was "darker and less suffused with rufous" than the similar-looking Wagner's mustached bat (*P. rubiginosa*) from South America.

Distribution. Allen's mustached bat occurs in Colombia, Venezuela, northwestern Guyana, and on several islands off the coast of South America, including Trinidad and Tobago. Although not currently known from Grenada or the Grenadine Islands, this bat inhabits Saint Vincent.

Measurements. Total length: 90–99 mm; tail length: 20–27 mm; hindfoot length: 12–14 mm; ear height: 20–25 mm; forearm length: 58–63 mm; body mass: 14–20 g. Individuals from Saint Vincent are similar to the Trinidadian population in external and cranial measurements, whereas bats from northern Venezuela and Margarita

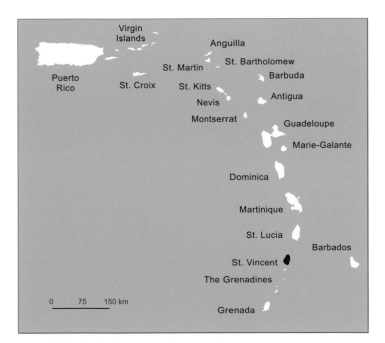

Island are noticeably smaller. On Saint Vincent, males are larger, on average, than females in some cranial measurements but not in length of forearm.

Description. The fur of Allen's mustached bat is short and dense. Dorsal hairs are usually dark grayish-brown, whereas ventral hairs have whitish or buffy tips, giving the hairs an overall paler appearance. The pointed ears connect by low and inconspicuous ridges that fuse on top of the naked snout and form a prominent midline tubercle. The tragus is simple and lanceolate, and has a small secondary fold. Tufts of stiff hairs sprout from the upper lip, forming the mustache, whereas the broad lower lip contains many small wartlike papillae. As in all members of the Mormoopidae, the tail is well developed but shorter than the interfemoral membrane, and about half the tail protrudes from the top of the membrane when the animal is at rest. The only other mormoopid in the southern West Indies is Davy's naked-backed bat, but that species is easy to recognize by its smaller size and hairless wings completely covering the back.

Natural History. In the laboratory, this tropical bat endures cold well and maintains its body temperature (36.4 °C) even at air temperatures as low as 8 °C. In the field, it also seems more tolerant of lower temperatures than other mormoopids and frequently roosts in sites where the ambient temperature hovers near 26 °C. Allen's mustached bat typically gathers in mixed-sex assemblages inside extensive caverns, but this bat also accepts small caves and occasionally abandoned buildings or hollow trees for roosting. On Saint Vincent, the only known dayroost is a shallow cave with a low ceiling that these mammals share with the Antillean fruit-eating bat (*Brachyphylla cavernarum*); the two species, though, do not intermingle.

On the mainland, this bat occurs in an array of habitats, including agricultural zones, tropical dry forest, evergreen forest, and cloud forest, up to 1500 m of elevation. Biologists have captured this species on Saint Vincent at sites ranging from the mouth of the

Buccament River up to 646 m on the southeast flank of La Soufrière Volcano. The species probably forages in cluttered forest patches, as do many other mustached bats, but it is also caught in mistnets set in house yards and banana plantations, over trails through secondary forest, across streams, and next to ponds.

Although no specific information is available concerning this bat's diet, similar mustached bats in the Greater Antilles eat large numbers of moths and flying ants. Echolocation calls emitted by this species are unique among bats on Saint Vincent. The second harmonic contains the most energy and includes a long-duration component of constant frequency near 62 kHz and two shorter elements—an upward frequency-modulated sweep at the beginning of the pulse and a downward one at the end.

In South America, Allen's mustached bat typically gives birth to a single young during the transition from the dry season to the rainy season (May–June). The same appears true on Saint Vincent, where biologists report pregnant females between 27 May and 5 June and lactating individuals between 27 July and 5 August. Adult males and females roost together during the mating period, but sexual segregation occurs for parturition and throughout lactation, as the females adopt warmer sites for use as nurseries.

Status and Conservation. Until recently, taxonomists considered this bat a subspecies of Parnell's mustached bat, and the IUCN has not yet evaluated the status of Allen's mustached bat as a separate taxon. Given its distribution in South America, though, the species is likely secure. However, the isolated population inhabiting the small (389 km²) island of Saint Vincent appears vulnerable to catastrophic storms, and the effects of the eruption of La Soufrière Volcano in spring 2021 are unknown.

Selected References. Goodwin and Greenhall 1961; Handley 1976; Vásquez-Parra et al. 2015; Kwiecinski et al. 2018; Pavan 2019.

Pteronotus macleayii

Macleay's Mustached Bat, Murciélago Bigotudo de Macleay, Ptéronote de Macleay

Name. *Pteronotus* is a combination of the Greek nouns *pteron* and *notos* and means "winged back." The first species in the genus was Davy's naked-backed bat (*P. davyi*), in which the hairless wings unite along the dorsal midline, thus hiding the fur of the back. The specific name honors William S. Macleay, who collected the specimens on which John E. Gray based his original description of this particular mustached bat.

Distribution. Macleay's mustached bat currently inhabits Jamaica, Isla de la Juventud, Cuba, and a few keys off the north coast of Cuba, such as Santa María and Romano. However, paleontologists have unearthed fossils of this bat from the late Quaternary period on the Bahamian island of New Providence and discovered other remains in Parc National La Visite, Massif de la Selle, in Haiti; the latter bones are the only records of this species from Hispaniola.

Measurements. Total length: 72–78 mm; tail length: 19–25 mm; hindfoot length: 9–10 mm; ear height: 16–18 mm; forearm length: 39–44 mm; body mass: 4–7.5 g. Males are larger than females in most measurements. The animals on Cuba and Jamaica represent different subspecies, and those from Jamaica are typically larger than their Cuban counterparts.

Description. Unlike Davy's naked-backed bat, the hairless wings of Macleay's mustached bat independently meet the body at the sides. The color of the dorsal fur usually ranges from grayish-brown to yellowish-brown, although some individuals appear orange-brown. Dorsal hairs are tricolored—dark at the base and tips but gray or white in the middle. Ventral pelage is grayish-white with a pale buffy tinge. The muzzle is short, and the lower lip has multiple wartlike bumps. A V-shaped notch separates the nostrils, and a small rectangular flap appears above each opening. Ears are narrow, pointed, and separated above the head; a deep cleft occurs in the lateral edge of each pinna, and the base of the ear appears continuous with the lower lip. The tall tragus is broad at the base, rounded at the tip, and has a prominent secondary fold. The tail emerges from the center, not at the end, of the well-formed uropatagium, and from seven to ten short, parallel ridges occur on the underside of this membrane, perpendicular to its posterior margin (see Fig. 6.2).

Macleay's mustached bat is morphologically similar to but slightly larger than the sooty mustached bat (*P. quadridens*). The sooty mustached bat, though, lacks the ridges on the tail membrane, the V-shaped notch between the nostrils, and the rectangular pads above them. Furthermore, the sooty mustached bat has from three to five small bumps above each nostril that are missing in Macleay's mustached bat.

Natural History. This species roosts exclusively in caves, where it forms colonies of several thousand individuals. Macleay's mustached bat frequently occupies high-domed chambers, located far from the entrance, where environmental conditions are hot, moist, and constant. Macleay's mustached bat often associates with other denizens of hot caves, including the sooty mustached bat and Parnell's mustached bat (*P. parnellii*), as well as the Antillean ghost-faced bat (*Mormoops blainvillei*), Cuban flower bat (*Phyllonycteris poeyi*), and Greater Antillean long-tongued bat (*Monophyllus redmani*).

Macleay's mustached bat begins leaving its cave by 8 to 29 minutes after sunset, and returns for the day by 28 to 46 minutes before sunrise. Although ultrasonic devices detect these bats foraging throughout the evening, activity peaks late in the night, between 8 and 10 hours after sunset, precisely when the sooty mustached bat is least active. Macleay's mustached bat uses broadband echolocation calls with high intensity and short duration, and they include up to four harmonics, although the second harmonic contains the most energy. Each pulse begins with a short constant-frequency component followed by a downward sweep, from about 70 to 55 kHz. Calls last 4 milliseconds. The structure of the calls, along with wings that have a high aspect ratio and low wing loading, indicates an animal equipped for hunting insects by slow aerial hawking in background-cluttered habitats, such as forest edges or around the crowns of trees. Analyses of stomach contents suggest that Macleay's mustached bat preys only on flying insects, including flies, beetles, orthopterans, and homopterans.

This small mammal reproduces once per year and bears a single young each time. Testicles enlarge in January and February, suggesting that copulation may begin early in the year. Nevertheless, reports of gravid females are restricted to March through July, with a peak in May; lactation occurs during June and July and climaxes in August. Males and females spend most of the year roosting in the same cave, but during pregnancy and lactation, either sex may leave and roost elsewhere.

Status and Conservation. Macleay's mustached bat is widespread and abundant on Cuba and Juventud but less common on Jamaica. Although the IUCN classifies the species as Least Concern overall, the Jamaican population, in particular, is threatened by habitat loss and disturbance to its caves through guano extraction and increasing visitation by tourists and spelunkers.

Selected References. Silva-Taboada 1979; Genoways et al. 2005; Macías et al. 2006a; Mancina et al. 2012; Emrich et al. 2014.

Pteronotus parnellii

Parnell's Mustached Bat, Murciélago Bigotudo Mayor, Ptéronote de Parnell

Name. In Greek, *pteron* means "wing," and *notos* indicates "rear" or "back." This combination refers to the unusual way that the wing membranes connect along the midline of the back in Davy's naked-backed bat (*P. davyi*), which was the first species placed in *Pteronotus*. The specific epithet *parnellii* honors Dr. Richard Parnell, an ichthyologist who collected specimens of the bat from Jamaica and sent them to John E. Gray, who named the species in 1843.

Distribution. For many decades, taxonomists considered Parnell's mustached bat a wide-ranging species that occurred on all the Greater Antilles and on the mainland, from Mexico to South America. However, based on molecular evidence, evolutionary biologists have split the group into multiple species, and the name

Pteronotus parnellii now refers only to the animals residing on Jamaica and Cuba, including some of the keys in the Camagüey Archipelago. However, fossils from Grand Cayman Island, Isla de la Juventud, and Abaco and New Providence islands in the Bahamas, as well as mainland Florida, suggest a much wider distribution for this species during the Pleistocene epoch.

Measurements. Total length: 76–86 mm; tail length: 18–22 mm; hindfoot length: 10–13 mm; ear height: 19–22 mm; forearm length: 50–55 mm; body mass: 10–16 g. Individuals from Jamaica are larger than the bats from Cuba. Mean length of forearm is slightly greater in females, although males weigh a little more.

Description. Parnell's mustached bat is a medium-sized bat, with a slender body and short, dense fur that appears olive-brown to grayish-brown on the back and lighter on the belly. Dorsal hairs are weakly tricolored, darker at the base and tip and somewhat lighter in the middle of the shaft; ventral hairs, in contrast, are distinctly bicolored, with brown on the basal half and whitish-grey at the

outer end. Inconspicuous ridges connect the pointed ears on top of the muzzle to form a prominent fleshy bump, and a distinct tuft of stiff hairs, the mustache, occurs on the upper lip between the nostril and eye. Unlike Davy's naked-backed bat, the dark-colored wings of Parnell's mustached bat attach to the side of the body. As in all members of the Mormoopidae, the interfemoral membrane is well developed, and its posterior margin is perpendicular to the axis of the body and not pointed; the tail itself is shorter than the membrane and emerges from the center of the uropatagium. This mustached bat is the largest mormoopid in the Greater Antilles, and size alone is sufficient to distinguish it from Macleay's mustached bat (*P. macleayii*) or the sooty mustached bat (*P. quadridens*), which also live on Cuba and Jamaica.

Natural History. Parnell's mustached bat is common and widespread on both Cuba and Jamaica, occurring from sea level to at least 1300 m in elevation. These bats are obligate cave dwellers that gather in colonies consisting of a few individuals up to a few thousand animals. During the day, they frequently form compact clusters and hide inside cavities in the ceiling or hang on walls or beneath ledges, sometimes as low as 2.5 m above the floor. This species usually roosts in huge multispecies aggregations, often within the most expansive chambers of hot caves. Like all mormoopids, these bats generally roost apart from the other species that use the same cave, although Parnell's mustached bats do occasionally mingle with Greater Antillean long-tongued bats. Parnell's mustached bat is also more eclectic in its choice of roosting environment than the sooty mustached bat or Macleay's mustached bat. For instance, Parnell's mustached bat accepts smaller rooms and passageways and, in general, is more tolerant of cooler and drier conditions, either within a hot cave or in other underground retreats.

These animals usually begin emerging from their roost between 10 and 35 minutes after sunset, at the same time as Greater Antillean long-tongued bats. Individual mustached bats come and go at the cave throughout the night, and they apparently do not use separate nightroosts. In the surrounding forest, foraging activity is greatest within 1 or 2 hours after sundown and then decreases each hour until midnight, before continuing at low levels until about 30 minutes before sunrise. Mammalogists occasionally capture Parnell's mustached bat while it flies above pastures, in plantations of banana and coconut, along streams, and over ponds. However, this mammal spends most of its time hunting within the cluttered interior of secondary or old-growth forests. Typical flight speed in the open is about 25 km/hr. Parnell's mustached bat remains active even as nighttime air temperature falls to 8 °C.

On Cuba, 93% of these bats include at least some moths in their diet. However, an array of insect types completes the menu, including flies, dragonflies, crickets, small cockroaches, and beetles, such as weevils and June bugs (*Phyllophaga*). The weight of insects consumed in one foraging bout can equal 21% of the animal's body mass. Flying insects are less plentiful during the dry period of the year, and these mammals store fat late in the rainy period to tide them over the time of reduced abundance. As the dry season progresses, males lose 25% of their body fat, whereas females lose 52%.

Most echolocating bats use calls with a low "duty cycle," meaning that the duration of the sound is short relative to the total time between the start of one signal and the beginning of the next. However, Parnell's mustached bat and other members of the subgenus *Phyllodia*, such as the Puerto Rican and Hispaniolan mustached bats (*P. portoricensis* and *P. pusillus,* respectively), are unusual among New World bats in using long-duration echolocation calls having a high duty cycle, with comparatively little time beween pulses. In addition, these bats emit sounds of mostly constant frequency and are capable of detecting slight differences in frequency (i.e., a Doppler shift) between the outgoing pulse and returning echo caused by movement of an insect's body or the flapping of its wings.

For Parnell's mustached bat, the second harmonic is the strongest, and it usually begins with a brief upward change in frequency, followed by a long segment of constant frequency at about 61 kHz, and ends with a downward sweep of intermediate duration to 58 kHz. Total call duration is 16 to 29 milliseconds, and 75 to 90% of that time involves the constant-frequency component; overall, these pulses are about three to seven times longer than the calls of other insectivorous bats on the same islands. Calls emitted by this mustached bat are well suited for detecting the fluttering wings of moths amid the twigs and leaves of a complex forest environment. In addition to echolocation sounds, biologists have identified thirty-three distinct syllables that these bats emit, singly or in combination, for communication purposes within the roost; among mammals, only a few primates have such a rich collection of vocalizations.

Parnell's mustached bat is a highly synchronous breeder that produces one pup per year. On Cuba, copulation begins in January or February, the first neonates appear in mid-June, and all mothers wean their young by late September. Births may occur a month or more earlier on Jamaica. Newborn are hairless, except the vibrissae and some bristles on the snout and feet, and weigh as much as 38% of the mother's mass. Adult males and females segregate during the reproductive season and roost in different caves.

Status and Conservation. The IUCN has not reclassified Parnell's mustached bat since systematists broke this once wide-ranging taxon into nine separate species in the early 2000s. Nevertheless, this mammal appears common and its population seems stable on Cuba and Jamaica.

Selected References. Silva-Taboada 1979; Genoways et al. 2005; Fenton et al. 2012; Mancina et al. 2012; Pavan 2019.

Pteronotus portoricensis

Puerto Rican Mustached Bat, Murciélago Bigotudo de Puerto Rico, Ptéronote de Porto Rico

Name. The combination of *pteron* ("wing") and *notus* ("back") pertains to Davy's naked-backed (*P. davyi*) bat, the first species placed in the genus *Pteronotus*; in that unusual animal, the bare wings attach to the dorsal midline and not to each side of the body, as they do in the Puerto Rican mustached bat and most other species. The specific epithet *portoricensis*, of course, denotes Puerto Rico, where this mammal lives.

Distribution. This mammal is endemic to Puerto Rico and the smaller, adjacent island of Mona.

Measurements. Total length: 78–83 mm; tail length: 13–22 mm; hindfoot length: 8–13 mm; ear height: 16–21 mm; forearm length: 50–53 mm; body mass: 11–16 g. Individuals from Mona Island weigh 12% less and have forearms that are 8% shorter, on average, than do their counterparts from the main island of Puerto Rico.

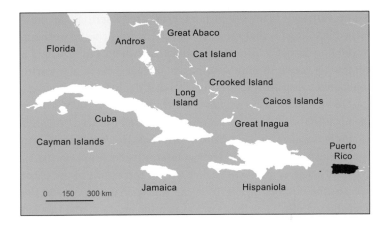

Description. The Puerto Rican mustached bat has a slender body, with short fur densely distributed over the body. Overall color ranges from dark gray to grayish-brown, with ventral hairs that are paler than those on the back. In the light of a headlamp, though, these bats often have a golden hue. Belly hairs are bicolored, with dark bases and lighter tips, whereas hairs from the back are weakly tricolored, with a lighter zone in the middle of the shaft. Ears and membranes are dark and hairless. Tufts of stiff hairs occur on the upper lip to either side of the nostrils, and the lower lip has numerous small bumps. The pointed ears connect by subtle ridges that join on top of the snout, forming a prominent fleshy protuberance. As in all mormoopids, the interfemoral membrane is extensive, and its back margin is horizontal and not V-shaped. The tail is shorter than the membrane; the vertebrae slide into the uropatagium when the bat is in flight, but they project from the middle of the membrane when the animal is inactive or crawling. It is remarkably similar to Parnell's mustached bat (*P. parnellii*) living on Cuba and Jamaica but slightly smaller on average. On its home island, the Puerto Rican mustached bat could be confused only with the sooty mustached bat (*P. quadridens*); that animal, however, is considerably smaller, lacks the fleshy knob above the snout, and has a row of three to five small bumps above each nostril.

Natural History. Mammalogists have captured Puerto Rican mustached bats at sites that range in elevation from coastal areas up to 700 m. This bat is intimately associated with hot caves, where it commonly roosts in multispecific assemblages, particularly with the Greater Antillean long-tongued bat (*Monophyllus redmani*), the brown flower bat (*Erophylla bombifrons*), and the island's other members of the Mormoopidae, the Antillean ghost-faced bat (*Mormoops blainvillei*) and sooty mustached bat. However, the Puerto Rican mustached bat is the least common mormoopid on the island. Although some caves shelter tens of thousands of the other species, the number of Puerto Rican mustached bats is typically a few hundred to a few thousand at any one location, and this bat occupies only about 15% of the island's hot caves. On Mona Island, though, the species seems more abundant.

The Puerto Rican mustached bat feeds heavily on beetles, moths, and flying ants, along with smaller amounts of bugs, flies, and dragonflies. Females consume more beetles and ants and fewer moths than do the males. Compared with the other two mormoopids on the island, the large-bodied Puerto Rican mustached bat has a more diverse diet, although some overlap occurs; foods eaten by the three species are more similar during the wet season, when insects are most abundant, and more alike in the mesic north of the island compared with the dry forests in the south.

The Puerto Rican mustached bat produces multiharmonic echolocation calls of high frequency. The second harmonic is the strongest, and it starts with a long segment of constant frequency at about 56 kHz, followed by a short downward sweep that ends at 47 kHz. The bat sends out calls every 30 to 99 milliseconds, and each pulse lasts 15 to 32 milliseconds, with the initial constant-frequency component accounting for up to 90% of call duration. The structure of its echolocation calls, combined with wings that provide good maneuverability and an ability to hover, suggest that this animal forages for flying insects in cluttered areas inside the forest, as does Parnell's mustached bat on Cuba.

Like other mormoopids, this bat probably produces a single offspring each year, but the timing of reproductive events is unclear. Wildlife biologists have captured pregnant females in May,

June, and August and discovered lactating females in September on Puerto Rico, as well as a maternity colony with fully furred pups during late November on Mona Island. Colonies fluctuate in size at different times of the year, which suggests sexual segregation during the breeding season.

Worldwide, owls account for 90% of all predation on bats, and these birds frequently attack other types of mustached bats on Cuba and Hispaniola. However, such incidents generally involve barn owls, which do not occur on Puerto Rico, and no reports of predation by owls exist for the Puerto Rican mustached bat or any other bat on the island. House cats and boas (*Chilabothrus inornatus*) presumably capture some of these mammals, although direct observations are lacking. Similar to the sooty mustached bat, the Puerto Rican mustached bat is frequently infested with cylindrical-shaped mites (*Lawrenceocarpus micropilus*) that cling to hairs in the bat's ears; however, the Puerto Rican mustached bat also consistently hosts bat flies (Streblidae), which do not afflict sooty mustached bats. Longevity is unknown.

Status and Conservation. Until 2016, taxonomists considered this bat a subspecies of Parnell's mustached bat, and the IUCN has not yet determined the status of the Puerto Rican mustached bat as a separate taxon. However, given its small populations in a limited number of caves on mostly one island, its conservation status needs examination.

Selected References. Gannon et al. 2005; Rodríguez-Durán and Padilla-Rodríguez 2010; Rolfe and Kurta 2012; Pavan 2019.

Pteronotus pusillus

Hispaniolan Mustached Bat, Murciélago Bigotudo de La Española, Ptéronote de Hispaniola

Name. In Davy's naked-backed bat (*P. davyi*), the first member of the genus, the wings do not join the body at the sides but meet instead in the middle of the back, and the word *Pteronotus* combines

Greek words that mean "winged back." The specific epithet *pusil-lus* is a Latin term for "very small" and refers to the reduced size of this species compared with the closely related Parnell's and Puerto Rican mustached bats (*P. parnellii* and *P. portoricensis*, respectively) that occur on nearby Cuba and Puerto Rico, respectively.

Distribution. This bat inhabits only Hispaniola, although paleontologists have also recovered fossils on adjacent Gonave Island.

Measurements. Total length: 73–82 mm; tail length: 18–22 mm; hindfoot length: 11–13 mm; ear height: 19–21 mm; forearm length: 48–51 mm; body mass: 8–13 g.

Description. The Hispaniolan mustached bat is medium-sized, with a slender body and short dense fur. Overall, the back is dark brown, although somewhat paler on the head and nape, whereas the ventral pelage is a lighter brown with a drab wash on the chest and abdomen and paler on the throat. The pointed ears are broad

at the base and linked by thin bands of tissue that unite above the muzzle and form a wide fleshy knob. Numerous small papillae stud the center of the lower lip, and tufts of hair on either side of the nostrils create the mustache. As in all *Pteronotus*, the tragus is simple and pointed and has a small secondary fold on its anterior edge. The back boundary of the broad uropatagium is straight and not distinctly pointed. The tail is shorter than the uropatagium and projects from the middle of the membrane when the animal is motionless or climbing a cave wall. The peculiar arrangement of the tail identifies this mammal as a mormoopid, and on Hispaniola, it could be mistaken only for the sooty mustached bat (*P. quadridens*). However, the large size, presence of a fleshy bump behind the nostrils, and proportionally shorter ears readily separate the Hispaniolan mustached bat from its smaller cousin.

Natural History. The Hispaniolan mustached bat is somewhat common and occurs throughout the island, with captures occurring as high as 1211 m on the north slope of Pic Macaya. During the night, biologists often catch these bats along streams or in forested sites, especially dry woodlands. During the day, these mammals typically gather in small colonies containing up to a few hundred animals, although the population at Cueva Honda de Julián, in the Dominican Republic, may include as many as ten thousand of these creatures. They roost exclusively in caves, usually hot caves that also are occupied by other mormoopids, such as the Antillean ghost-faced bat (*Mormoops blainvillei*) and sooty mustached bat, and by phyllostomids, such as the Greater Antillean long-tongued bat (*Monophyllus redmani*) and brown flower bat (*Erophylla bombifrons*). The number of Hispaniolan mustached bats present at a cave often varies substantially, for unknown reasons, between the wet and dry seasons.

Although some of these bats begin leaving their cave near sunset, emergence does not peak until after dark. The Hispaniolan mustached bat is an aerial insectivore. No specific information on its diet is available, but it probably consumes large amounts of moths and beetles, which dominate the diet of the related Puerto Rican and Parnell's mustached bats. The Hispaniolan mustached bat produces multiharmonic echolocation calls while foraging that differ from the sounds made by other bats on the island. The pulses begin with an extended constant-frequency segment, followed by a brief downward sweep; the second harmonic is the strongest, and the constant frequency component is about 68 kHz. These bats most likely hunt in the interior of forests.

Data on reproduction are scant. Mammalogists have caught pregnant females in March and September, nursing mothers during September, and nonreproductive individuals in May and August. Other mormoopids in the Greater Antilles give birth once per year to a single offspring, and the same is likely true for the Hispaniolan mustached bat.

Status and Conservation. For more than a century, taxonomists thought that this bat was a subspecies of Parnell's mustached bat, and only in the last 10 years have biologists concluded that the two are distinct species. The IUCN has not evaluated the status of the Hispaniolan mustached bat, but more information on the life of this insular endemic will be required for a proper assessment of its conservation status.

Selected References. Tejedor et al. 2005a; Núñez-Novas et al. 2016, 2019; Pavan 2019.

Pteronotus quadridens

Sooty Mustached Bat, Murciélago Bigotudo Chico, Ptéronote Fuligeneux

Name. In Greek, *pteron* means "wing" and *notus* refers to "rear." These words denote the unusual appearance of Davy's naked-backed bat (*P. davyi*), which was the initial species described for the genus. In that animal, the hairless wings attach to the body at the center of the back, rather than to the side, as they do in all other bats in the West Indies. The specific epithet *quadridens* comes from the Latin nouns for the number "four" (*quattuor*) and "tooth" (*dens*), and indicates the presence of four small toothlike projections along the anterior rim of the ear in the original specimen.

Distribution. The sooty mustached bat resides only in the Greater Antilles—on Cuba, Jamaica, Hispaniola, and Puerto Rico, includ-

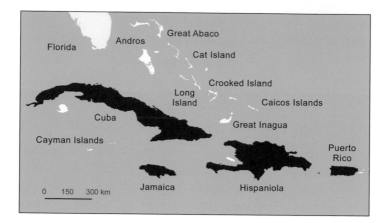

ing Mona. Fossil specimens, though, occur on South Abaco, North Andros, and New Providence islands in the Bahamas, suggesting a broader range in the past.

Measurements. Total length: 59–80 mm; tail length: 13–21 mm; hindfoot length: 6–11 mm; ear height: 13–19 mm; forearm length: 36–39 mm; body mass: 4–7 g. Animals from Cuba are a distinct subspecies and are smaller than the bats from the other main islands.

Description. Most of these bats appear sooty or grayish-brown from a distance, but individual hairs are tricolored, with dark sections at the base and tip of the shaft and a sandy-colored band in the center. The pelage of some individuals, though, appears slightly orange or yellow, especially in juveniles. Wings, tail membrane, and pinnae are naked and black. The tall pointed ears often angle toward the face and, when folded downward, extend past the animal's snout; a deep cleft occurs in the outer margin of the pinna. Short, bristly hairs (the "mustache") protrude from lips that appear permanently puckered, and wartlike tubercles and cutaneous flaps adorn the nose and mouth. A tail emerging from the center, rather than the end, of the broad uropatagium marks this animal as a member of the family Mormoopidae.

The sooty mustached bat is the smallest mormoopid living on

the Greater Antilles, although it overlaps slightly in size with the larger Macleay's mustached bat (*P. macleayii*) from Cuba. However, Macleay's mustached bat has from seven to ten short, parallel ridges or bands of tissue on the underside of the tail membrane that are perpendicular to the trailing edge (see Fig. 6.2); no such lines occur in the sooty mustached bat. In addition, the sooty mustached bat has from three to five tiny bumps above each nostril that are missing in Macleay's mustached bat.

Natural History. The sooty mustached bat is one of the most abundant species of bat on Cuba and Puerto Rico, but less so on Jamaica, where it is known from only a handful of locations. Biologists catch this species with mistnets in many types of habitats, including both xeric and moist landscapes. These animals are active at sites from sea level to at least 1250 m.

Like other members of its family, these bats are denizens of so-called hot caves, where the ambient temperature ranges from a "cool" 28 to a sweltering 40 °C, and relative humidity hovers around 90%. Such a warm environment results from the heat produced by thousands of tiny furnaces (the bats) and by the chemical breakdown of guano by millions of microorganisms. The small-bodied sooty mustached bat rapidly loses heat and moisture through the surface of its skin, but the seemingly severe environmental conditions in the cave greatly lessen these physiological problems.

Colonies typically consist of fewer than one thousand individuals on Jamaica, whereas populations on Cuba usually range between five and fifteen thousand bats. However, some hot caves on Puerto Rico, such as Cucaracha Cave, near Aguadilla, house more than 140 thousand sooty mustached bats. Although other species shelter in the same subterranean site, the sooty mustached bat keeps to itself, roosting in deep recesses within the cavern.

On Puerto Rico, the sooty mustached bat is usually the first species to leave the cave each night, beginning between 11 minutes before and 10 minutes after sunset. These bats exit en masse and fly in an undulating column that travels over hills and valleys, crossing roads and forests; some columns maintain cohesiveness for 9 km or more, before individuals scatter and begin hunting on their own.

Average flight speed is about 27 km/hr. A few individuals return to and leave the cave multiple times throughout the night, but most animals finally come home for the day between 17 minutes before and 10 minutes after sunrise. On Cuba, feeding activity gradually decreases as the ambient temperature falls below 26 °C, although some bats continue to forage even at temperatures as low as 8 °C.

Short, broad, and rounded wings suggest that the sooty mustached bat maneuvers well and possibly forages in somewhat cluttered environments, such as the forest understory. This mammal produces high-frequency echolocation calls that allow it to detect flying insects while avoiding leaves, branches, and other obstacles. Each pulse begins with a short stretch of quasi-constant frequency that transitions into a frequency-modulated sweep from 82 to 59 kHz. Average call duration is 5 milliseconds.

About 42% of the volume of this bat's diet is composed of beetles, whereas moths and flying ants combine to form another 50%. The remainder consists of flies, including mosquitos, spiders, bugs, and the occasional earwig or small roach. These tiny mammals can pack their stomachs with insects amounting to 25% of the bats' body mass during frenzied feeding right after sunset.

Females give birth to a single young once per year. Gravid individuals occur with increasing frequency from February through May and June, and almost all females are pregnant during those last 2 months. A few bats give birth in June, although the proportion of adults that is lactating does not peak until July and then decreases through September. Newborns are up to 30% of adult body mass at birth, and the mothers leave their large pups in dense clusters of 50 to 200 other youngsters while out foraging. Male and female adults often roost in separate caves during the maternity period on Puerto Rico, although such sexual segregation does not seem to occur on Cuba.

Birds, reptiles, and mammals prey on the sooty mustached bat. An early emergence makes this animal susceptible to predation by diurnal birds, such as American kestrels (*Falco sparverius*), red-tailed hawks (*Buteo jamaicensis*), and merlins (*Falco columbarius*); these avian predators snatch the bat out of the air, quickly land on a high perch, and consume the carcass after removing the wings.

Nocturnal avian predators include ashy-faced and American barn owls (*Tyto glaucops* and *T. furcata*, respectively). The Puerto Rican boa (*Chilabothrus inornatus*) consistently eats these bats, and endemic boas living on Cuba (*C. angulifer*), Hispaniola (*C. striatus*), and Jamaica (*C. subflavus*) may as well. Feral cats on Puerto Rico catch the sooty mustached bat as it emerges, and even the greater bulldog bat (*Noctilio leporinus*) occasionally attacks and eats its smaller cousin. Many (85%) individuals harbor at least one ectoparasite, with the two most common being chiggers attached to the wing and tail membranes and cigar-shaped mites (*Lawrenceocarpus micropilus*) that cling to hairs inside the pinna and on the tragus; as many as seventy-nine mites occur in the ears of just one bat. Females harbor more parasites than do males.

Status and Conservation. Although this bat is somewhat vulnerable because it relies on an uncommon type of roost, a hot cave, it is a very common species that inhabits multiple islands. The IUCN classifies the species as Least Concern.

Selected References. Kurta et al. 2007; Rodríguez-Durán et al. 2010; Mancina et al. 2012; Rolfe and Kurta 2012; Rolfe et al. 2014.

Funnel-Eared Bats

The family Natalidae includes only three genera and eleven species. These bats live throughout the Neotropics, from northern Mexico to northeast Paraguay, as well as the West Indies, including the Bahamas. In other families of bats, most species inhabit the mainland, and only a small proportion occupy Caribbean islands; however, the Natalidae shows the opposite trend, with eight of the eleven species restricted to the islands. Throughout its distribution, natalids dwell in tropical habitats ranging from semiarid scrublands to humid forests.

These mammals are characterized by broad, funnel-shaped ears and a short, twisted tragus. Hairs on the upper lip impart the appearance of a mustache, and the inconspicuous eye appears hidden between the ear and the mustache. The uropatagium totally encases the tail vertebrae, which extend to the back edge of the membrane (see Fig. 0.2). Proportionately, these bats have the lengthiest tails and hind legs of any bats in the West Indies; the tail is typically longer than the head and body combined, and the length of the tibia in most species is over half that of the forearm. The calcar is also extensive and, in most natalids, supports 50 to 65% of the trailing edge of the uropatagium. The fur is dense, long, and soft, with color that ranges from gray to yellowish-, orangish-, or reddish-brown. Wing and tail membranes are brown or black and often translucent. All natalids are small; the largest, the Cuban greater funnel-eared bat (*Natalus primus*), weighs at most only 13 g, whereas the most diminutive species, Gervais's funnel-eared bat (*Nyctiellus lepidus*), is one of the tiniest mammals in the world at a mere 2 to 3 g. The skull is elongate, with a narrow, dorsoventrally flattened rostrum, and the cheek teeth have W-shaped ridges (dilambdodont dentition) for shearing the exoskeletons of their insect prey.

Bats in this family spend the day in caves, cavities in trees, and occasionally buildings. In the West Indies, they often roost in caves that contain or are located near bodies of water, an association presumably related to the rapidity with which some natalids dehy-

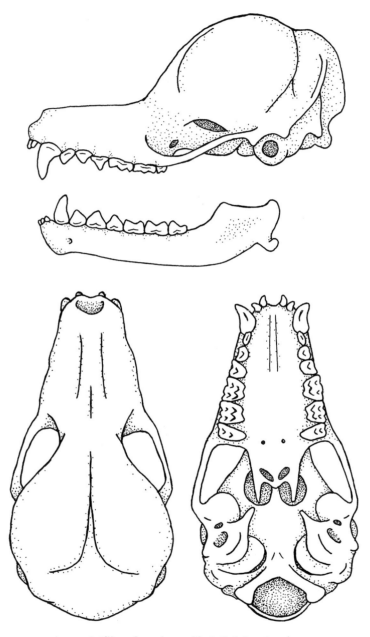

Lesser Antillean funnel-eared bat, *Natalus stramineus*

drate while outside their humid shelters. For instance, if Mexican greater funnel-eared bats (*Natalus mexicanus*) are removed from their daytime roost, their wings quickly become dry and crinkly, and the animals die within just 1 or 2 hours. Natalids gather in colonies that range from a few individuals to several hundred animals. Two of the eleven species, though, are apparently restricted to single caves, making them highly imperiled.

Natalids forage in the shrubby understory and within or directly over the forest canopy, feeding mostly on soft-bodied prey, such as moths, termites, and flies. Consequently, echolocation involves frequency-modulated calls of high frequency, short duration, and low intensity. The wings are broad, and the flight of these mammals is highly maneuverable and slow and often likened to the fluttering movements of moths. In addition to catching insects in midair, some natalids may be gleaners, plucking their prey from the surface of a leaf or tree trunk.

Courtship and copulation occur over the winter dry season, and parturition happens during the spring-summer rainy period, when insect activity is greatest. Gestation lasts 5 months or possibly longer, litter size is one, and births occur once per year. Adult males, nonreproductive females, and reproductive females usually segregate, with expectant and nursing mothers moving to the deepest parts of the cave or sometimes to a completely different roost. Adult males possess a unique domed structure above the snout, called the natalid organ, which is composed of sensory cells and glands that discharge a thick greenish substance (see Fig. 6.1c); the exact function of the organ is still a mystery, but the secretions presumably play a role in mating or territorial behavior.

Selected References. Tejedor et al. 2005b; Tejedor 2011, 2019.

Chilonatalus macer

Cuban Lesser Funnel-Eared Bat, Murciélago Oreja de Embudo Chico, Natalide de Cuba

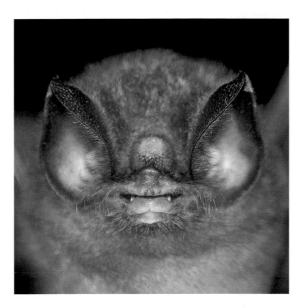

Name. Bats in the genus *Chilonatalus* have large, funnel-shaped ears, similar to bats within the genus *Natalus*, but also have broad lips with fleshy outgrowths, like those of mustached bats in the genus *Pteronotus*. An earlier synonym for *Pteronotus* was *Chilonycteris*, with *chilo* meaning "lip" in Greek. *Chilonatalus* is simply a combination of those two names. Gerrit S. Miller, who described this particular species in 1914, emphasized that it had a more elongate rostrum than the Caribbean lesser funnel-eared bat (*C. micropus*) from Jamaica, and the specific epithet *macer*, which is Latin for "skinny" or "thin," apparently refers to this difference.

Distribution. As its common name suggests, this species currently lives on Cuba and Isla de la Juventud, although fossil and subfossil remains from the late Quaternary period occur on Grand Cayman Island, as well as Cuba.

Measurements. Total length: 85–92 mm; tail length: 46–56 mm; hindfoot: 7–8 mm; ear height: 13–16 mm; forearm length: 32–34 mm; body mass: 2.5–4 g.

Description. This species is one of the smallest bats on Cuba. Pelage is dense and lax and overall color is orange-brown to light brown; the back is slightly darker than the belly. Individual hairs are from 5 to 8 mm long on the back but shorter on the underside, and each shaft is bicolored, with a dark tip and a lighter base. The legs are inordinately long compared with those of most bats yet typical of natalids; length of the tibia in this species is more than 50% that of the forearm. The posterior margin of the broad interfemoral membrane is V-shaped and fringed with fine hairs. The broad ears are funnel shaped and slightly pointed at the tip, and the short tragus forms a twisted helix. The elliptical nostrils open at the tip of the upper lip, and dense patches of curved hairs form a mustache on the back halves of the upper lip. A forward-projecting, lightly haired, fleshy protuberance occurs on the midline of the snout behind the nostrils, and from the side, this bump, along with the upper and lower lips and a dermal flap beneath the chin, give the appearance of four lips (see Fig. 6.1c). Funnel-shaped ears, hairy fringe on the back edge of a pointed tail membrane, and a double-lipped profile separate the Cuban lesser funnel-eared bat from all other species on Cuba and surrounding islands. The ovoid natalid organ of males begins on top of the forehead and extends along most of the snout so that the length of the organ equals half the length of the skull; the natalid organ is quite obvious in the accompanying photo.

Natural History. The Cuban lesser funnel-eared bat is widespread in lowland areas of the island, up to an elevation of 450 m. Although sometimes occurring in dry habitats, such as thorn scrub, this bat most often frequents moist environments, including evergreen and semideciduous woodlands. These mammals generally form small colonies of thirty to fifty individuals, although occasionally a few hundred animals may be present. The usual roosting site is a room with a low ceiling inside a warm and humid cave, where each bat hangs quietly by a single foot, well separated from others of its kind; solitary animals occasionally occupy cooler sites near the entrance. Different species of bats often occur in different parts of the same caverns, especially various mormoopids, as well as Gervais's funnel-eared bat (*Nyctiellus lepidus*), Waterhouse's leaf-nosed bat (*Macrotus waterhousii*), the Jamaican fruit-eating bat (*Artibeus jamaicensis*), and the big brown bat (*Eptesicus fuscus*); however, the Cuban lesser funnel-eared bat almost always roosts apart and does not interact with the others. In addition to subterranean sites, there are three old and very atypical reports of individuals of this species resting during the day inside a tobacco house, within a clump of sugarcane, and even inside a tent; most biologists assume that these seemingly aberrant sites represent emergency shelters adopted by animals unable to return to their home cave before sunrise.

The evening exodus from the cave begins about 15 minutes after sunset and continues sporadically for 30 to 45 minutes. These bats have a slow, fluttering flight, reminiscent of large moths. Low wing loading and broad wings allow these mammals to forage close to the ground and access dense vegetation, where they catch insects in midflight by scooping them out of the air with a wing or their expansive tail membrane. The diet is specialized, composed principally of moths, although the stomach of one bat also contained parts of a spider. Natalids, in general, dehydrate quickly when outside their humid caves, and rapid water loss combined with slow, low-level flight suggest that these bats do not forage far from home.

Reproductive data are sparse, but the Cuban lesser funnel-eared bat apparently has a single annual estrus, giving birth to one young per litter at the end of the rainy season. Naturalists have reported Cuban lesser funnel-eared bats in early pregnancy during May, births at the end of June, and nursing mothers as late as 10 August. Only nonbreeding females are known between December and March. The Cuban lesser funnel-eared bat may be similar to Gervais's funnel-eared bat in which the sexes segregate during gestation and lactation, with females moving to warmer and deeper chambers of a cave or migrating a few kilometers to a separate maternity site. The American barn owl (*Tyto furcata*) and the bare-legged owl (*Margarobyas lawrencii*) are the only documented predators.

Status and Conservation. The Cuban funnel-eared bat was considered a subspecies of the Caribbean funnel-eared bat (*Chilonatalus micropus*) until 2011. The IUCN currently classifies the Cuban funnel-eared bat as Data Deficient, indicating that additional research is needed to determine its conservation status.

Selected References. Silva-Taboada 1979; Tejedor 2011, 2019.

Chilonatalus micropus

Caribbean Lesser Funnel-Eared Bat, Murciélago Orejón Chico, Natalide à Pattes Courtes

Name. Members of *Chilonatalus* have physical characteristics of bats in the genus *Chilonycteris* (now called *Pteronotus*) and also in the genus *Natalus*. To reflect this similarity, Gerrit S. Miller simply merged the two names when describing the new taxon *Chilonatalus* in 1898. The word *micropus* is a combination of Greek words meaning "small foot."

Distribution. This species inhabits Jamaica and Hispaniola, as well as the isolated Caribbean islands of Saint Andrew and Providence;

the latter islands, which are governed by Colombia, are small (17–26 km²) and located more than 625 km southeast of Jamaica and about 200 km east of Nicaragua.

Measurements. Total length: 80–89 mm; tail length: 45–48 mm; hindfoot length: 7–8 mm; ear height: 13–16 mm; forearm length: 31–35 mm; body mass: 2.6–5 g. Females have slightly longer forearms than do the males. Two subspecies exist that differ in size, and the animals from Hispaniola and Jamaica are larger than are those living on Providence.

Description. The body appears light grayish-brown to yellowish- or reddish-brown, with the belly lighter than the head or back. Hairs along the back are from 4 to 7 mm in length and are bicolored, dark at the tips and lighter at the base. As in all members of the Natalidae, the distinctive pinnae are tall, very wide, and funnel shaped. The nostrils are elliptical in outline and open at the end of short tubular projections on the outer margin of the upper lip. A thin flap of hairy skin occurs just below the upper lip, and a forward-

projecting, sparsely haired, fleshy bump sits on the snout behind the nostrils; from the side, this lower flap and upper protuberance, combined with the true lips, give an impression of two pairs of lips. In Caribbean lesser funnel-eared bats, the natalid organ of adult males is hemispherical in shape, mostly hairless, and located at the junction of the rostrum and cranium. Despite the animal's small body, the prominent penis is longer than that of any other member of the family; the structure varies from 4 to 6 mm long when flaccid, which amounts to about 15% of the length of the forearm. No other bat on Hispaniola or Jamaica displays the double-lipped appearance, which is unique to the genus *Chilonatalus*, and although other funnel-eared bats are native to these two islands, the forearm and foot of the Caribbean species are distinctly smaller.

Natural History. Throughout its range, the Caribbean lesser funnel-eared bat is an uncommon species, and most knowledge of its ecology and behavior comes from Jamaica, where biologists have captured this animal from less than twenty-five sites. The bat appears to frequent moist forests that receive as much as 2900 mm of rain per year and are located from sea level up to 400 m in elevation. This funnel-eared bat is an obligate cave-roosting species, and most selected caverns are the so-called hot caves, which accumulate metabolic heat produced by the bats and by microbes in the guano and result in a roosting environment that is consistently warmer than on the surface. The caves also are very humid, and some have high concentrations of carbon dioxide that make breathing difficult for humans yet seemingly have no impact on the bats. This mammal potentially shares its caves with nine other species of bats, but it tends to roost separately from the others.

Presence of W-shaped ridges on the cheek teeth (dilambdodont dentition) indicate that this species is insectivorous, but no specific dietary information exists. Based on the structure of the external ear, biologists predict that this bat hunts its prey using frequency-modulated echolocation calls between 40 and 80 kHz. Caribbean lesser funnel-eared bats probably intercept their prey while the mammal flies along forest edges, around shrubs and the crowns of trees, and perhaps in the interior of forests. As do other natalids, these animals fly very slowly and have high maneuverability, which allows them generally to avoid the mistnets that field biologists use to capture bats.

Data on reproduction are very fragmentary. To date, mammalogists have not captured a single pregnant individual, although up to 90% of females are lactating in mid-July. However, none produces milk in late July, when adult males have regressed testes, suggesting that the annual reproductive cycle ends by August. The mating system is unknown, but the very lengthy penis suggests female promiscuity and a need for deep delivery of sperm that the next partner is less likely to dislodge.

Status and Conservation. The Caribbean lesser funnel-eared bat is widespread on Jamaica, with records from most parishes; nevertheless, this animal is mostly uncommon and locally abundant only at a few caves, such as Oxford, St. Clair, and Windsor. It is known from just one cave on Hispaniola, and little information is available concerning the bats on Saint Andrew and Providence islands, which also harbor dense human populations. The IUCN ranks this species as Vulnerable.

Selected References. Kerridge and Baker 1978; Tejedor 2011.

Chilonatalus tumidifrons

Bahamian Lesser Funnel-Eared Bat, Murciélago Cabezón, Natalide des Bahamas

Name. Bats in the genus *Chilonatalus* have fleshy outgrowths near the lips, like animals in the genus *Chilonycteris* (now called *Pteronotus*), and cone-shaped ears, like members of the genus *Natalus*. *Chilonatalus* is simply a combination of those two names. *Tumidifrons* likely derives from the Latin words *tumere*, meaning "to swell," and *frons*, meaning "front," and references the prominent natalid organ on the head of adult males.

Distribution. This animal is known from only its namesake islands, the Bahamas. It currently inhabits Great Abaco, Andros, and San Salvador islands, although fossils indicate that these bats also once lived on Cat, Great Exuma, and New Providence islands.

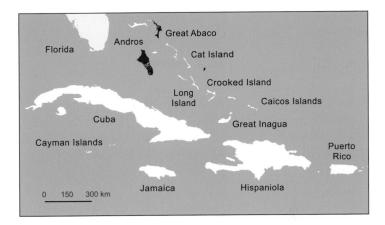

Measurements. Total length: 85–93 mm; tail length: 47–48 mm; hindfoot length: 7–8 mm; ear height: 15–17 mm; forearm length: 32–36 mm; body mass: 3–3.5 g.

Description. These mammals are some of the smallest bats in the West Indies and weigh less than two American pennies. Hairs on the back of this diminutive mammal are bicolored, dark at the tip and lighter at the base, and overall the body appears light brown to orangish-brown. The belly, in contrast, is a pale buff or light yellow. Dorsal hairs are from 5 to 9 mm long, but ventral hairs are shorter at 4 to 7 mm. The ears are light yellow, with dark edges, and the wing membranes are dark brown. Ears are tall, broad, and shaped somewhat like a funnel. The natalid organ is essentially naked and large, covering about half the skull between the ears. Elliptical nostrils open at the end of tubular projections on the margin of the upper lip, and a lightly haired, fleshy bump projects forward above the lip. If you view the bat from the side, this fleshy projection, the upper and lower lips, and the flap of skin below the mouth combine to give the impression of a "double-lipped" face, which is characteristic of all lesser funnel-eared bats. A thin calcar supports the well-developed tail membrane for two-thirds of its width, and short hairs line the trailing edge of the membrane. The only other funnel-eared bat in the Bahamas is Gervais's funnel-eared bat (*Nyctiellus lepidus*), but that species has a shorter forearm, lacks the double-lipped appearance, and is missing the fringe of hairs on the tail membrane.

Natural History. The Bahamian lesser funnel-eared bat frequents deciduous forests with medium rainfall (1000–1300 mm annually). This species roosts exclusively in humid caves, both large and small, that have constant but moderate internal temperatures of 23–24 °C. Inside the caves, the bats roost solitarily, above water and over dry areas, typically hanging from the ceiling or a ledge by a single foot. Different caves contain from ten to five hundred of these mammals, but the population seems to fluctuate, even on a daily basis. At Alter Cave, on San Salvador Island, for example, the number of these bats varies unpredictably from twenty to two hundred individuals, with some animals moving overnight, more than 1 km, to a different cave. They occasionally share their underground home with the big brown bat (*Eptesicus fuscus*), Waterhouse's leaf-nosed bat (*Macrotus waterhousii*), or the buffy flower bat (*Erophylla sezekorni*), although the various species roost apart while inside the cavern. On one occasion, a colony containing about three hundred Bahamian lesser funnel-eared bats apparently abandoned their home after one to two hundred boisterous buffy flower bats occupied the site.

Bahamian lesser funnel-eared bats begin leaving their cave about 10 minutes after sunset, and emergence of the colony is complete within 35 minutes or so. As in other natalids, flight is slow and maneuverable and often appears mothlike. Their flight style is suitable for foraging around or in the dense vegetation of the local forests, known as coppice, and for avoiding mistnets set by biologists intent on capturing these bats. Nothing, though, is known concerning their use of echolocation or their specific diet. Nonetheless, other small species of funnel-eared bats on Cuba eat moths, flies, and leafhoppers, as well as flying termites and ants, and dietary choices by the Bahamian lesser funnel-eared bat are probably comparable.

There is essentially no knowledge of this bat's reproductive cycle or social system, although there are reports of exclusively male colonies forming in July. In Gervais's funnel-eared bat, males and females roost separately during the summer reproductive season, and in that species, females are pregnant from March to July and lactating from July through September. Reproduction by the Bahamian lesser funnel-eared bat may follow the same schedule.

Status and Conservation. According to the IUCN, the Bahamian lesser funnel-eared bat is Near Threatened. Populations are fragmented and scattered across three small islands, and the size and trend of these groups are unknown. Threats include human activities inside roosting caves and natural disasters, including hurricanes and flooding of low-lying underground shelters.

Selected References. Tejedor 2011; Speer et al. 2015.

Natalus jamaicensis

Jamaican Greater Funnel-Eared Bat, Murciélago Orejón Jamaiquino, Natalide de Jamaïque

Name. Taxonomists are not certain, but the name *Natalus* may be derived from the Latin word *natalis*, meaning "of one's birth." They speculate that the term refers either to the unknown "birthplace" (collection locality) of the original specimen or to the small and newborn-like appearance of the Lesser Antillean funnel-eared bat (*N. stramineus*), which was the original species placed into the genus. *Jamaicensis* indicates the homeland of the Jamaican greater funnel-eared bat.

Distribution. This rare mammal is endemic to the island of Jamaica, where biologists have found living specimens in just a single location—St. Clair Cave, in St. Catherine Parish. The only other documented occurrence is from Wallingford Cave, in St. Elizabeth Parish, where Harold E. Anthony discovered a single fossilized mandible about 100 years ago.

Measurements. Total length: 105–112; tail length: 57–60 mm; ear height: 18 mm; forearm length: 44–46 mm; body mass: 5.9–7.3 g.

Description. The overall color of the lax pelage varies from yellowish-brown to brownish-orange. Dorsal hairs are yellowish-brown at the base and darker at the tips; ventral shafts are yellowish-brown at the bottom, as well, but pinkish toward the end.

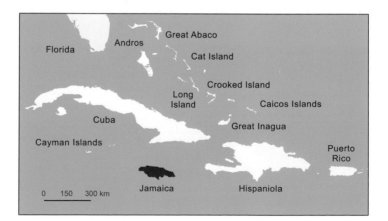

Hairs from the back are about 9 to 11 mm long, while those from the belly are conspicuously shorter, about 6 to 7 mm in length. The skin of the funnel-shaped ears appears light brown, and in lateral view, the broad pinnae obscure the black beady eyes. Mustache-like hairs coat the upper lip and the top of the flattened snout. The lengthy hind legs support a well-developed interfemoral membrane that forms a backward-directed V and has a fringe of scant hairs along the trailing edge; the tail is as long or longer than the body and extends to the point of the V but not beyond. The natalid organ of males is elliptical in outline and covers the forehead, from the back of the rostrum to the top of the skull.

Mustached bats (*Pteronotus*) have similar facial hair, but their tail is short and does not span the interfemoral membrane. The comparable-sized Jamaican brown bat (*Eptesicus lynni*) also has a pointed uropatagium; however, its membrane does not sport a hairy fringe, and its blackish ears are not funnel shaped. The only other bat on the island that has a mustache, long tail, and V-shaped membrane is the Caribbean lesser funnel-eared bat (*Chilonatalus micropus*); nevertheless, that species is considerably smaller, with a forearm length of less than 36 mm, and it too lacks a furred border on the tail membrane.

Natural History. St. Clair Cave is located in a landscape covered by semideciduous forest at about 100 m above sea level. This cavern is a hot cave, and the Jamaican greater funnel-eared bat roosts in a small chamber adjacent to the warmest room, which local speleologists have dubbed the "Inferno Passage." Air temperature is about 30 °C, relative humidity approaches 100%, and levels of carbon dioxide are elevated, making breathing a challenge for human explorers. Donald McFarlane (1985, 17), who studied caves, bats, and other organisms in the Greater Antilles for many years, provides an apt description of the conditions within the Inferno Passage: "The floor of the passage is largely occupied by thigh-deep pools of fetid water and liquid guano. The walls are a living carpet of scavenging invertebrates—cockroaches, cave crickets, millipedes, and innumerable smaller forms. The air, vibrating with the beat of a myriad of unseen wings and raised some five or six degrees by the heat of the bats' bodies, resembles some kind of Stygian sauna bath and is filled with a rain of bat urine and excrement."

Approximately 50 Jamaican greater funnel-eared bats once shared the side chamber with 150 Caribbean lesser funnel-eared bats. Individuals of the larger species roosted closer to the entrance of the caldarium, and they usually hung by one foot so that even the slightest movement of air caused the animal to swing to and fro. While roosting, each animal remained distant from others of its kind by about 7 to 10 cm, but when disturbed, these bats flew about the cave in a compact mass that resembled a flock of fluttering moths. Other species found in St. Clair Cave include three species of mustached bats, the buffy flower bat (*Erophylla sezekorni*), the Greater Antillean long-tongued bat (*Monophyllus redmani*), and the endangered Jamaican flower bat (*Phyllonycteris aphylla*), among others.

Although specific dietary information is unavailable, this mammal is certainly insectivorous. Broad wings and slow maneuverable flight are typical of other members of the family, allowing them to forage in dense vegetation, and the Jamaican greater funnel-eared bat likely does the same. No data exist concerning reproductive patterns, longevity, or mortality factors. Litter size is probably one, as it is in all other members of the family.

Status and Conservation. The Jamaican greater funnel-eared bat is one of the most endangered species of mammals, having one small population restricted to a single known cave that has no official protection. This mammal is considered Critically Endangered, based on its restricted distribution, combined with habitat fragmentation and introduced predators, especially feral house cats, that are negatively affecting the remaining bats.

Selected References. Goodwin 1970; McFarlane 1985; Genoways et al. 2005; Tejedor 2011.

Natalus major

Hispaniolan Greater Funnel-Eared Bat, Murciélago Orejón de La Española, Natalide d'Hispaniola

Name. The generic name is apparently from the Latin adjective *natalis*, which means "related to birth." Use of this word might reference the small head and body (4–5 g) of the Lesser Antillean funnel-eared bat (*N. stramineus*), which was the first species placed into the genus *Natalus*. *Major* is a Latin adjective for "greater." When preparing the description of this species in 1902, Gerrit S. Miller remarked that it was much larger than the Lesser Antillean funnel-eared bat.

Distribution. This bat resides only in the Dominican Republic and Haiti on the island of Hispaniola in the Greater Antilles.

Measurements. Total length: 100–112 mm; tail length: 50–59 mm; hindfoot length: 8–11mm; ear height: 13–19 mm; forearm length: 41–45 mm; body mass: 5.5–10 g.

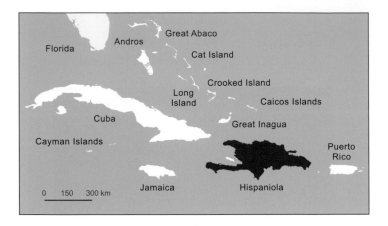

Description. There are two main patterns of coloration—an overall grayish-tan or an orangish- to yellowish-brown—although intermediate shades occur in some individuals. Hairs along the back are dense, long (about 7 mm), and bicolored, with tips darker than the bases; ventral hairs also are weakly two-toned or unicolor. As in all *Natalus*, Hispaniolan greater funnel-eared bats have small eyes virtually enclosed within the wide, conical pinnae that are swirled into a funnel shape. The short muzzle appears flattened; stiff hairs project from the top of the snout, and others protrude from the upper lip and provide the look of a disorderly mustache. The legs are long, with the length of the tibia about equal to or more than half the size of the forearm. The tail extends to the edge of the pointed tail membrane, which terminates with a fringe of short hairs. The elliptical natalid organ of males extends from the back of the rostrum to the top of the head.

Vespertilionid bats, such as the big brown bat (*Eptesicus fuscus*), are somewhat similar in size and also have a V-shaped tail membrane. However, the wide funnel-shaped ears, long legs, and hairy edge of the membrane clearly separate the Hispaniolan funnel-eared bat from any vespertilionid in the Greater Antilles. The only other natalid on Hispaniola is the Caribbean lesser funnel-eared bat (*Chilonatalus micropus*), but that species is much smaller, with a forearm length less than 36 mm.

Natural History. The Hispaniolan greater funnel-eared bat is widespread on the island, although not very abundant. This mammal is active in various types of forest, including dry, semiarid, and moist woodlands, where annual precipitation ranges from 700 to 2800 mm. Wildlife biologists have captured it at sites from sea level to about 1000 m in elevation.

Like other members of the family, the Hispaniolan greater funnel-eared bat generally roosts underground, in both small and large caves. It often spends the day in hot caves, especially those with pools of water, and these bats apparently select chambers or corridors next to, but not in, the hottest room (caldarium). The bats hang by one or two feet, separated by 10 cm or so from others of their species, on the walls or from low ceilings. When disturbed, these animals tend to fly away close to the ground and near the walls of the cavern. The number of Hispaniolan greater funnel-eared bats in any single site is typically small, and the largest known colony, in Cueva Honda de Julián, in the Dominican Republic, comprises only one hundred individuals. Hot caves, with their wide range of temperatures, are popular with many types of bats, and as many as ten other species may share a subterranean roost with the Hispaniolan greater funnel-eared bat, including the Jamaican fruit-eating bat (*Artibeus jamaicensis*), Greater Antillean long-tongued bat (*Monophyllus redmani*), Cuban flower bat (*Phyllonycteris poeyi*), and sooty mustached bat (*Pteronotus quadridens*). Despite the predilection for caves, a group of nine Hispaniolan greater funnel-eared bats once sheltered inside a large, hollow tree; the cavity was big enough to accommodate a human, and the bats clung to the top of the void, 3 m above the ground. In addition, a colony of fifty of these mammals roosted within a brick-lined tunnel, 1.5 m high and open at both ends, at an abandoned sugar mill in Santo Domingo.

These anaimals are insectivorous, although no specific information on their diet is available. They apparently begin foraging just after twilight. Like other species in the family, Hispaniolan greater funnel-eared bats have broad wings and a slow fluttering flight, yet they are highly maneuverable, which suggests that they hunt within thick vegetation. In general, biologists believe that natalids are prone to desiccation when outside their humid roosts, and consequently, these bats likely do not travel far from the moist environment of their home cave.

The seasonal timing of pregnancy and lactation is totally unknown. Births probably occur once per year, near the end of the dry season or the start of the rainy period, and litter size is one, as in other members of the family. The only reported predator is the American barn owl (*Tyto furcata*).

Status and Conservation. Although the Cuban and Jamaican greater funnel-eared bats (*N. primus* and *N. jamaicensis*, respectively) are known from only single caves, the Hispaniolan species occurs in at least thirty localities. Nonetheless, the Hispaniolan greater funnel-eared bat seems mostly dependent on hot, moist caves, which are not abundant and not protected on their island. Overall, in the Caribbean, there is a clear need to conserve hot caves to protect endemic fauna—both vertebrates and troglobitic invertebrates. The IUCN currently considers this bat as Near Threatened.

Selected References. Timm and Genoways 2003; Tejedor et al. 2004, 2005a; Tejedor 2011.

Natalus primus

Cuban Greater Funnel-Eared Bat, Murciélago Oreja de Embudo, Grande Natalide

Name. The generic name may be from the Latin word *natalis*, which means "related to birth," and some biologists speculate that the word references the small size of most natalids. *Primus* is the Latin word for "first." Harold E. Anthony, who described this particular funnel-eared bat in 1917, did not explicitly state why he selected *primus* as the specific epithet, but he did emphasize that this species appeared larger than all other members of the family Natalidae.

Distribution. The single extant population identified by biologists inhabits Cueva La Barca, in the Guanahacabibes Peninsula, at the

westernmost tip of Cuba. In the past, though, the species was more widespread. Fossil and subfossil remains occur in caves throughout Cuba, as well as on Isla de la Juventud, Grand Cayman, and the Bahamas (Abaco, Andros, New Providence, and Eleuthera islands).

Measurements. Few external measurements of this rare mammal are available. Ear height: 20–22 mm; forearm length: 46–52; tibia length: 25–29 mm; and body mass: 6–13 g. Males are heavier and have longer tibias than females.

Description. Two color morphs exist, one yellowish and the other grayish. Individual hairs are about 8 mm long; dorsal hairs are bicolored, with bases lighter than tips, whereas ventral hairs are monocolored. The muzzle is long and dorsoventrally flattened. The tip of the muzzle projects slightly beyond the upper lip, and the small nostrils open downward and to the side, rather than forward. A tuft of dense hairs on either side of the upper lip gives the impression of a mustache. The lower lip is thick and has numerous

transverse grooves. The funnel-shaped ears are broad, and in lateral view, the outer margin of the pinna curves forward and hides the eye and the short, twisted tragus; the inner and outer margins of the ear are straight and contribute to a squarish outline. The legs are distinctly lengthy and include a tibia that is more than half as long as the forearm. A thin calcar supports a large uropatagium for half its width, and short hairs line the trailing edge of the membrane, which forms a backward-directed V. The combination of large forearm, long legs, funnel-shaped ear, and V-shaped membrane distinguishes the Cuban greater funnel-eared bat from any other species on the island.

Natural History. Cueva La Barca is located near sea level, only 800 m from the coast, and surrounded by semideciduous forest, in a region receiving 1400 mm of rainfall annually. The cavern is a phreatic cave with five rooms, including a hot room or caldarium, several small entrances, and two pools of water ("guano swamps"). This cave is home to fourteen species of bats, which is the highest species richness in any subterranean site of the West Indies. Cuban greater funnel-eared bats generally occupy well-ventilated corridors and the darkest chambers near, but not in, the hot room; air temperature at roosting sites is about 24 or 25 °C, with relative humidity close to 100%. Individuals typically roost along the walls, about 1 m from the ground. The animals dangle freely from rocky protrusions or ledges, by one or both feet and without their belly contacting the wall. The bats roost in small groups of a few dozen or more animals, although they remain at least 10 cm from each other. These mammals are skittish, and after the smallest disturbance, they flee, usually flying in a constant stream, one by one, within 1 to 2 m of the wall and no more than 1.5 m above the floor.

Most Cuban greater funnel-eared bats leave their roost each night between 30 and 50 minutes after sunset. No data exist concerning home range, although field biologists captured one individual about 3.5 km from its cave. Echolocation pulses as the bats fly along a passage inside Cueva La Barca contain two or three harmonics, with most energy in the second. A typical second harmonic begins near 114 kHz and then rapidly decreases to 64 kHz over 1.7 milliseconds. As do other natalids, this species flies slowly and is highly maneuverable, and it probably forages in the cluttered understory of the surrounding forest or around the canopy of trees. Analyses of stomach contents indicate that these mammals feed mostly on moths, crickets, and beetles and less frequently on ants, flies, lacewings, and true bugs. In captivity, these bats are aggressive and often attack other species, even those that are larger.

The Cuban greater funnel-eared bat has a 6-month-long reproductive season. Biologists have watched multiple pairs copulating inside Cueva La Barca in mid-April and have captured pregnant females, each containing a single embryo, as early as 1 May. However, scientists have never captured mothers with dependent young, suggesting that expectant females move past a guano swamp seldom crossed by humans and into the darkest and hottest areas of the cave before giving birth. Records of lactating females exist as late as September, but only nonreproductive individuals are known from November and December.

Status and Conservation. The Cuban greater funnel-eared bat is one of the rarest and most threatened of Caribbean bats. Taxonomists originally described this bat from fossilized remains, and the species was believed extinct until biologists located the colony in Cueva La Barca in 1992. When rediscovered, the population comprised a few thousand individuals, but less than 750 bats remain. Small populations of any species can lead to inbreeding and increased risk of disease and can potentially lessen the group's resilience to natural or human-driven calamities. Cueva La Barca, the only known roost for this bat, is vulnerable to ongoing ceiling collapse, potential flooding from severe storms, and increased visitation by tourists and cavers. An additional area of concern is the unknown impact of forest fragmentation, caused by logging in the surrounding woodlands, where these bats presumably forage. The IUCN lists the species as Vulnerable; however, this classification is too conservative and seemingly belies the urgent need of conservation actions to preserve this insular endemic.

Selected References. Tejedor et al. 2004; Tejedor 2011; Sánchez et al. 2017; De la Cruz Mora and García Padrón 2019.

Natalus stramineus

Lesser Antillean Funnel-Eared Bat, Murciélago Orejón de las Antillas Menores, Natalide Pailée

Name. In his original description of the genus in 1838, John E. Gray offered no explanation for the choice of the name *Natalus*. Later authors suggested that the word came from the Latin *natalis*, meaning "of one's birth," and they surmised that Gray was referring to the unknown capture locality ("birthplace") of the type specimen or to the animal's small and delicate appearance. *Stramineus* is the Latin adjective for "straw-colored" and describes the overall pale yellow coloration of the first specimen. At one time, the name *Natalus stramineus* applied to virtually all continental forms of *Natalus*, as well as to all populations in the West Indies, but today taxonomists restrict its use to those animals living in the Lesser Antilles, north of Saint Vincent.

Distribution. The Lesser Antillean funnel-eared bat roams over most major islands from Martinique in the south to Anguilla in the north. This mammal inhabits Anguilla, Antigua, Barbuda,

Dominica, Martinique, Montserrat, Nevis, Saba, and the islands of Basse-Terre, Grande-Terre, Marie-Galante, and Terre-de-Bas in the Guadeloupean Archiplego. There is also an unverified report of a skull from this species in a cave on Saint Martin.

Measurements. Total length: 98–105 mm; tail length: 49–57 mm; hindfoot length: 7–10 mm; ear height: 12–16 mm; forearm length: 37–42 mm; body mass: 4.3–6.5 g.

Description. At 5 g, the Lesser Antillean funnel-eared bat is small for a bat but medium-sized for a natalid. Fur is long and lax, with an overall color of grayish-, yellowish-, or orangish-brown. Dorsal hairs are 8 or 9 mm long and darker at the tip than the base, whereas ventral hairs are lighter, unicolored, and from 6 to 7 mm long. The muzzle is wide and flat, and tufts of hairs on the upper lip and snout form an obvious mustache. The diminutive eyes and twisted tragus are hidden in lateral view by broad, funnel-shaped

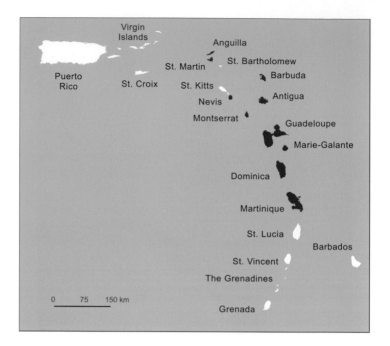

ears; the pinnae are dark colored on their posterior surface, but from the front, they appear light colored and rimmed in dark brown or black. Hind legs are quite long, about as long as the head and body combined. The tail vertebrae continue all the way to the end of a V-shaped interfemoral membrane, which terminates in a fringe of fine hairs. The male natalid organ ranges from elliptical to wedge-shaped and covers the forehead from the back of the rostrum to the top of the head.

Vespertilionid bats, such as the Guadeloupean big brown bat (*Eptesicus guadeloupensis*) and Schwartz's myotis (*Myotis schwartzi*), also have a wide, pointed tail membrane, but they lack the flattened muzzle, mustache, and broad ears of a Lesser Antillean funnel-eared bat. No other member of the family Natalidae occurs on the same isles as this funnel-eared bat, which simplifies identification.

Natural History. The Lesser Antillean funnel-eared bat is most often encountered at intermediate elevations, but it does roam as high as 770 m on the slopes of Piton Boucher on Martinique. Ecologists have captured this species in moist forests and even rain forests, but surprisingly, it also frequents dry habitats. For example, acoustic detections and mistnetting captures occur most frequently in dry coastal forests on Guadeloupe, and this bat appears common on xeric islands such as Anguilla, Antigua, and Barbuda.

This mammal typically spends the day hidden inside a humid cave, but it may roost in a rock crevice, among boulders, or within human-made tunnels and tarrish pits, which are shallow holes excavated by early European colonizers for building materials. On many islands, such as Antigua, Barbuda, Nevis, and Saint Martin, the Lesser Antillean funnel-eared bat occupies the side tunnels of small caves, while other species, such as the Antillean fruit-eating bat (*Brachyphylla cavernarum*), inhabit the more spacious chambers and large domes in main rooms and corridors. Biologists have often speculated that natalids desiccate rapidly in dry atmospheres, and the side tunnels presumably have less air movement and higher humidity. Other species that may live in the same cave include the Jamaican fruit-eating bat (*Artibeus jamaicensis*), insular single-leaf bat (*Monophyllus plethodon*), and Davy's naked-backed bat (*Pteronotus davyi*).

Population size varies from a few animals at multiple caves to several hundred in Deer Cave on Barbuda, Le Grand Trou à Diable on Marie-Galante, and La Grotte de Courcelles on Guadeloupe. The number of animals within a subterranean site seems to fluctuate within and between years, suggesting that these bats move among caves. Regardless of population size, though, individual Lesser Antillean funnel-eared bats generally hang alone and maintain a distance of 5 to 10 cm from each other.

This bat has broad wings and low wing loading, suggesting that it is capable of hovering and slow maneuverable flight within structurally complex habitats. When flying in open forest on Guadeloupe and Martinique, a Lesser Antillean funnel-eared bat emits echolocation sounds of low intensity that decrease almost linearly from 103 to 54 kHz, over 3 milliseconds, with time between pulses averaging 202 milliseconds. However, as the bat begins to hunt in the crowded understory, calls become multiharmonic, which allows the animal to discern more details of what lies ahead, and the pulse interval shortens to just 23 milliseconds, which gives the animal faster updates on the position of obstacles or prey. This mammal probably includes moths in its diet, as do other natalids, and supplements those insects with leafhoppers, flies, and perhaps small beetles, among others.

In general, natalids give birth once per year to a single offspring, but reproductive events for the Lesser Antillean funnel-eared bat are apparently not tightly synchronized among islands. Births occur on Dominica, for example, as early as 14 April, yet females are still pregnant on Guadeloupe as late as 18 August, even though less than 50 km separate the two islands. Additional dates for pregnant females are 19 April on Dominica, 19 May on Anguilla, 25–26 May on Saba, and 3 June on Barbuda; records for nursing mothers include 19 April on Dominica, 19 May on Anguilla, and 30 June on Saba.

One of these bats ventured into a building, flew into the whirling blades of a ceiling fan, and eventually died from its injuries. Another collided in midair, inside a cave, with an Antillean fruit-eating bat, which weighs almost ten times as much as a Lesser Antillean funnel-eared bat, and the impact broke the arm of the smaller animal. The Dominican boa (*Boa nebulosa*) is the sole documented predator.

Status and Conservation. The IUCN classifies the Lesser Antillean funnel-eared bat as a species of Least Concern. However, populations of this mammal are heterogeneously distributed in the Lesser Antilles; the species is apparently common on some islands (e.g., Dominica) but rare on others (e.g., Nevis). Throughout their range, these bats seem dependent on humid caves for dayroosting, and woodlands surrounding each cave for foraging; however, such sites are not abundant, and continuing expansion of human activities is a major threat. On Guadeloupe, for instance, developers bulldozed the land around La Grotte de Courcelles, removing the natural vegetation and causing a partial collapse of the cave.

Selected References. Genoways et al. 2007a; Ibéné et al. 2007; Pedersen et al. 2007; Tejedor 2009; Barataud et al. 2017.

Nyctiellus lepidus

Gervais's Funnel-Eared Bat, Murciélago Mariposa, Chauve-souris Papillon

Name. The generic name is derived from the Latinized diminutive of *nyx*, which is Greek for "night." The specific epithet *lepidus* refers to the animal's small size and flight behavior, which is similar to those of a lepidopteran (butterfly or moth).

Distribution. This bat currently lives on Cuba and nearby Isla de la Juventud, as well as several islands of the Bahamas, including Cat, Eleuthera, Great Exuma, Little Exuma, and Long. Fossils also occur on Andros and New Providence islands.

Measurements. Total length: 63–70 mm; tail length: 25–35 mm; hindfoot length unavailable; ear height: 8–14 mm; forearm length: 26–32 mm; body mass: 2–3 g. Females are larger and heavier than males.

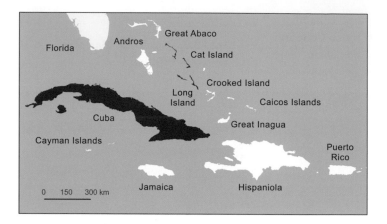

Description. Gervais's funnel-eared bat is one of the smallest mammals in the world and the only member of the genus *Nyctiellus*. The dense pelage varies from grayish-brown to yellowish-brown to reddish-brown. Individual hairs are about 5 mm in length and bicolored, with tips darker than bases. The snout is broad and dorsoventrally flattened. The nostrils open on a rounded skin pad on the margin of the upper lip, which includes a dense coating of parallel ventrally curving hairs that form a mustache. Unlike the ears of other members of the family, the distal third of the pinnae in a Gervais's funnel-eared bat is narrow with a slightly rounded tip. The small natalid organ of the male is squarish in outline, centered on the snout, and not in contact with the forehead. The naked interfemoral membrane is V-shaped at the back and lacks a fringe of hairs. The Bahamian lesser funnel-eared bat (*Chilonatalus tumidifrons*) and the Cuban lesser funnel-eared bat (*C. macer*) are similar in appearance, but both are slightly larger than Gervais's funnel-eared bat and have a border of short hairs on the back edge of the tail membrane. In addition, the length of the tibia exceeds half the length of the forearm in most natalids, but in Gervais's funnel-eared bat, the size of this leg bone is less than half that of the forearm.

Natural History. Gervais's funnel-eared bat is a cave-dwelling species, although one colony occupied an abandoned cistern on Exuma in the Bahamian archipelago. This bat dwells in at least fifty-one caves on Cuba and twelve in the Bahamas that vary in length from 10 to 1700 m. Most diurnal roosts are located less than 100 m above sea level and less than 10 km from the coast.

Within these subterranean retreats, Gervais's funnel-eared bat selects humid and warm rooms, but it typically avoids the hottest parts of hot caves. The bats roost in small groups of ten to fifteen individuals within solution cavities in the ceiling, or up to several thousand animals might occupy the same large chamber. Regardless of population size, individuals are always widely spaced and not in contact with each other. The Jamaican fruit-eating bat (*Artibeus jamaicensis*), buffy flower bat (*Erophylla sezekorni*), Waterhouse's leaf-nosed bat (*Macrotus waterhousii*), Greater Antillean long-tongued bat (*Monophyllus redmani*), Cuban lesser funnel-eared bat, and big brown bat (*Eptesicus fuscus*) often roost within different sections of the same cave. However, similar to most cave-dwelling species in the West Indies, Gervais's funnel-eared bat typically does not interact with the other bats while in its underground retreat.

Outside caves, biologists have collected Gervais's funnel-eared bat in moist and dry habitats, including evergreen, semideciduous, and scrub forests, as well as mangroves. The small body and short broad wings of Gervais's funnel-eared bat facilitate a slow and very maneuverable flight pattern that allows the use of cluttered habitats, such as the forest understory. This bat feeds totally on insects, including leafhoppers, flies, moths, and termites; a full stomach contains finely chopped insect remains that may amount to 25 to 30% of the bat's weight. Foraging activity occurs in short bouts of 45 minutes or less, near dusk and dawn, when many small insects are more active; the bats spend most of the night, though, quietly roosting within the home cave. While leaving its roost, this mammal uses low-intensity frequency-modulated echolocation calls with multiple harmonics. The second harmonic usually has the most energy and sweeps from 114 to 71 kHz over 2.7 milliseconds; average time between calls is 32 milliseconds.

Births occur once each year. Mammalogists report finding pregnant individuals between March and July, and nursing females from May to September, although lactation peaks in July and August. Nevertheless, nonreproducing females occur year-round, and even during July, up to 10% are not pregnant or lactating, which suggests that some bats may not become sexually mature in their first year. The single offspring has bluish-gray skin, is mostly hairless, and weighs up to 41% of its mother's mass. Not surprisingly, the females do not habitually carry such huge pups outside the cave on foraging trips, although earlier reports erroneously suggested otherwise.

Adult males and females comingle in the same cave in greatest proportions during September and October, when testicular development is maximum, suggesting that copulation occurs in those months. During the maternity season, though, the sexes roost apart. Sometimes the females simply shift to the deepest part of a cave, where they give birth and raise the young, although expectant mothers often move to a different underground roost that the females use as a nursery. Recaptures of banded animals indicate that some maternity centers are from 8 to 15 km distant from sites occupied during the nonbreeding period. Maternity caves shelter as many as thirty thousand individuals, whereas caves used exclusively by males house just a few hundred bats or less. Air temperature inside caves used by males is often less than 26 °C, whereas the larger number of tiny bodies at maternity sites heats the air to over 28 °C.

In 1995, researchers banded 2494 Gervais's funnel-eared bats at three caves in central Cuba. In 2008, about 17 years after marking, two of these bats, both males, were recaptured in their original cave, and both individuals appeared healthy at the time. This is a phenomenal lifespan for such a tiny mammal, and the longest reported for bats living in the West Indies.

Status and Conservation. Biologists commonly encounter this species on Cuba and the Bahamas, and the IUCN considers it a species of Least Concern.

Selected References. Silva-Taboada 1979; Tejedor 2011; Vela Rodríguez et al. 2019; Vela Rodríguez and Mancina 2020.

Bulldog Bats

On the mainland, this Neotropical family occurs from central Mexico to northeast Argentina. It also inhabits the Caribbean, where its range includes the Greater and Lesser Antilles, the Bahamas, and many continental islands off the coast of Central and South America. The family has a single genus, *Noctilio*, which includes two species, the lesser bulldog bat (*Noctilio albiventris*) and the greater bulldog bat (*N. leporinus*), although only the latter mammal inhabits the West Indies.

In general, both species appear similar, but as the common names suggest, a greater bulldog bat typically weighs about twice as much as a lesser bulldog bat. Both have a distinctive, hairless face and a doglike snout, with the upper lip split in the center. Short, stiff, gray or orange hairs cover their bodies, and a pale stripe runs along the back. These animals also come equipped with large internal cheek pouches, pointed and well-separated ears, long claws, and a wide interfemoral membrane that stretches from ankle to ankle and extends back as far as the claws when spread. The tail is about half the length of the membrane, and the last few vertebrae protrude from its dorsal surface.

The rostrum is short, only half as long as the domed braincase. The mandible is stout, and the sagittal crest is well developed, suggesting a strong bite. The canines are large, pointed, and slightly curved, whereas the cheek teeth have distinct W-shaped ridges.

Noctilionids occur in both wet and dry forests and savannahs, but these bats are usually associated with coastal areas, rivers, and lakes at low elevations. Each species regularly roosts in caves, rock crevices, hollow trees, and occasionally abandoned buildings. The greater bulldog bat becomes active later each night than its smaller cousin, and the two species emphasize different foods. The lesser bulldog bat preys primarily on insects that it plucks from the surface of a water body with its large hindfeet, although this mammal occasionally takes small fish and consumes some fruit. The greater

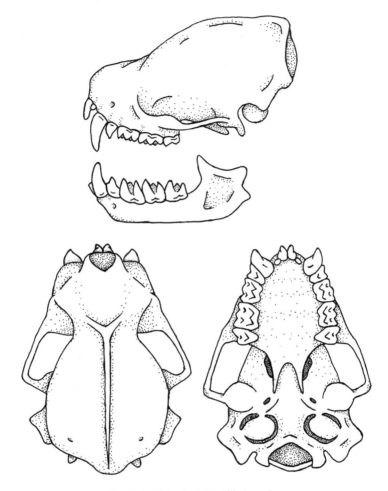

Greater bulldog bat, *Noctilio leporinus*

bulldog bat, in contrast, concentrates on catching fish and supplements those creatures with large insects and crustaceans.

Both species form small colonies, typically consisting of less than fifty individuals but sometimes including as many as several hundred animals. Within each colony, the bats segregate into harems of two to ten adult females that a lone reproductive male watches over. Females annually produce a single pup that develops slowly and subsists on milk for 2 to 3 months, much longer than most other kinds of bats. As her offspring approaches weaning, the mother feeds it masticated food that she brings home in her cheek pouches. Newly volant young follow their parent and maintain verbal contact with her, performing duets, with one individual vocalizing and then the other, back and forth.

Selected Reference. Medellín 2019.

Noctilio leporinus

Greater Bulldog Bat, Murciélago Pescador, Noctilion Pêcheur

Name. *Noctilio* is derived from the Latin word *noctis*, meaning "of the night." The specific epithet *leporinus* is also Latin and translates as "like a hare," thus describing the somewhat harelike appearance imparted by the elongate ears and split upper lip.

Distribution. The greater bulldog bat occurs throughout the Greater and most of the Lesser Antilles, as well as on Great Inagua

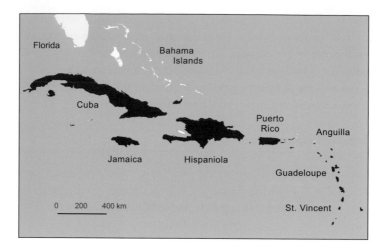

in the southern Bahamas. On the mainland, this species ranges from northern Mexico to northern Argentina.

Measurements. Total length: 76–130 mm; tail length: 23–35 mm; hindfoot length: 25–34 mm; ear height: 16–31 mm; forearm length: 81–92 mm; body mass: 48–87 g. Size varies among islands, with no evident geographic pattern. Males are up to 15 to 20% heavier than females, depending on the island.

Description. This beast is unmistakable in appearance. It is the largest West Indian bat, with long and narrow wings that span over a half meter. The fur is distinctively short with a velvety look, and the color ranges from rufous-brown to gray to yellowish-brown on the back but lighter on the belly. A pale line, almost white in some populations, extends along the center of the back, from head to rump. Ears, snout, wings, and tail membrane are dark brown to grayish-brown. The ears are tall, slender, and significantly pointed, and the distinctive tragus has multiple projections. The muzzle is slightly elongated, with naked cheeks and a pointed nose, and the upper lips are split and form a hood over the mouth, thus simultaneously conveying the look of a harelip and the snout of a bulldog. This bat has a short, well-developed tail, and its tip protrudes from the dorsal portion of the uropatagium rather than the posterior margin. The hind legs are massive and sport elongate toes tipped with large, curved claws. The cheek teeth have W-shaped ridges (dilambdodont dentition) similar to those of strictly insectivorous bats.

Natural History. The greater bulldog bat inhabits dry and humid tropical forests, mangroves, agricultural areas, and large urban parks. Wherever this mammal lives, though, it is never far from standing or slow-moving water, such as bays, estuaries, lagoons, wide sluggish rivers, and freshwater lakes and ponds. Consequently, the greater bulldog bat frequents lowlands along the coast and rarely ventures above 500 m in elevation.

These bats seldom roost in buildings and most often choose hollow trees, rock crevices, or caves. The animals generally congregate in small colonies containing up to thirty individuals, although a few populations, like the one in Cueva Grande de Caguanes on Cuba, include over two hundred bats. In underground sites, greater bulldog bats typically occupy deep solution cavities in the ceiling, always high above the floor, in areas of complete darkness or in dimly lit passages near the entrance. When approached by humans, this bat is quickly on alert but does not take flight; instead, the animal hunkers down and creeps backward, deeper into cracks or holes.

Inside a roosting cavity, greater bulldog bats typically form a harem, with one adult male and from two to ten females, whereas bachelors roost alone or with a handful of other males. All greater bulldog bats possess a gland just beneath the wings, near the armpit, that oozes a yellowish-colored, odiferous substance that is powerful enough that humans can detect the scent as the animal flies by in the darkness. Females coat the head with their own secretions and frequently nuzzle the armpits of harem mates. The chemical composition of the discharge differs among individuals and presumably conveys information concerning a bat's identity, sex, and reproductive status.

Greater bulldog bats are active at any time of night, although less so when the moon is bright. They forage alone or in small groups of up to fifteen individuals, beginning at dusk and sometimes in the late afternoon. Each foraging bout lasts from 0.75 to 2.5 hours, but the animal spends most of the night resting and socializing. Nocturnal roosts are usually located away from the day-roost and closer to the foraging grounds, and individuals from different diurnal sites may occupy the same nightroost.

Worldwide, seven species of bats include fish in their diet, but the greater bulldog bat does so on a more consistent basis than do any of the others. Species of fish that are eaten undoubtedly vary with location, but some of those taken on Culebra, which lies east of Puerto Rico, include tilapia (*Orochromis mossambicus*), silversides (*Atherinomorus stipes*), and ballyhoo (*Hemiramphs brasiliensis*). The greater bulldog bat has two main strategies for capturing underwater prey. In the first technique, the mammal flies within 4 to 50 cm of the surface until its echolocation registers a splash or ripple made by a small fish, about 25 to 75 mm in length; the bat then swoops to the specific site, snaps the hind legs downward and forward into the water, and gaffs its prey using the formidable claws. The second strategy is more opportunistic and involves "trawling" or "raking," whereby the bat simply drags its feet through the water, for up to 10 m, until it fortuitously contacts a fish. Regardless of technique, the bat quickly transfers its catch to the mouth; the fish is chewed and swallowed, or the remains are temporarily stored in the mammal's expansive cheek pouches and transported to a nightroost for further processing. Like their human counterparts, fishing bats are mostly unsuccessful, making from fifty to two hundred passes over their preferred hunting area until they capture just one fish; nevertheless, a greater bulldog bat manages to eat up to forty of these aquatic vertebrates every night. A fishy diet results in smelly feces, and sometimes biologists locate the roosts of these bats by the odor of the rotting material and the resulting swarm of flies.

Although fish dominate their diet, these mammals also hunt for small crabs and various insects, such as cockroaches, crickets, beetles, true bugs, and winged ants. Insects are gathered in midair with the wing or tail membrane or gleaned from surfaces using the hind feet. When foraging on insects, these bats hunt over fields, in lighted areas, and along roads. On Puerto Rico, greater bulldog bats occasionally munch on smaller species of bats, such as Pallas's mastiff bat (*Molossus molossus*), the sooty mustached bat (*Pteronotus quadridens*), and flightless pups of the Antillean fruit-eating bat (*Brachyphylla cavernarum*).

When flying within 4 to 10 cm of the water's surface, just "looking" for fish, the bat emits echolocation calls that begin with a prolonged constant-frequency component at about 52 to 56 kHz, before rapidly sweeping down to approximately 30 kHz. Total pulse duration averages 9 milliseconds, with 20 milliseconds between calls. The greater bulldog bat and its cousin, the lesser bulldog bat, emit the most intense (loud) sounds ever recorded from a bat at 140 dB; in contrast, a smoke alarm positioned 10 cm from a human ear registers only 105 dB. A greater bulldog bat apparently eavesdrops on its colony mates and alters its foraging path in response to successful feeding sounds produced by others.

Litter size is one, and a full-term fetus weighs 12 or 13 g, or about 21% of the weight of the mother. Pregnant or lactating females occur in every month, but most births occur between February and June, depending on island. Reproductive events are prolonged; gestation lasts 5 months or more, and lactation requires another 2 or 3 months.

The most commonly reported mortality factor, surprisingly, is collision with automobiles, presumably occurring as the bats hunt for insects over land, and Gilberto Silva-Taboada even describes one greater bulldog bat flying through the window of a moving bus and striking a passenger in the chest. In addition, the spinning blades of wind turbines occasionally kill these mammals. A wild-caught individual lived for 11.8 years in captivity.

Status and Conservation. The greater bulldog bat is common throughout its wide distribution and appropriately classified as Least Concern.

Selected References. Silva-Taboada 1979; Gannon et al. 2005; Übernickel et al. 2013; Aizpurua and Alberdi 2018; Rodríguez-Durán and Rosa 2020.

New World Leaf-Nosed Bats

Phyllostomids originated about 35 million years ago and had a relatively short and explosive evolutionary history concentrated in the Neotropics. In terms of number of species, New World leaf-nosed bats currently constitute the most diverse group of bats in the Americas and the second largest family of bats in the world, with 11 subfamilies, 60 genera, and about 260 species. Members of the Phyllostomidae inhabit mainland America, from the southern United States to central Chile and northern Argentina. They are the most diverse group of mammals in the West Indies, where twenty-two species from fourteen genera are distributed among more than sixty-five different islands; seventeen of the species are endemic to the Antilles.

Phyllostomids range in size from the 4 g Brosset's big-eared bat (*Micronycteris brosseti*) of French Guiana, to over 225 g for the carnivorous spectral bat (*Vampyrum spectrum*) of South America, which is the largest bat in the New World, with a wingspan of almost 1 m. Most species are various shades of gray, brown, black, or red; however, some are completely white, and many have facial lines, body stripes, or colored patches about the shoulders. Some leaf-nosed bats, such as the Honduran white bat (*Ectophylla alba*), are unusual among mammals in that they absorb pigments (carotenoids) from their fruity diet and concentrate them in the skin of their ear, tragus, and nose leaf, resulting in a yellowish color. Wings are usually broad. Tails, though, are short, rudimentary, and frequently absent, and the uropatagium is often nothing more than a narrow band of skin paralleling the legs (see Fig. 0.2).

In Greek, *phyllon* means "leaf," and *stoma* refers to the "mouth." Biologists, however, now use the latter term to denote any anatomical "opening," and in this case, the opening is the nostril. Hence, the name Phyllostomidae refers to a fleshy appendage surrounding the nasal openings, called a "nose leaf" (see Fig. 0.5). The nose leaf has two general components—a U-shaped "horseshoe," which lies below and to the sides of the nostrils, and an upright "lancet," which sits dorsal to the openings. Unlike other New World bats, most phyllostomids emit some or all their echolocation pulses through their nostrils, rather than an open mouth, and the nose leaf apparently helps focus the outgoing sounds. Both structures, though, especially the lancet, vary in size and shape among species. For example, in the common sword-nosed bat (*Lonchorhina aurita*), the thin, blade-shaped lancet is taller than the very large ears, whereas in vampires and some nectar-feeding species, it becomes a simple fold of tissue.

Echolocation calls are brief, very directional, high in frequency, and often lower in intensity (quieter) than pulses emitted by mormoopids, natalids, or vespertilionids, making leaf-nosed bats difficult to detect by naturalists even with appropriate equipment. Consequently, biologists historically have called these mammals the "whispering bats." Nonetheless, phyllostomids combine their other senses with echolocation to take advantage of an amazing array of foods, including flowers, pollen, nectar, leaves, fruit, blood, insects, frogs, lizards, birds, bats, and mice. The Jamaican fruit-eating bat (*Artibeus jamaicensis*), for instance, relies on olfaction to guide it close to a new fruiting tree and then switches to echolocation to determine the precise position of a single ripe fruit. Pallas's long-tongued bat (*Glossophaga soricina*), in contrast, often uses vision for initial detection of nectar-producing flowers that strongly reflect ultraviolet light, whereas the common vampire (*Desmodus rotundus*) detects infrared radiation (heat) given off by its mammalian prey. Many phyllostomids return to forest patches, productive trees, or even specific flowers, night after night, relying on a well-developed spatial memory. Such dependence on multiple sensory pathways and memories correlates with a larger brain in phyllostomids compared with other bats that simply echolocate.

The enormous variation in diet results in immense variation in shape of the skull, teeth, and other food-gathering structures. To illustrate, a frugivorous species, such as the Jamaican fruit-eating bat, generally has a short rostrum that can develop a strong bite, well-developed canines for securely holding fruits during transport, and

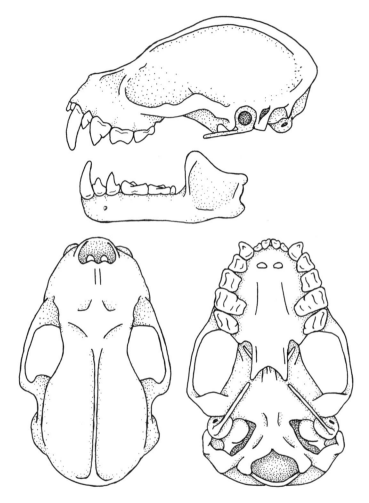

Jamaican fruit-eating bat, *Artibeus jamaicensis*

cheek teeth with low rounded cusps for breaking up fleshy, fibrous, or meaty material. In contrast, the liquid diet of nectar-feeding bats, such as the Greater Antillean long-tongued bat (*Monophyllus redmani*), does not require robust teeth, but their snouts are long, and their tongues are even longer and highly protrusible, allowing the mammal to reach deeply into flowers. In the tube-lipped tailless

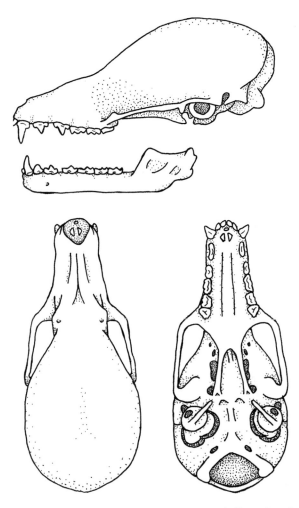

Greater Antillean long-tongued bat, *Monophyllus redmani*

bat (*Anoura fistulata*), from Ecuador, the tongue protrudes 85 mm, which is 1.5 times as long as the animal's entire body; among vertebrates, only chameleons have a longer tongue.

Other facets of of the lives of phyllostomids also vary greatly among species. They generally take shelter in caves; however, many occupy tree hollows or buildings, and some simply rest hidden among the leaves on a tall tree. A few species, like the common tent-making bat (*Uroderma bilobatum*), use their teeth to modify large leaves of various palms, philodendrons (*Philodendron*), or heliconias (*Heliconia*) and construct a "tent" that protects the animals from the elements. Most reproduce more than once annually, in synchrony with food availability, but some apparently have a single reproductive event per year. The mating system of a few species is promiscuous, others are monogamous, and many phyllostomids form harems, consisting of a single adult male and several receptive females.

Selected References. Solari et al. 2019; Fleming et al. 2020; Leiser-Miller et al. 2020; Gessinger et al. 2021; González-Gutiérrez et al. 2022.

Ardops nichollsi

Lesser Antillean Tree Bat, Ardops de Dominica, Ardops des Peites Antilles

Name. The author of the name *Ardops* did not indicate its meaning, but at least two possibilities exist. *Ardops* may stem from the Greek terms *ard*, meaning "heat," "glow," or "passion," and *ops*, indicating the eyes or referring to the face. Hence, this name may suggest the "fiery eyes" or "passionate face" of this handsome mammal. Alternatively, *Ardops* may mean "plow face." An ard is a type of Neolithic plow, consisting of two pieces of wood, one vertical and the other horizontal, and the term could be a reference to the well-formed nose leaf of this bat. The species name *nichollsi* is in honor of Dr. H. A. A. Nicholls, who collected the original specimen under the auspices of the West Indian Exploration Committee that was established by the Royal Society and the British Association for the Advancement of Science.

Distribution. This medium-sized bat occurs on most major islands in the Lesser Antilles, from Saint Martin in the north to Saint Vincent in the south. The major occupied isles include Antigua, Dom

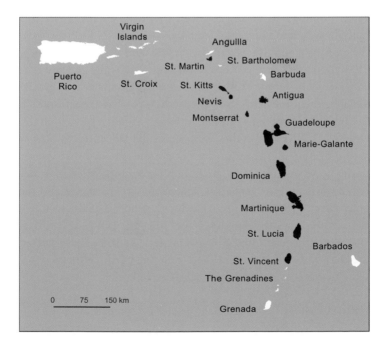

inica, Guadeloupe, Marie-Galante, Nevis, Martinique, Montserrat, Saba, Saint Eustatius, Saint Kitts, and Saint Lucia, as well as Saint Martin and Saint Vincent.

Measurements. Total length: 58–73 mm; tail length: 0 mm; hind-foot length: 12–18 mm; ear height: 12–18 mm; forearm length: 41–56 mm; body mass: 13–30 g. Measurements overlap between the sexes, but females tend to be larger than males. There are five distinct subpopulations. One of these groups occurs from Guadeloupe and Marie-Galante northward to Saint Martin, whereas the other four occur, one each, on the more isolated islands of Dominica, Martinique, Saint Lucia, and Saint Vincent. Although Lesser Antillean tree bats are similar in external appearance throughout their range, these five subpopulations differ genetically and often in physical dimensions. For example, animals from Dominica and Saint Vincent are consistently smaller than are those living on the other islands.

Description. The color of the back varies from a dark brown to a buffy-brown, and individual hairs are tricolored—dark brown at the base, pale buffy-brown in the middle, and tipped with brown. The belly is a rich brown washed with a grayish-white and lacks the tricolored hairs of the back. Dorsal hairs are from 9 to 11 mm long but shorter on the belly. Prominent white patches appear on each shoulder where the wings attach. Although shoulder spots occur on other bats in the West Indies, such as the red fig-eating bat (*Stenoderma rufum*), these patches are not present on any other species in the Lesser Antilles. The wings and interfemoral membrane are dark brown, and the ears are a lighter brown. The tragus and the outer edges of the pinna are tinged with yellow, and the inside of the ear may appear yellowish as well. Lesser Antillean tree bats have a prominent pointed nose leaf but lack a tail. The front part of the face, or rostrum, is quite short, and the upper teeth occur in a semicircular arc, rather than a nearly straight line.

Natural History. Extended field studies on the natural history of the Lesser Antillean tree bat are lacking, so much information about the species comes from the anecdotal notes of field biologists, with some observations over 100 years old. For example, Oldfield Thomas, who first described the species in 1891, noted that "this bat is said to hang all day under the branches of trees, and not take refuge in holes and crannies as most other species do." Similarly, G. K. Noble indicated in 1914 how he found an adult female and a fully grown youngster hanging together from a tree, directly over a path through a forest on Guadeloupe. After assembling these and other observations, biologists now believe that the Lesser Antillean tree bat spends the day concealed in the dense foliage of trees, rather than in caves or tree hollows. Hiding among leaves is a roosting habit that this species shares with its relatives on the Greater Antilles, including the Cuban fig-eating bat (*Phyllops falcatus*) and red fig-eating bat.

The timing of nightly activity is largely unknown, but mammalogists capture these bats in mistnets as early as 20 to 30 minutes after sunset. The Lesser Antillean tree bat is primarily a frugivore that often consumes the fruit of higuillo (*Piper dilatatum*), hogplum (*Spondias mombin*), monkey paws (*Marcgravia umbellata*), pitch apple (*Clusia rosea*), shortleaf fig (*Ficus citrifolia*), and trumpet tree (*Cecropia schreberiana*). These bats are occasionally captured in or near cultivated plantations of banana (*Musa*), breadfruit (*Artocarpus altilis*), and mango (*Mangifera indica*), but it is unclear whether the bats are just passing by or are actually eating these fruits.

Ecologically, the Lesser Antillean tree bat is most often associated with closed-canopy evergreen and seasonal forest. Extensive stands of these forests still blanket portions of the Lesser Antilles, from Montserrat southward to Saint Vincent, primarily because of conservation efforts designed to preserve the watershed of each island and maintain a supply of drinking water for the human population. North of Montserrat, however, only remnants of these woodlands persist. The dearth of forests on some islands means that the Lesser Antillean tree bat has very little habitat available in much of its geographic range. For example, on Antigua, this mammal is limited to approximately 22 km² in the southwestern corner of the island, which is less than 8% of the total surface area. Similarly, on tiny Saba, just 4 km² of effective habitat remain on the northwestern quarter of the isle, and on Saint Martin, appropriate woodlands cover perhaps 1.5 km² along the western slope of Pic du Paradis. In general, this bat seems more abundant at higher elevations, rather than coastal areas, perhaps because high-elevation forests are less modified by humans. The abundance of the Lesser Antillean tree bat, though, varies from island to island; on Saint Kitts, for instance, this species represents 49% of the fruit-eating bats captured in mistnets, whereas on Saint Martin, these mammals amount to only 3% of the frugivores.

Records of pregnant individuals exist for March, April, late May, June, July, and August. Many females are simultaneously lactating and pregnant, especially in June and July, indicating that these bats often experience a postpartum estrus and are capable of reproducing twice per year. Litter size is one, as in most bats.

Since the eruption of the Soufrière Hills Volcano on Montserrat beginning in 1995, the teeth of many Lesser Antillean tree bats, as well as those of other fruit-eating species, have become greatly worn, with only the roots remaining in some individuals. Such extreme dental wear results from chewing fresh fruit coated with volcanic ash, and bats living on Saint Vincent may be facing a similar challenge after the volcanic eruption on that island in 2021. The Lesser Antillean barn owl (*Tyto insularis*) is the lone reported predator.

Status and Conservation. The IUCN is unable to classify the Lesser Antillean tree bat, because so little is known about this mammal. Nevertheless, life on tiny islands is always precarious, and the species faces continuing challenges from increasing development for tourism and the heightened frequency and intensity of tropical storms associated with global warming. Both factors threaten the remaining forests on which these bats depend for shelter and food.

Selected References. Jones and Genoways 1973; Pedersen et al. 2005; Larsen et al. 2017; Kwiecinski et al. 2018.

Ariteus flavescens

Jamaican Fig-Eating Bat, Murciélago Leonado, Arite des Figuiers

Name. In Greek, *Ariteus* means "warlike" or "pugnacious." The reason for assigning this name is unknown. It may refer to the disposition of this mammal or to its formidable appearance caused by having a large nose leaf attached to a small head. The specific epithet *flavescens* is the Latin word for "yellowish," which denotes the yellow tint to the fur and ears of many individuals. This animal sometimes is called the naseberry bat.

Distribution. This species occurs on Jamaica, where it is locally abundant and found in woodlands throughout the island.

Measurements. Total length: 50–69 mm; tail length: 0 mm; hindfoot length: 11–14 mm; ear height: 13–18; forearm length: 36–44 mm; body mass: 9–16 g. Sexual dimorphism is common, with adult females typically larger than males.

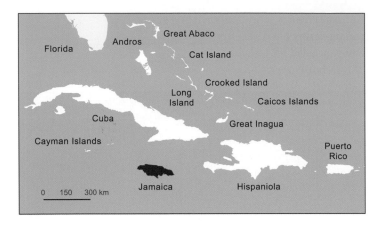

Description. The Jamaican fig-eating bat is one of four species of so-called short-faced bats that live in the West Indies; these animals are characterized by a small rostrum, relative to the size of the head, and upper teeth that are arranged in an almost semicircular arc. The overall color of a Jamaican fig-eating bat is a buffy-brown, although the head tends to be more yellowish and the belly is lighter. Individual hairs are long, wavy, and silky to the touch, and they become noticeably paler from the tip to the base of the shaft. The only noticeable markings are white patches on each shoulder where the wings attach. The tragus and the edges of the ear appear pale yellow in some individuals. Tail vertebrae are absent, and the furred interfemoral membrane is narrow and confined to the margins of the legs. The nose leaf is proportionally large and prominent; it is lanceolate in shape, simple behind, and rounded at the base in front. The wings are short and broad, a feature that provides increased maneuverability in complex forested habitats. The white shoulder spots distinguish this bat from all others on Jamaica, although such epaulettes exist on other short-faced species, such as the red fig-eating bat (*Stenoderma rufum*) on Puerto Rico and the Lesser Antillean tree bat (*Ardops nichollsi*), which dwells on numerous islands from Saint Martin to Saint Vincent.

Natural History. Radio-tracking studies indicate that the Jamaican fig-eating bat is a tree-roosting species that usually spends the day concealed within dense foliage. The bats roost in many types of tree, both native and introduced, including West Indian almond (*Terminalia catappa*), West Indian elm (*Gauzuma ulmifolia*), Malay apple (*Syzygium malaccense*), Panama rubber (*Castilla elastica*), and various other figs. Selected trees are large, with an average diameter of 82 cm. The bats typically hang about 15 m above the ground and never lower than 8 m. Each bat roosts alone, although multiple Jamaican fig-eating bats sometimes spend the day in the same tree. Individuals are not loyal to a single roost and typically move to another tree every 2 to 3 days. However, each bat remains faithful to a small area of the forest.

These mammals live from near sea level to as high as 1400 m in elevation. Naturalists have captured them in mistnets set in primary forest, remnants of old forest with open understory, fruit plantations, pasture-dominated landscapes, and regenerating woodlands. In addition, roost trees have been located next to roadways and active farm fields. Hence, this species does not seem to have a preferred habitat, and it appears tolerant of human-mediated disturbance.

As its common name suggests, the Jamaican fig-eating bat is primarily a fruit-eater. Observations of foraging individuals show that these bats consume the fruit of cashew (*Anacardium occidentale*), cacao (*Theobroma cacao*), fustic tree (*Maclura tinctoria*), naseberry (*Manilkara zapota*), rose apple (*Syzygium jambos*), and several species of figs, such as Jamaican cherry (*Ficus americana*), shortleaf fig (*Ficus citrifolia*), and white fig (*Ficus carica*). By extracting plant DNA from the feces of these mammals, biologists also know that banana (*Musa*), breadnut (*Brosimum alicastrum*), and trumpet tree (*Cecropia schreberiana*) are acceptable foods. The diet overlaps broadly with that of other frugivores on the island, such as the Jamaican fruit-eating bat (*Artibeus jamaicensis*), and individuals from as many as six species may flit around the same fruiting tree. Jamaican fig-eating bats begin feeding promptly at dusk and are active at any time of night.

The Jamaican fig-eating bat has a stomach that is radically dif-

ferent from other bats, and in fact, different from that of all studied mammals. The stomach of most bats consists of a single open
chamber, but in this species, a wall of tissue divides the stomach
into two seemingly identical compartments connected by a small
passage. In addition, the wall of the stomach and the partition between the chambers contain skeletal muscle instead of the usual
smooth muscle found in mammalian digestive tracts. The reasons
for these unusual anatomical traits are unknown, but they suggest
that the Jamaican fig-eating bat is internally processing food in a
manner different from other species.

Mammalogists have captured pregnant individuals in the dry
and wet seasons, between early April and late July. However, the
fetuses recovered in late July were small, which indicates that lactation continues through August and may extend into September.
During June and July, many females are nursing youngsters while
concurrently supporting an embryo, suggesting that at least some
adults mate soon after giving birth and are able to produce young
twice each year. Litter size is one, and the offspring is hairless and
pink-colored at birth.

Known predators include the American barn owl (*Tyto furcata*)
and the Jamaican boa (*Chilabothrus subflavus*). A field investigator
once detected a radio-tracked boa high in a tree, precisely where
the signal of a radio-tagged bat was located; the snake had a noticeable bat-sized bulge in its belly, and over time, the two signals
slowly moved through the tree in concert. American barn owls
often nest near the entrance of a cave, and although fresh pellets
produced by this bird on Jamaica contain mostly rodents, 27% of
the nonrodent remains consist of Jamaican fig-eating bats. Paleontologists have frequently recovered fossil bones of these mammals from shallow caves on the island and once believed that the
animals had died while roosting underground; however, these old
bones almost certainly represent the degraded remains of regurgitated owl pellets, rather than indicate an alternative roosting site
for this tree-dwelling bat.

Status and Conservation. The IUCN labels this common species
as Least Concern.

Selected References. Genoways 2001; Genoways et al. 2005; Sherwin and Gannon 2005; Hayward 2013.

Artibeus jamaicensis

**Jamaican Fruit-Eating Bat, Murciélago Frutero Común,
Artibée de la Jamaïque**

Name. The generic name combines two Greek words meaning "presence of facial lines." The specific epithet indicates that the type specimen came from Jamaica.

Distribution. This species occurs from northern Colombia to southern Mexico and throughout the Greater Antilles and the Lesser Antilles, as far south as Martinique and Barbados. The Jamaican fruit-eating bat is also reported from Mayaguana and both Little and Great Inagua in the southern Bahama Islands and from Providenciales, which is part of the Turks and Caicos Islands. This large bat occasionally appears in the Florida Keys, especially Key West, but it is not certain whether those animals represent a res-

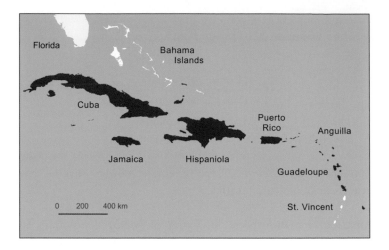

ident breeding population. On Saint Lucia, the Jamaican fruit-eating bat and Schwartz's fruit-eating bat have interbred to produce a population of hybrid animals referred to as *Artibeus jamaicensis* x *schwartzi*.

Measurements. Total length: 78–89 mm; tail length: 0 mm; hind-foot length: 16–18 mm; ear height: 20–27 mm; forearm length: 52–67; body mass: 36–48 g.

Description. Overall, the fur varies from ashy gray to ashy brown and is darker on the back and lighter on the belly. Individual hairs on the back are whitish at the base. Juveniles, in contrast, appear mousey gray all over. Light-colored stripes occur on either side of the adult face, but they are often indistinct; one line travels from the base of the ear to the angle of the jaw, and the other passes across the forehead to the back of the nose leaf. Ears, tragus, and membranes are naked and dark brown to dark gray in color. The lance-shaped nose leaf is broad at the base and comes to a distinct pointed tip. Wartlike papillae line both lips. The uropatagium is narrow, naked, and supported at the ankle by a short calcar. There is no tail.

A large body size and prominent nose leaf distinguish this animal from most other bats throughout the Greater Antilles and on most of the Lesser Antilles. On Saint Vincent, though, the related great fruit-eating bat (*Artibeus lituratus*) has the same body form but is typically bigger, and has distinct facial stripes and a yellow tinge to the ears and tragus. On Guadeloupe and Montserrat, the Guadeloupean big-eyed bat (*Chiroderma improvisum*) is similar in size, but that animal has a hairy tail membrane and a white line on its back that readily separate it from a Jamaican fruit-eating bat.

Natural History. The Jamaican fruit-eating bat lives at elevations from sea level to 2135 m. This mammal occupies a range of habitats, including humid evergreen forests, semideciduous woodlands, dry scrub, agricultural areas, and even urban parkland. It roosts within thick foliage, under leaves, and in cool caves, hollow trees, and human structures. In Central and South America, this mammal sometimes occupies "tents" made by other species of bats from the broad leaves of palms and philodendrons (*Philodendron*), but tent roosting is unknown in the West Indies.

Within a roost, the most common social group for the Jamaican fruit-eating bat consists of a single breeding male and a harem of six adult females (range: 1–14), along with their dependent young. Occasionally harems exceed fourteen females, and in those situations, two adult males are typically part of the group, with the second being the son of the harem master. Bachelor males, nonreproductive females, and juveniles older than 2 or 3 months roost alone or in small groups.

In terms of diet, the Jamaican fruit-eating bat is a generalist frugivore, although it obtains additional energy and nutrients from flowers, pollen, nectar, leaves, and insects. The fruits of various trumpet trees (*Cecropia*) and figs (*Ficus*) are available year-round in many areas, and they typically dominate the menu; however, the bat supplements these common foods with other fruits as they become seasonally available. Across its range, the Jamaican fruit-eating bat includes the products of more than two hundred plant species in its diet, and such dietary flexibility is one characteristic that allows this mammal to exploit diverse habitats and to tolerate human disturbance. Nonetheless, only fifteen to twenty types of plant are typically on the menu at any given locality.

A Jamaican fruit-eating bat begins its nightly search for food shortly after sunset, when it flies directly to a fruiting tree that it visited the night before. Using vision, olfaction, and echolocation, the animal searches for ripe fruit and may hover briefly while making a selection. Large fruits, such as mango (*Mangifera indica*), are consumed on the spot, but most items are carried 25 to 250 m to a feeding roost in a different tree. As it chews, the bat swallows the resulting juice and many seeds. However, fibrous material, as well as some seeds, become packed into an indigestible wad in the mouth, called a "spat," which the animal simply spits to the forest floor. The bat makes repeated trips between the fruiting tree and the feeding roost, and over the course of an evening, it may process double its own weight in fruit. By expelling spats and defecating throughout the night, the mammal effectively disperses the seeds of the plants on which it feeds. These bats obtain sufficient water from their juicy diet alone and apparently do not need to drink. On the mainland, they are much less active whenever the moon is bright, but such lunar phobia does not seem to occur on the islands.

The Jamaican fruit-eating bat continuously emits faint echolocation calls during flight. The sounds are frequency-modulated pulses that contain up to five harmonics, with the second, or occasionally the third, being the most intense. The second harmonic starts near 73 kHz and declines to 40 kHz; signal duration is 2.6 milliseconds, and time between pulses averages 73 milliseconds. On average, the third harmonic is about 17 kHz higher for the beginning and ending frequencies. These calls allow a Jamaican fruit-eating bat to detect an object the size of a fig from a distance of 3 to 5 m.

The Jamaican fruit-eating bat has a bimodal breeding pattern. In general, many births occur between February and April. Just 2 to 4 days after giving birth, a mother comes into heat and mates again so that she becomes simultaneously pregnant and lactating; her second offspring is typically born between June and August. After the second birth, she may copulate once more; however, the resulting embryo pauses development while still inside the uterus and becomes dormant until mid-November, when normal growth resumes. The exact timing of offspring production differs on the different islands of the West Indies, and possibly from year to year, probably in relation to fruit availability. Males with full sexual activity occur in all seasons.

Females typically give birth to a single young. Gestation usually takes from 3.5 to 4 months but can be as long as 7 months when delayed embryonic development occurs. The female gives birth while roosting head down, and the newborn emerges unaided, headfirst. Mothers eat the placenta. A newborn weighs 27% of maternal mass and is partly furred, with ears and nose leaf erect and eyes open. A youngster has a complete set of adult teeth after 40 days, begins to fly at 50 days, is weaned by 60 days, and reaches adult mass by day 80. After 8 months, young females become sexually mature, but the males do not do so until 12 months of age.

Some bat-human conflict develops when a fruit tree or a feeding roost grows next to a human dwelling; the stream of spats and feces that emanates from the tree on a nightly basis can "paint" the outer walls of the house, much to the dismay of the homeowner. Predators from the West Indies include ashy-faced (*Tyto glaucops*), American (*T. furcata*), and Lesser Antillean barn owls (*T. insularis*), and the Cuban and Jamaican boas (*Chilabothrus angulifer* and *C. subflavus*, respectively). The fast-moving blades of wind turbines on Puerto Rico killed four of these bats, even though the turbines were located far from any fruiting trees. Ash produced by active volcanos may coat fruits and lead to extreme tooth wear as the animal feeds or cause lung damage after inhaling the particles. One Jamaican fruit-eating bat survived for 9 years in the wild.

Status and Conservation. The Jamaican fruit-eating bat is widespread and abundant, constituting more than 50% of the nightly catch in mistnets at most sites in the Caribbean basin. This adaptable animal is a species of Least Concern.

Selected References. Ortega and Castro-Arellano 2001; Ortega et al. 2003; Gannon et al. 2005; Lobova et al. 2009.

Artibeus lituratus

Great Fruit-Eating Bat, Murciélago Frutero Grande, Grande Artibée

Name. The generic name comes from the Greek words *arti*, meaning "facial lines," and *beus*, referring to the "presence of" something. The specific name *lituratus*, in contrast, is from a Latin adjective meaning "branded" or "marked." Both parts of the scientific name indicate the well-defined white facial stripes in this species and other members of the genus.

Distribution. In the West Indies, the great fruit-eating bat lives only on Saint Vincent. However, this species also inhabits other islands to the south, including Grenada and the Grenadine Islands of Bequia, Union, and Carriacou, as well as the continental islands of Trinidad, Tobago, and Isla de Margarita. This bat is widespread on the mainland and occurs from southern Mexico to Paraguay and southern Brazil.

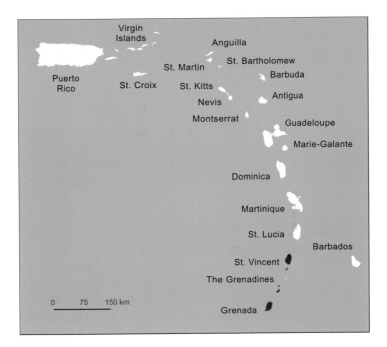

Measurements. Total length: 91–107 mm; tail length: 0 mm; hind-foot length: 15–21 mm; ear height: 20–27 mm; forearm length: 66–76 mm; body mass: 61–73 g. These ranges include measurements from animals on Saint Vincent and from the Grenadines.

Description. The great fruit-eating bat possesses orange-brown pelage, a pair of white facial stripes on each side, and a yellow-tinged tragus and outer ear. The upper facial stripe begins at the base of the nose leaf, runs above the eye and over the back of the head, and ends at the posterior edge of the pinna. The lower stripe begins at the back margin of the mouth and extends to the base of the external ear. The tail membrane has become a narrow strip of skin that is furred at its base on the dorsal side, and a slight fringe of hair lines the posterior edge near the midline. A small calcar is present. The prominent nose leaf is well developed and shaped like an equilateral triangle. Wartlike bumps are present on both top and bottom lips.

Compared with Schwartz's fruit-eating bat (*A. schwartzi*) on Saint Vincent, the great fruit-eating bat has more distinct facial stripes and ventral fur that is darker brown and without white tips. Individual species of bats have different personalities and react differently when captured. The great fruit-eating bat is aggressive and belligerent, which also helps distinguish it from the more abundant and calmer Schwartz's fruit-eating bat.

Natural History. Biologists have captured only four individuals, all males, at just two locations on Saint Vincent. One site was northeast of Kingston, in an area of second-growth forest and banana trees, along a ridge above agricultural lands that included several banana plantations; other fruits growing nearby were breadfruit (*Artocarpus altilis*), cacao (*Theobroma cacao*), nutmeg (*Myristica*), and soursop (*Annona muricata*). The second locale was at Old Sandy Bay in Charlotte Parish, along a rough road leading uphill to farmland and pastures. On the Grenadines, these bats also are uncommon, accounting for only 2 to 14% of the fruit bats captured on the three islands.

Elsewhere in its range, the great fruit-eating bat is abundant and occupies diverse natural habitats, from sea level to over 2600 m in elevation. It frequents tropical rain forests but also inhabits deciduous woodlands, seasonally dry forests, and savannahs. Urbanized neighborhoods with a suitable density of mature trees are acceptable too. Dayroosts are usually in tall trees, where the bats rest unseen, typically at least 3 to 6 m above the ground, among dense palm fronds or other thick foliage. The bats have multiple such roosts and will change every few days, sometimes returning to a previously occupied site and sometimes going to a new tree. Tree hollows and shallow caves occasionally provide shelter. These bats typically hang in small social units containing three to fifteen animals, and many of these groups consist of a single breeding male and a number of adult females.

The great fruit-eating bat is an adaptable species that consumes an array of native and exotic fruits throughout its geographic range. Wherever it lives, though, this mammal mostly relies on fruits that are available year-round and supplements those dependable resources with additional species that are sporadically abundant. In many parts of this bat's distribution, fruits of local trumpet trees (*Cecropia*) and figs (*Ficus*) dominate the menu. Nevertheless, flowers, pollen, leaves, and insects also provide additional calories and some nutrients.

This bat leaves the roost just after sundown and flies a few hundred meters to a fruiting tree that the animal likely discovered the previous night. At the tree, the mammal circles and hovers above the canopy or around exposed crowns, searching for ripe fruits among the branches. After landing and securing a fruit in its jaws, the bat moves to a different tree, 50 to 200 m away, in which the animal rests and eats. It rips off a bit of fruit with its teeth, munches a little, and then presses the partly chewed food against the roof of its mouth using the tongue. This process expels the juice, which the bat promptly swallows. However, the pulp and most seeds remain in the mouth as a fibrous wad, or "spat," that the bat eventually spits to the ground below. Some of these expelled seeds eventually germinate, and thus, the bat plays an important role in seed dispersal in tropical forests. Individuals may remain near a particular fruiting tree for hours, making repeated trips back and forth, before going to another foraging site 1 to 2 km away and repeating the procedure. Over the course of a single night, a lone great fruit-eating bat may pluck, carry, and process fruits that, in total, weigh twice as much as the animal itself. Foraging can occur at any time of night, but these bats seem most active 1 to 2 hours after sunset and again before dawn.

Although no information on echolocation calls is available for the great fruit-eating bat on Saint Vincent, some data exist for hand-released animals from Trinidad. Those bats emit steep frequency-modulated calls that vary in structure but, on average, sweep from 80 to 51 kHz. Average interpulse interval of the strongest harmonic is 69 milliseconds, with a pulse duration of 2.3 milliseconds.

Five females caught on Carriacou were nursing in late May and early June, and males from Saint Vincent and from the Grenadines were in reproductive condition at that time. Births in Trinidad occur primarily in March–April and again in August–September, and great fruit-eating bats on Saint Vincent may follow a similar timeline. Gestation ranges from 3.5 to 4 months, and litter size is almost always one.

Status and Conservation. The great fruit-eating bat is regarded as Least Concern by the IUCN because of the broad distribution of the species on the mainland. However, the status of the small population of great fruit-eating bats on Saint Vincent is unknown, especially after the eruption of the La Soufrière Volcano in the spring of 2021. Similar volcanic events that began on Montserrat in the 1990s severely affected local populations of frugivorous bats, with many animals probably killed outright and the survivors suffering from hair loss, respiratory problems, and severe tooth wear caused by eating fruit covered in volcanic ash.

Selected References. Larsen et al. 2007; Muñoz-Romo et al. 2008; Lobova et al. 2009; Kwiecinski et al. 2018.

Artibeus schwartzi

Schwartz's Fruit-Eating Bat, Murciélago Frutero de Schwartz, Artibée de Schwartz

Name. *Artibeus* originates from the Greek words *arti*, signifying "facial lines," and *beus*, indicating "presence of"; hence, the generic name references the white or gray stripes on the face of bats in this genus. The specific name *schwartzi* honors zoologist Albert Schwartz, who actively collected specimens of bats and other creatures in the West Indies during the mid-twentieth century.

Distribution. Schwartz's fruit-eating bat inhabits Saint Vincent and the Grenadine Islands of Bequia, Mustique, Canouan, Union, and Carriacou.

Measurements. Total length: 90–107 mm; tail length: 0 mm; hindfoot length: 17–21 mm; ear height: 19–23 mm; forearm length: 56–69 mm; body mass: 56–73 g.

Description. Pelage appears dark brownish, grayish, or black; ventral fur is lighter than the dorsum and has a grayish wash caused

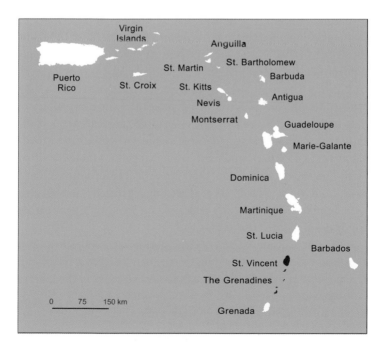

by hairs that are pale at the base, dark in the middle, and gray at the tips. The base of the hairs on the shoulder are white or pale and provide an indistinct, pale collar. A white stripe extends from the angle of the jaw toward the ear, and another line leads from the back of the nose leaf to the top of the head. The wings are dark brown or gray and become paler at the tip. A tail is absent, and the narrow uropatagium is naked and runs along the interior of the hind legs. Compared with the great fruit-eating bat (*A. lituratus*), Schwartz's fruit-eating bat lacks a fringe of hairs along the border of the tail membrane and has darker brown pelage, less distinct facial stripes, smaller overall body size, and less aggressive behavior when handled.

Natural History. Over evolutionary time, most animals develop from a single ancestral type. However, Schwartz's fruit-eating bat represents a fascinating case of what biologists call "reticulate evolution," in which a new form of animal develops by merging the DNA from two or more ancestors. The modern Schwartz's fruit-eating bat is actually a hybrid that combines characteristics of the Jamaican fruit-eating bat (*A. jamaicensis*), the flat-faced fruit-eating bat (*A. planirostris*) from South America, and a third, yet-unidentified ancestor. On Saint Vincent, populations of Schwartz's fruit-eating bat are reproductively isolated and have become a fully self-sustaining species. To the south, molecular and morphological data indicate populations on the Grenadines are Schwartz's fruit-eating bat, although some genetic relatedness to the Jamaican fruit-eating bat and the flat-faced fruit-eating bat exists. Finally, north of Saint Vincent, on Saint Lucia, there appears to be an active "hybrid swarm," in which hybrid bats are breeding with each other but also with animals that genetically appear to be Schwartz's fruit-eating bat or the Jamaican fruit-eating bat.

On Saint Vincent, Schwartz's fruit-eating bat is widespread and constitutes about half of the frugivorous bats caught on the island. A major collecting site is the 8 ha Botanical Gardens in the Liberty Lodge section of Kingstown. This site possesses a variety of buildings, manicured lawns, and a collection of exotic and native Caribbean flowers and trees. Other capture locations include numerous areas with fruiting trees of banana, cacao (*Theobroma cacao*), mango (*Mangifera indica*), or wild red ginger (*Alpinia purpurata*).

This bat is undoubtedly frugivorous, although specific information on which fruits contribute to the diet is lacking. The wings have an average aspect ratio, low tip index, and high wing loading that allow slow flight and good turning ability, especially within complex environments. As in most phyllostomids, echolocation calls contain multiple harmonics, with most energy in the second harmonic. Individual pulses are low in intensity and decrease steeply from an average of 74 to 37 kHz; mean duration of individual calls is 2.7 milliseconds, and time between calls is 97 milliseconds.

Known roosting sites of Schwartz's fruit-eating bat on Saint Vincent include Cane Garden Point Cave, Dennis' Cave, and Wynne Cave, all of which are sea-level cavities created by waves pounding into volcanic bluffs. Within these shallow caves, the bats occupy holes in the ceiling near the entrance, with a Saint Vincent big-eared bat (*Micronycteris buriri*) occasionally roosting nearby. Schwartz's fruit-eating bat also occupies the concrete ruins of the Mount Pleasant Resort, along with Miller's long-tongued bat (*Glossophaga longirostris*). The caves and the ruins are brightly lit compared with the roosting sites of most other species of bats.

In late May and early June, as the dry season ends and the wet season starts, 27% of the adult females are pregnant, 28% are nursing, and 13% are concurrently pregnant and lactating. In late July and early August, at the peak of the rainy season, 28% are lactating; however, the proportion of gravid individuals is only 12%, and none is both pregnant and nursing. Apparently some, but not all, females mate quickly after giving birth near the end of the dry season and produce a second offspring later in the year, but this postpartum estrus does not occur after the second birth. Moreover, the July–August slowdown in reproductive activity suggests that breeding does not occur year-round; parturition probably begins in April and ends in June, with a second birthing period from late July into August. Litter size is one.

Status and Conservation. Biologists know little concerning the abundance of this mammal, and consequently, the IUCN lists the species simply as Data Deficient. Nevertheless, its conservation concerns are similar to those of other endemic species in the West Indies and include tropical storms, climate change, continued loss of habitat due to human activities, and the occasional explosion of Saint Vincent's La Soufrière Volcano.

Selected References. Jones 1978; Larsen et al. 2010; Kwiecinski et al. 2018.

Brachyphylla cavernarum

Antillean Fruit-Eating Bat, Murciélago Hocico de Cerdo, Brachyphylle des Petites Antilles

Name. The generic epithet comes from the Greek words *brachys*, indicating "short," and *phyllon*, meaning "leaf"; together they refer to this bat's small, broad nose leaf. The specific name *cavernarum* is from Latin and denotes the caves in which these animals typically reside.

Distribution. The range of this mammal extends from Puerto Rico, eastward across the Virgin Islands, and southward through the Lesser Antilles. In the U.S. Virgin Islands, the Antillean fruit-eating bat definitely occupies Saint Croix, Saint John, and Saint Thomas, as well as Grass Cay and Thatch Cay, and in the British Virgin Islands, it inhabits the islands of Guana, Jost Van Dyke, Norman, and Tortola. This mammal is a resident of all major islands of the Lesser Antilles, from Anguilla in the north to Barbados and Saint Vincent in the south.

Measurements. Total length: 80–103 mm; tail: 0 mm; hindfoot length: 18–23 mm; ear height: 18–24 mm; forearm length: 59–71; body mass: 29–54 g. Males are usually larger than females. Individuals are smallest on Barbados and largest on the other islands of the Lesser Antilles, whereas animals from Puerto Rico and the Virgin Islands are intermediate in size.

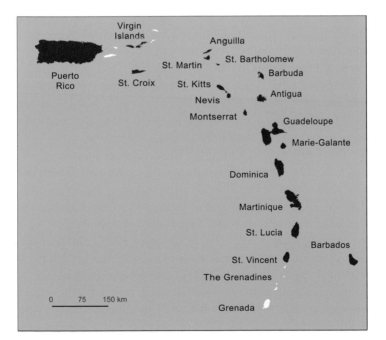

Description. Hairs are white to yellowish at the base with brown tips. Pelage color is often not even; darker patches of fur occur at various sites on the head and back, and a V-shaped mantle is frequently evident, starting near the shoulders and joining in the middle of the back. The flanks are typically lighter in color. Ears are short and separate, and the tragus is triangular in outline. The rudimentary nose leaf is just a slight ridge about the nostrils that gives the animal a piglike appearance. Several papillae dot the lower lips and form a V-shaped pattern on the chin. The uropatagium is narrow and naked, the minute tail is hidden within the membrane, and a supporting calcar is absent or very small. Large body size, stocky build, and piggish snout separate the Antillean fruit-eating bat from other species throughout its broad range in the West Indies.

Natural History. This adaptable species occurs in a myriad of habitats, including coastal plains, dry scrub, desert-like environments, and lowland and upland forests, although urban zones with highly fragmented woodlands are not commonly used. The Antillean fruit-eating bat typically spends the day in a well-ventilated cave that is somewhat cool (<25 °C), but some occasionally roost in hot caves, although never in the deepest, warmest rooms. This mammal prefers total darkness, but sometimes it uses well-lighted areas near the underground entrance. Other acceptable resting sites are leafy treetops, rocky crevices, and unused buildings, especially in areas where caves or other dark locations are lacking.

The Antillean fruit-eating bat forms colonies containing several hundred to several thousand animals; for example, at least five thousand of these mammals occupy an old sugar factory on Guadaloupe, and twenty thousand of them roost in Bats Cave on Antigua. On some of the smaller islands of the Lesser Antilles, only a single large colony may exist. Within each cave, Antillean fruit-eating bats are gregarious, generally roosting in multiple groups of various sizes, but the composition of each group is fluid, with individuals constantly joining or leaving. Caves are frequently a scarce resource, so different species often share an underground site with Antillean fruit-eating bats, although they typically hang in rooms or corridors apart from the others. Despite its social nature, biologists describe the Antillean fruit-eating bat as quarrelsome, petulant, belligerent, and pugnacious toward members of the same or different species, as well as any human who attempts to disentangle one from a mistnet.

When flying, the Antillean fruit-eating bat produces ultrasonic calls, with up to four harmonics, although the second harmonic is generally the most intense. Average signal duration is 2.6 milliseconds, and the interpulse interval is 76 milliseconds; mean starting and ending frequencies of the second harmonic are 67 and 38 kHz, respectively. The Antillean fruit-eating bat has an average aspect ratio (6.36), high wing loading (13.6 Newtons/m^2), and average wing-tip index (1.43), which allow this mammal to be maneuverable and agile at slow flight speeds. The structure of the wings and echolocation calls suggests this bat forages along the edge of vegetation, such as above and around the canopy of trees.

The Antillean fruit-eating bat is really an omnivore, with the diet consisting of an array of fruits, flowers (pollen and nectar), and insects that vary with time of year and island. Commonly eaten fruits include cacao (*Theobroma cacao*), guava (*Psidium guajava*), false tamarind (*Lysiloma latisiliquum*), mango (*Mangifera indica*), papaya (*Carica papaya*), manjack (*Cordia*), royal palm (*Roystonea borinquena*), sapodilla (*Manalkira sapota*), sour orange (*Citrus aurantium*), and West Indian almond (*Terminalia catappa*). Many of these bats are coated with pollen when captured, and they apparently eat the flowers of royal palm, West Indian almond, portia (*Thespesia populnea*), sausage tree (*Kigelia africana*), silk cotton (*Ceiba pentandra*), and West Indian locust (*Hymenaea courbaril*), among others. These animals also consume substantial quantities of insects, including many scarab beetles (*Phyllophaga*), especially during the dry season, when fruits and flowers are less abundant. However, it is unknown whether the bats take the beetles in flight or whether the insects are simply attracted to the same flowers or fruits that the bats seek during that time of year.

These mammals are one of the last to leave their cave each night, with most departing an hour or more after sunset. Although biologists occasionally capture an individual carrying fruit, these animals do not appear to use specific feeding roosts, as do Jamaican fruit-eating bats (*Artibeus jamaicensis*). When feeding in a tree, an Antillean fruit-eating bat tolerates conspecifics but shows hostility toward the similar-sized Jamaican fruit-eating bat, forcing it to leave the tree and forage elsewhere. Radio-tracking indicates that Antillean fruit-eating bats roam as far as 22 km from their home cave during the night.

The Antillean fruit-eating bat generally produces a single offspring once per year, although the specific timing of reproductive events varies slightly among islands, due to differences in climate and local abundance of food. Throughout most of the bat's range, mammalogists report pregnant females from January to June and nursing mothers from April through July. On Saint Croix, as a specific example, copulation occurs in December and January, followed by gestation from January through April during the dry season; parturition then begins at the onset of the rainy period in May. Births are highly synchronous on most islands, and at one colony on Saint Croix, all females gave birth within a 3-week period during late May and early June. On Puerto Rico, though, reproductive events appear less coordinated, with captures of lactating individuals occurring between January and August, which suggests that these bats may give birth twice in 1 year. During the maternity period, adult males and females often segregate, sometimes roosting in separate locations within the same cave or sometimes choosing completely different underground retreats.

Predators include the Lesser Antillean barn owl (*Tyto insularis*) and the house cat. Endemic boas on Puerto Rico (*Chilabothrus inornatus*), Dominica (*Boa nebulosa*), and Saint Lucia (*Boa orophias*) take a few of these robust bats, either as they emerge in the evening or while roosting during the day, and require up to 41 minutes to dispatch and swallow just one individal after capture. Furthermore, the greater bulldog bat (*Noctilio leporinus*), the largest bat in the West Indies, occasionally enters a cave while adult Antillean fruit-eating bats are foraging and snacks on hairless pups that are awaiting their mothers' return. Modern wind turbines cause additional fatalities. After the eruption of the Soufrière Hills Volcano on Montserrat, the teeth of many fruit-eating bats became polished, rounded, and often worn to the gumline after months of processing fruits covered in volcanic ash, and the same may occur on other islands with active volcanoes. Nonetheless, the diverse diet of these bats and their preference for caves make them less susceptible to tropical storms than other species, such as the Jamaican fruit-eating bat or the Lesser Antillean tree bat (*Ardops nichollsi*); for instance, the Antillean fruit-eating bat increased from 4% of the frugivores captured on Montserrat before Hurricane Hugo to 45% after the storm.

Status and Conservation. Although the Antillean fruit-eating bat does not frequent highly disturbed habitats, it is one of the most common frugivorous species in much of the Caribbean, and the IUCN rightfully regards this mammal as a species of Least Concern.

Selected References. Vaughan Jennings et al. 2004; Bacle et al. 2007; Pedersen et al. 2012; Lenoble et al. 2014a.

Brachyphylla nana

Cuban Fruit-Eating Bat, Murciélago Hocico de Cerdo Chico, Brachyphylle de Cuba

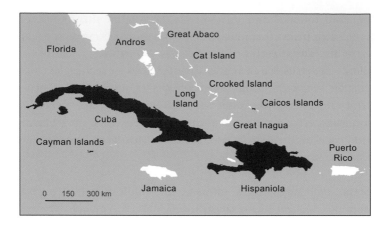

Name. The name *Brachyphylla* stems from two Greek words, *bracyhys* and *phyllon*, meaning "short leaf" and refers to the greatly reduced nose leaf of this member of the Phyllostomidae. The specific epithet *nana* is also Greek and means "dwarf," denoting the small size of this species compared with the only other member of the genus, the Antillean fruit-eating bat (*B. cavernarum*).

Distribution. The Cuban fruit-eating bat occurs on Cuba, including Isla de la Juventud and a few keys of the northern archipelago, such as Romano and Salinas, and this bat also currently inhabits Grand Cayman, Middle Caicos, and Hispaniola. In addition, Pleistocene and subrecent fossils are known from Cayman Brac and Jamaica, as well as Andros and New Providence in the Bahamas, indicating a broader distribution in the past.

Measurements. Total length: 69–102 mm; tail length: 0 mm; hindfoot length: 13–20 mm; ear height: 12–22 mm; forearm length: 52–63 mm; body mass: 23–65 g (upper limit corresponds to pregnant females).

Description. This is a large West Indian bat, with short, broad wings that often span more than 40 cm. Overall, the animal appears grayish-brown across the head and back and somewhat lighter on the belly. However, animals feeding on flowers frequently become coated with pollen that imparts a yellowish or even greenish cast to the fur. Dorsal hairs of adults are distinctly bicolored, with light bases versus brown tips, but ventral hairs vary little in color along the shaft. Ears, snout, wings, and tail membrane are grayish-brown. The ears are small, broad, rounded, and separate. The muzzle is short, with a dermal fold behind the nostrils, and the snout is topped by a thick vestigial nose leaf, which appears to have a shallow notch at its dorsal midline. The upper lips, together with the dermal fold and small nose leaf, convey the look of a pig. The tail is rudimentary or absent. Legs are muscular, and the uropatagium is just a narrow band of skin—a combination that allows greater agility when scrambling through trees. This species could be confused with the Cuban flower bat (*Phyllonycteris poeyi*) or the buffy flower bat (*Erophylla sezekorni*), which also have piglike snouts; however, both those mammals have a longer tail, a shorter forearm (≤ 51 mm), and a relatively more elongated snout than the Cuban fruit-eating bat. The cheek teeth are broad, with well-developed rounded cusps and steep, narrow valleys between adjacent cusps.

Natural History. These bats forage in dry or humid tropical forests at mostly low-to-moderate elevations. However, Cuban fruit-eating bats shelter exclusively in caves, and the abundance of suitable underground sites may explain why this mammal is common in some areas but not others. Although these bats most frequently roost in hot caves, typically outside the caldarium, they occasionally opt for smaller, cooler, and less climatically stable caves, especially in regions where hot caves do not occur. Cuban fruit-eating bats gather in colonies of tens to tens of thousands of individuals, with fewer bats inhabiting cooler, more exposed sites. Each animal typically hangs by itself from the ceiling and sometimes from the wall of the cavern. Although these mammals occasionally form mixed groups with Cuban flower bats, they generally roost apart from other species. At some locations, such as Cueva La Chepa and Cueva Los Patos in the Dominican Republic, sharp declines in captures occur during the dry season, suggesting strong seasonal shifts in cave occupancy, perhaps related to food availability or reproductive events.

This bat quickly flees when approached in its subterranean haunt, but if captured and held in hand, the animal relentlessly struggles to break free and continuously emits strident squeaks and audible cries reminiscent of human laughter.

Cuban fruit-eating bats are nocturnal, starting activity late and retreating into their underground shelter earlier than most species, typically before dawn. On Hispaniola, the bulk of the colony exits the dayroost between 1 and 2 hours after sunset, with 75% of the animals departing within 2.5 hours of sundown, and on Cuba, activity at foraging sites peaks in hours 3 to 5 after sunset. A few bats enter and leave the home cave throughout the night, but most gather in small groups at other nearby caves for sporadic periods of nightroosting. These animals fly low and slowly.

The name "fruit-eating" bat is somewhat of a misnomer, because this animal is actually omnivorous. On Cuba, for example, stomach analyses show that 68% of the bats consume some pollen each night, whereas 48% ingest only pollen; similarly, 52% of the animals feed at least partly on insects, but 32% eat insects exclusively. Seeds, indicating fruit consumption, appear in just 1% of the samples. Types of pollen present reveal that the bats feed at the flowers of unidentified cacti and palms, as well as various legumes, such as false tamarind (*Lysiloma latisiliquum*), sicklebush (*Dichrostachys cinerea*), withe (*Bigonia diversifolia*), woman's tongue (*Acacia farnesiana*), and yellow flamboyan (*Peltophorum adnatum*). Most insects in the diet are beetle larvae and lepidopteran caterpillars, but adult flies and small airborne roaches also are prey. Although some insects are probably ingested accidentally as the mammal feeds on flowers, the high prevalence of insects in the diet suggests that Cuban fruit-eating bats actively seek these invertebrates. Scientists believe that this species evolved from a nectar-feeding ancestor, though, and some modern behaviors support such a hypothesis; for example, in captivity, Cuban fruit-eating bats eat peeled bananas mostly by repeatedly lapping at the soft fruit with their tongue, rather than using their teeth. Foul-smelling flatulence often emanates from these creatures and presumably results from digestive processing of common items in the diet.

During the reproductive period, females separate from males

and form maternity groups, usually within the same cave. Observations at a Cuban colony suggest promiscuous mating, seasonally concentrated in November and coinciding with the largest testicular size. Reproductive events are prolonged; gestation lasts at least 5 months, and lactation requires another 2 or 3 months. On Cuba, biologists encounter gravid females between December and May, with the highest frequency from February to April, and lactating individuals from May to August. On Hispaniola, reproductive events occur slightly later, with pregnant bats reported as late as June and nursing females present into September. Litter size is one, and a full-term fetus weighs about 10 g or 30% of the weight of the mother.

Gerrit S. Miller based his original 1902 description of the Cuban fruit-eating bat on a single skull recovered from an owl pellet. Since that time, more than 225 pellets produced by the American barn owl (*Tyto furcata*) and ashy-faced owl (*T. glaucops*), from Cuba, Grand Cayman, and Hispaniola, have yielded remains of this mammal, indicating that nocturnal raptors are common predators. In addition, the Cuban boa (*Chilabothrus angulifer*) successfully attacks this bat on occasion. There are no data on longevity.

Status and Conservation. Conservationists know of a single active roost on Middle Caicos Island, and only a few records exist for Haiti, which may reflect the sampling effort in that nation. Nevertheless, this species is abundant and widespread in the Dominican Republic and on Cuba and Juventud. Consequently, the IUCN deems it a species of Least Concern.

Taxonomists currently classify living animals from Hispaniola and Middle Caicos versus those from Grand Cayman and the Cuban isles as separate subspecies, but genetic data suggest that the two groups may represent distinct species. If this hypothesis gains support, then the bats on Cuba and Grand Cayman would retain the name *Brachyphylla nana*, and those associated with Hispaniola and Middle Caicos would become *Brachyphylla pumila*.

Selected References. Silva-Taboada 1979; Mancina and Castro-Arellano 2013; Núñez-Novas et al. 2014, 2016, 2019.

Chiroderma improvisum

Guadeloupean Big-Eyed Bat, Murciélago Volandero, Chiroderme de la Guadeloupe

Name. In Greek, *chiro* is "hand" and *derma* means "skin"; together, these terms refer to the wing of the bat. In Latin, *improvisum* signifies "unforeseen" or "unexpected." Biologists were surprised to discover this animal on Guadeloupe, because the closest record of any member of the genus *Chiroderma* was from the island country of Trinidad and Tobago, 550 km to the south. The common name, big-eyed bat, indicates that the eyes of animals in this genus appear larger, relative to the size of the head, than those of most other species in the family Phyllostomidae.

Distribution. This mammal lives in the northern Lesser Antilles—on the islands of Guadeloupe, Montserrat, Nevis, and Saint Kitts. Nevertheless, archeologists uncovered a jawbone that was at least 800 years old at Folle Anse on Marie-Galante, suggesting that the range of the species once included that island too.

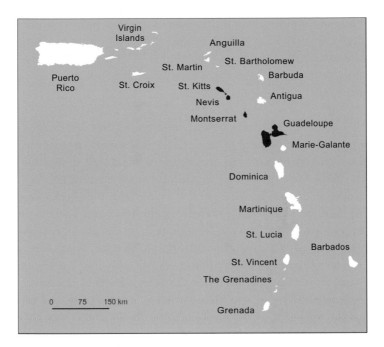

Measurements. Total length: 70–87 mm; hindfoot length: 15–19 mm; ear height: 21–22 mm; forearm length: 57–60 mm; body mass: 35–60 g. However, sample size for each measurement is only from two to four individuals.

Description. Overall, the fur on the back appears grayish-brown, with a thin, central, white stripe extending from below the shoulders to the rump; the fur on the belly is slightly paler. Dorsal hairs are tricolored, with white tips that give a frosted appearance; the middle of the shaft is buff and its base is gray. Indistinct white lines occur above the eyes, and even more obscure lines are below the eyes. Ears are light brown with yellowish edges. The spear-shaped nose leaf is well developed and also light brown in color, whereas the wing membranes are somewhat darker. The narrow membrane on the inside of the legs extends past the knees and is well haired on its dorsal surface. There is no tail. The Guadeloupean big-eyed bat is the largest member of its genus, and in the central and northern Lesser Antilles, this species could be confused only with other large leaf-nosed bats, such as the Jamaican fruit-eating bat (*Artibeus jamaicensis*) or Schwartz's fruit bat (*A. schwartzi*); however, those mammals lack the mid-dorsal stripe and have a naked tail membrane.

Natural History. Mammalogists have captured the Guadeloupean big-eyed bat on only ten occasions, with just a single animal each time, making this mammal one of the rarest and least-studied species in the West Indies. The Guadeloupean big-eyed bat flies high in open areas, as indicated by its first capture near the top of a large net (30 m long and 6 m high) that was set in a pasture about 20 m from gallery forest. Other captures have occurred along dry riverbeds bordered by trees, over tracks through secondary forest, in nets set next to a pond, in swamp forests, and near a house. The animal roams the islands from sea level up to at least 350 m in elevation.

Echolocation calls are multiharmonic, low intensity, and frequency modulated, with most energy from 75 to 85 kHz and a downward concave curve in the spectrogram. The Guadeloupean big-eyed bat is presumably a fruit-eater, and although specific dietary information is not available, this animal probably consumes figs (*Ficus*), as do other species of *Chiroderma* in South and Central America. Sparse data on the timing of reproduction include a pregnant bat from Guadeloupe caught on 7 April, a lactating individual from Montserrat netted on 12 July, and a nonreproductive female from Saint Kitts captured on 28 November. The mandible found on Marie-Galante came from a kitchen midden associated with a pre-Columbian village, suggesting that these bats may have been a source of food for the human residents.

Status and Conservation. The Guadeloupean big-eyed bat is rare throughout its distribution and is classified as Endangered, although the conservation assessment occurred before the discovery of this bat on the islands of Saint Kitts and Nevis.

Selected References. Baker and Genoways 1978; Ibéné et al. 2007; Beck et al. 2016; Lenoble 2019; Lim et al. 2020.

Erophylla bombifrons

Brown Flower Bat, Murciélago de las Flores, Érophylle de Porto Rico

Name. Biological historians speculate that *Erophylla* combines the name of the Greek god of love, Eros, and the Greek word *phyllon*, meaning "leaf," because the small pointed nose leaf of this animal supposedly resembles an arrow tip used by Eros. In 1899, Gerrit S. Miller described two new species, giving one the specific epithet *planifrons* and the other *bombifrons*. *Frons* refers to the "forehead" in Latin, and Miller was contrasting the relatively flat (*plani*) transition from the rostrum to the braincase in *planifrons* with the more rounded braincase of *bombifrons*. *Bombi* apparently originates from *bombus*, a Latin term for a "booming sound," which became the basis for the word "bomb," an exploding projectile. However, in the late 1800s, *bombé* entered the English dictionary and was applied to anything "with outward curving lines," such as the domed shape of a bomb, a piece of furniture, or the skull of a brown flower bat.

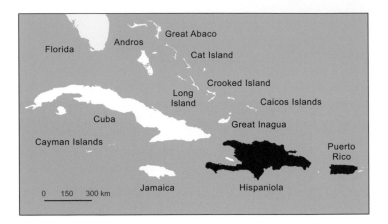

Distribution. This species inhabits the large islands of Hispaniola and Puerto Rico.

Measurements. Total length: 74–77 mm; tail length: 11–14 mm; hindfoot length: 13–15 mm; ear height: 15–19 mm; forearm length: 42–49 mm; body mass: 13–19 g.

Description. The fur is short and silky. Dorsal hairs of the brown flower bat are snow-white at their base and chestnut-brown at the tips, whereas hairs on the head and face are lighter and unicolored. The ears and membranes are dark brown. The rostrum is elongate, and the tip of the snout sports a short triangular nose leaf. The lower lip is dotted with small papillae and bears a shallow groove at its midline that allows the tongue to slip and slide back and forth. The uropatagium is V-shaped and quite narrow at the ankle, where a short calcar is present. The tail is shorter than the femur, but the vertebrae continue past the back margin of the even-shorter membrane.

The brown flower bat is distinctly darker than its blond relative, the buffy flower bat (*E. sezekorni*), which is found on the Bahamas, Cuba, Cayman Islands, and Jamaica. Moreover, the forehead abruptly rises from the snout in the brown flower bat but is more gently sloped in the buffy flower bat. On Hispaniola and Puerto Rico, the brown flower bat might be confused with other species that have light-colored pelage and small-to-nonexistent nose leafs—the Cuban fruit-eating bat (*Brachyphylla nana*), Antillean fruit-eating bat (*B. cavernarum*), and Cuban flower bat (*Phyllonycteris poeyi*). However, the Cuban flower bat lacks a calcar, and the two fruit-eating bats are larger, with forearm lengths greater than 50 mm. Furthermore, the triangular nose leaf of the brown flower bat, though small, is distinctive.

Natural History. Although the brown flower bat typically does not occur at high elevations (>1200 m), fossils of this species have been uncovered at Trouing Jean Paul, a limestone cave in southeastern Haiti, at an elevation of 1825 m. Modern field biologists capture this species in various habitats ranging from intact forests to highly fragmented woodlands in areas of anthropogenic disturbance. Thousands of these bats spend the day in cooler portions of hot caves, where air temperatures hover between 25 and 28 °C, and they occupy both dark and well-lit chambers. Females and males share roosts throughout the year. Although this mammal often occurs in subterranean sites with other types of bats, such as the sooty mustached bat (*Pteronotus quadridens*), Antillean ghost-faced bat (*Mormoops blainvillei*), and Greater Antillean long-tongued bat (*Monophyllus redmani*), the various species generally separate within the cavern according to temperature preferences and do not form mixed-species groups.

The brown flower bat is one of the last species to leave a cave each night, with peak emergence delayed until total darkness, about 40 to 60 minutes after sunset, and this bat returns home well before dawn. Echolocation calls are low intensity, low duty cycle, and typically consist of two or three harmonics. In hand-released animals, the fundamental has the greatest energy and typically begins at a frequency of 54 kHz and falls to about 27 kHz over 4.7 milliseconds, with a long interpulse interval of 107 milliseconds.

These bats are omnivorous, and on any given night, 85% of brown flower bats consume fruits, 76% capture insects, and 75% visit flowers for nectar or pollen. Beetles are the most common prey, but moths, flies, and flying ants also contribute to the animal portion of the diet. Panama berry (*Mutingia calabura*), elder (*Piper aduncum*), and turkey berry (*Solanum torvum*) are the most popular fruits, whereas nectar or pollen comes primarily from blossoms of banana, guava (*Psidium guajava*), Panama berry, and wild tamarind (*Leucaena leucocephala*). Lactation is an energetically stressful period for any female mammal, and these bats apparently lessen their intake of insects and increase consumption of high-caloric plant products while nursing.

To attract a mate, male buffy flower bats in the Bahamas use wing-flapping displays, a garlic-scented secretion from a gland above the eye, and specialized vocalizations, and perhaps brown flower bats do the same. Brown flower bats mate in December and January, parturition follows in late spring, and nursing mothers are present from May through September. Females typically give birth to one pup per year. Neonates weigh 25% of the mother's mass, and the forearm is half as long as that of the adult. Young bats must cling to their mother or to the roost while she forages, and consequently, the hindfeet of newborns are 83 to 92% of adult size. At birth, pups are hairless, ears are partly erect, eyes are open, and wings are translucent. Birth is breech, as it is in many bats.

Predators of the brown flower bat include house cats, American barn owls (*Tyto furcata*), and the Puerto Rican boa (*Chilabothrus inornatus*). After sunset, as the bats crowd together and leave the cave, one or more snakes dangle from rocky projections, tree branches, or exposed roots that surround the entrance. The reptile uses its jaws to grab any flying bat that strays too close and quickly throws two coils around the mammal's body to suffocate it; total processing time, from snatch to swallow, is only about 12 minutes. Hurricanes also negatively affect brown flower bats. After Hurricane Georges hit Puerto Rico in 1998, the brown flower bat went from being one of the most common species mistnetted on the island to one of the rarest for 10 months following the storm. Wind farms, though, likely have a minor impact on brown flower bats; a 2-year study on Puerto Rico found only one brown flower bat killed by a turbine.

Status and Conservation. The brown flower bat appears abundant and common on Hispaniola and Puerto Rico, and the IUCN classifies this species as Least Concern. Populations of brown flower bats on the two islands are genetically distinguishable but not isolated from each other, possibly because storms occasionally help transport animals from one island to the other. Having a larger genetic pool of individuals may protect this species from the negative impacts of inbreeding caused by small population sizes on single islands.

Selected References. Rodríguez-Durán and Soto-Centeno 2003; Soto-Centeno and Kurta 2003, 2006; Soto-Centeno et al. 2017.

Erophylla sezekorni
Buffy Flower Bat, Murciélago Cubano de las Flores, Érophylle de Sezekorn

Name. The generic term combines the name of the Greek god of love, Eros, and the word *phyllon,* meaning "leaf," because the nose leaf of these bats presumably resembles the shape of an arrow tip used by Eros. In 1861, Juan Cristobal Gundlach, the renowned

Cuban naturalist, named the species after Eduard Sezekorn, a German ornithologist.

Distribution. This species is restricted to the West Indies, where it lives on the Bahamas (excluding Little Inagua Island), Cayman Islands, Cuba (including Juventud), Jamaica, and the Turks and Caicos Islands.

Measurements. Total length: 69–84 mm; tail length: 11–16 mm; hindfoot length: 13–17 mm; ear height: 17–22 mm; forearm length: 43–51 mm; body mass: 13–21 g.

Description. The fur is short and silky. Overall, the back appears beige or chestnut-brown, but individual hairs are distinctly bicolored, white at the base and brownish at the tip. Hairs on the belly, face, and head are lighter and uniform in color. The elongate snout is tipped with a tiny triangular nose leaf that is only a few millimeters tall. To some people, when viewing the animal straight on, the tip of the snout and its small nose leaf vaguely resemble an upside-down heart. The bottom lip possesses a number of wartlike protuberances and is cleft at the midline, thus allowing the long protrusible tongue, which is studded with tiny papillae, to slide in and out. Ears and membranes are usually dark brown. The uropatagium is narrow, and about half the length of the short tail projects behind the membrane; a thin strip of the uropatagium extends to the ankle, where the tissue is supported by a calcar that is just 1 to 2 mm long.

Color of fur distinguishes this bat from most other species within its range, except the Cuban fruit-eating bat (*Brachyphylla nana*), Cuban flower bat (*Phyllonycteris poeyi*), and Jamaican flower bat (*P. aphylla*). However, the Cuban fruit-eating bat has a larger body (forearm length >51 mm) and lacks an obvious tail. In both flower bats, the uropatagium does not continue to the ankle, and consequently, neither species has a calcar.

Natural History. The buffy flower bat is active in diverse natural and disturbed habitats, and ecologists have captured it at elevations from sea level to 1300 m. On Cuba and Jamaica, this animal always seeks daytime shelter in underground locations. The bats prefer moderate air temperatures, avoiding the warmest sections of hot caves but occasionally occupying bell holes or cavities in cooler caves; similarly, they tolerate different levels of brightness, generally roosting in total darkness but occasionally in sites with diffuse lighting. On Grand Bahama, buffy flower bats sometimes dwell in buildings, such as a windowless cement-walled storage shed and an abandoned hotel.

Colonies range in size from a few individuals to several hundred animals, although populations may fluctuate. On Jamaica, for instance, buffy flower bats apparently abandon caves when nearby plants are not fruiting or in bloom. Nonetheless, these animals consistently use the same underground locales from year to year and generally return to specific sites within a cavern. Both sexes roost together throughout the year, but buffy flower bats rarely intermingle with different species that might share the same section of a cave.

The buffy flower bat is a late flyer, leaving its diurnal retreat well after dark. When traveling through a forest, this mammal produces low-intensity multiharmonic echolocation calls. Either the fundamental or the second harmonic has the most energy. The frequency-modulated pulses of the fundamental last 2.3 milliseconds and decrease from an average of 60 to 32 kHz, with a 40-millisecond interval between sounds. The second harmonic, in contrast, begins at 90 kHz and ends at 52 kHz. This mammal forages at low height, using a slow, maneuverable flight pattern, and may take breaks throughout the night, resting at sites not used during the day.

Buffy flower bats have an omnivorous diet containing fruit, nectar, pollen, and insects. On Jamaica, these animals feed on the fruit of breadnut (*Brosimum alicastrum*) and clammy cherry (*Cordia collococca*), and they are frequently mistnetted in plantations of banana and near fruiting fustic trees (*Maclura tinctoria*), suggesting that these plants provide food as well. On Cuba, stomach contents indicate that the menu includes the pollen of the sausage tree (*Kigelia pinnata*) and fruits of an endemic bromeliad, the Cuban hohenbergia (*Hohenbergia penduliflora*), along with beetles, moths, flies, and small cockroaches. These bats also visit flowering century plants (*Agave americana*) and sisals (*Agave sisalana*), in the Cayman Islands and Bahamas, respectively.

During the mating season, adult males occupy specific areas of the cave where they perform displays, and females visit these sites to copulate, although they often mate with nondisplaying males as well. The exhibitions involve wing flapping, audible and ultrasonic acoustic signals, and use of a garlicy scent from secretions produced by a gland above the eye (supraorbital gland). Furthermore, buffy flower bats are unusual among mammals in having a salivary gland that is sexually dimorphic; although its role in mating is unknown, the gland is enlarged and green colored in adult males but tiny and beige in females and immature individuals. Mating occurs in December and January, gestation follows during February to May, and lactating bats are evident from June to September. Females give birth to a single naked and blind pup per year. Known predators include the American barn owl (*Tyto furcata*) and the Cuban boa (*Chilabothrus angulifer*).

Status and Conservation. The buffy flower bat is somewhat uncommon on Jamaica and the Cayman Islands, but it appears abundant on the Bahamas and Cuba. Consequently, the IUCN classifies this mammal as a species of Least Concern. Although buffy flower

bats are widespread, populations from different islands display remarkably high genetic similarity. These animals are not strong fly-
ers capable of long-distance dispersal, but presumably, storms, like
hurricanes, help carry individuals and their genes between islands
on a frequent basis, thus reducing the negative impacts of insular
isolation on genetic variation.

Selected References. Murray and Fleming 2008; Fleming and Murray 2009; Murray et al. 2009; Muscarella et al. 2011; Speer et al.
2019.

Glossophaga longirostris

Miller's Long-Tongued Bat, Murciélago Hocicudo, Glossophage de Miller

Name. The generic name combines the Greek words *glossa*
("tongue") and *phaga* ("to eat"), whereas the specific name is a
merging of the Latin terms *longus* ("long") and *rostrum* ("snout").
Hence, this animal is the "long-snouted tongue feeder," which is a
wonderfully descriptive name for a nectar-loving bat.

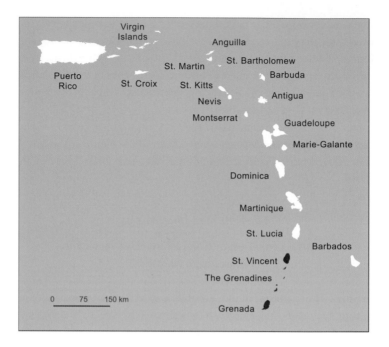

Distribution. Miller's long-tongued bats are widely distributed in northern South America and on several continental islands in the southern Caribbean, such as Aruba, Bonaire, Curaçao, Margarita, and Trinidad. Saint Vincent is the sole island in the West Indies occupied by this species, although it also occurs on neighboring islands to the south, such as Grenada and many of the Grenadines.

Measurements. Total length: 57–80 mm; tail length: 4–17 mm; hindfoot length: 5–14 mm; ear height: 7–17 mm; forearm length: 33–43 mm; body mass: 6–15 g.

Description. The dorsal pelage appears dark brown, and overall, the ventral fur is a lighter brown that turns darker on the chest and throat. Body hairs are pale at the base but dark at the tip. The short tail stretches only to the middle of the uropatagium and is totally wrapped in the membrane. The tip of the highly protrusible tongue is covered with numerous bristle-like papillae that aid in extracting sugary nectar from flowers. A well-formed spear-shaped nose leaf tops the narrow snout and marks this bat as a member of the family Phyllostomidae. As is typical of many nectar-feeding bats, the cheek teeth of Miller's long-tongued bat are thin and elongate.

On Saint Vincent, Miller's long-tongued bat is similar in size to two other phyllostomids—the Saint Vincent big-eared bat (*Micronycteris buriri*) and the insular single-leaf bat (*Monophyllus plethodon*). However, Miller's long-tongued bat has a narrower rostrum and ears that are only about half as tall as are those of a Saint Vincent big-eared bat. The insular single-leaf bat also has a narrow snout, but its tail is longer and obviously extends behind the narrow uropatagium.

Natural History. Miller's long-tongued bat lives in an array of natural habitats below 650 m of elevation. It occurs in deciduous and evergreen woodlands but also in arid and semiarid environments dominated by spiny scrub and thorn forests, as well as savannahs and intertwined gallery forests. This animal frequently inhabits suburban areas and visits fruit plantations; almost half the bats caught in banana plantations on Saint Vincent are this species.

Daytime roosts include caves, rock crevices, culverts, hollow trees, and abandoned houses and factories. In caves, this species often hangs from the walls in illuminated and well-ventilated rooms near the entrance, whereas in human-made structures, these mammals typically cling to the ceiling. In general, they congregate in small groups, usually containing from a few to less than twenty individuals and rarely involving above one hundred bats. The largest colony on Saint Vincent, though, consists of about three hundred animals sheltering inside a 40-meter-long tunnel near Grand Sable. This bat frequently shares its daytime retreat with other species of bats, but like most species in the Antilles, Miller's long-tongued bat typically does not intermingle with the others.

These bats leave their diurnal roosts at dusk or soon thereafter, periodically reenter for short rests between foraging bouts, and finally return for the day near dawn. Data on movements and home range are lacking for this particular mammal, but they likely vary depending on the quality of the habitat. To illustrate, the related brown long-tongued bat (*Glossophaga commissarisi*) has a home range amounting to 13 ha and individual foraging areas covering 3 ha in intact lowland forests of Costa Rica. Pallas's long-tongued bat (*Glossophaga soricina*), in contrast, has a home range of 430 to 890 ha and performs long commuting flights of 2 to 3 km in fragmented landscapes of the savannah-like Cerrado of Brazil.

Echolocation calls of Miller's long-tongued bat are frequency modulated, low intensity, and contain at least three harmonics, although the fundamental is very weak. The highest energy is in the third harmonic, which steeply sweeps through a broad range of frequencies, from 119 to 72 kHz, in just 1.6 milliseconds. Average time between pulses is 48 milliseconds.

This bat flies slowly, maneuvers through complex environments, and is capable of hovering while foraging. Throughout its range, Miller's long-tongued bat feeds mainly on nectar, pollen, and fruit from numerous species of plants that vary with climate, season, and location. In arid and semiarid environments, Miller's long-tongued bat is strongly dependent on columnar cacti and agaves throughout the year. In dry lands of Venezuela, for example, up to 85% of its plant-derived food comes from succulent plants. Similarly, on Curaçao, dietary samples from 85% of these mammals contained pollen or seeds from columnar cacti, and 43% of the samples consisted exclusively of cacti products. Although specific dietary information is unavailable for Saint Vincent or nearby islands, Curtis' tree cactus (*Pilosocereus curtisii*), a night-flowering species, occurs in the dry, southwestern coastal areas of Saint Vincent and possibly provides nectar, pollen, and fruit for Miller's long-tongued bats, and the apple cactus (*Cereus repandus*), a well-known bat-pollinated species, grows on Grenada. Other local plants that likely yield nectar and fruit for this mammal include balata (*Manilkara bidentate*), locust (*Hymenaea courbaril*), and spiked pepper (*Piper aduncum*), as well as the introduced banana and guava (*Psidium guajava*).

Nectar is low in nitrogen, and some nectar-feeding bats, such as the Greater Antillean long-tongued bat (*Monophyllus redmani*) of Cuba and Puerto Rico, consistently seek out insects as a source of protein. However, only 1 to 4% of Miller's long-tongued bats ingest insects on any given night, which suggests that consumption of these invertebrates is incidental to feeding at flowers, rather than the result of active hunting. Although pollen also contains nitrogen, it has outer coatings of cellulose and other chemicals that most mammals have difficulty processing. Nonetheless, digestive enzymes secreted by Miller's long-tongued bat are capable of penetrating two-thirds of the pollen grains and extracting the nitrogen-rich interior.

Miller's long-tongued bats give birth to a single offspring per reproductive event. Nonetheless, the timing and frequency of reproduction varies considerably in different parts of the animal's range. On Aruba, Bonaire, and Curaçao, these bats reproduce once each year, with pregnancy and lactation occurring mostly between March and August. In continental Venezuela, two peaks of reproductive activity occur, the first from December to April, and the second from June to October. However, on Saint Vincent, Grenada, and the Grenadines, this species appears reproductively active throughout most of the year. On these latter islands, biologists have captured both pregnant individuals and nursing mothers in December at the start of the dry season, in May and June during the transition to the rainy period, and in July and August at the peak of the wet season. In other regions, pregnancy and lactation are tied to the seasonal availability of flowers and fruits; hence, year-round reproductive activity on Saint Vincent suggests a more constant food supply.

Longevity is unknown. Some of these bats fall prey to American barn owls (*Tyto furcata*) on Bonaire and Curaçao, and a few have died after entering buildings and colliding with fast-spinning fans on Aruba. Similar events probably occur in the Lesser Antilles.

Status and Conservation. Mistnetting surveys on Saint Vincent, the Grenadines, and Grenada indicate that Miller's long-tongued bat is common and widespread, often constituting 18% or more of the catch. However, the future of this bat on Saint Vincent is uncertain following the volcanic eruption of 2021. At a global level, though, the IUCN classifies this bat as a species of Least Concern, primarily because of its broad range and abundance on the mainland.

Selected References. Petit 1997; Nassar et al. 2003; Genoways et al. 2010; Kwiecinski et al. 2018.

Glossophaga soricina

Pallas's Long-Tongued Bat, Murciélago Musaraña, Glossophage Murin

Name. *Glossophaga* is from the Greek words *glossa* ("tongue") and *phaga* ("to eat") and indicates the nectar-lapping behavior of these bats. The specific epithet *soricina* combines the Latin suffix *-inus*, meaning "pertaining to," and the noun *sorex*, which denotes a "shrew"; together, they refer to the long, shrewlike snout of this bat.

Distribution. Pallas's long-tongued bat ranges from northern Mexico southeastward into South America, as far as Paraguay and northern Argentina. In the West Indies, though, this species occurs only on Jamaica.

Measurements. Total length: 58–79 mm; tail length: 5–10 mm; hindfoot length: 10–17 mm; ear height: 10–16 mm; forearm length:

36–38 mm; body mass: 10–14 g. Animals in Jamaica form a distinct subpopulation, and they are larger and paler in color than the bats in nearby parts of Mexico and South America.

Description. The fur of Pallas's long-tongued bat varies from grayish-brown to reddish-brown. This species has a moderately elongated snout that is mostly naked and topped by a well-formed spear-shaped nose leaf. The lower lip contains a shallow furrow at the midline that is bordered by a number of smooth wartlike bumps; the tongue is extruded through this groove by as much as 40 mm, and the tip of the tongue is coated with hundreds of hair-like papillae that act like a mop, soaking up nectar drawn into the mouth. The tail is totally wrapped in the uropatagium and very short, reaching only to the middle of the membrane.

Pallas's long-tongued bat is the smallest leaf-nosed bat on Jamaica and could be mistaken only for the Greater Antillean long-tongued bat (*Monophyllus redmani*), which is another tiny nectar-feeding species. However, that bat is slightly larger, with a forearm length greater than 38 mm, and has tail vertebrae that reach beyond the rear margin of the membrane. Furthermore, the upper incisors of Pallas's long-tongued bat are procumbent (noticeably pointing forward), whereas those of the Greater Antillean long-tongued bat are directed downward.

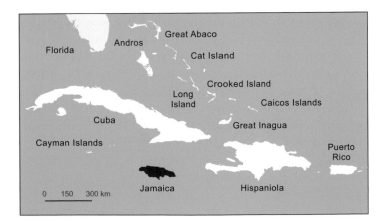

Natural History. On the mainland, Pallas's long-tongued bat is most common in lowlands, although it sometimes ranges as high as 2500 m in elevation. This mammal frequents various subtropical and tropical habitats, including primary and secondary vegetation, from rain forests to xeric woodlands to savannahs. Pallas's long-tongued bat is capable of surviving in human-modified landscapes dominated by farms, pastures, fruit plantations, and even urban developments. On Jamaica, this bat inhabits a similar diverse range of environments and occurs throughout most of the island, although the species appears uncommon above 700 m.

On the continents, this bat takes shelter in buildings, hollow trees, and underground locations, including caves, mines, and tunnels, but on Jamaica, all known roosts are in caves. Pallas's long-tongued bat apparently avoids extensive caverns and hot caves and usually occupies smaller and drier sites, where it typically hangs in compact clusters near the entrance, often in areas with diffuse light. Populations on the island are generally small, consisting of just a dozen bats up to a maximum of perhaps one hundred of these animals in a single cave. Naturalists sometimes find the Jamaican fruit-eating bat (*Artibeus jamaicensis*) or the Waterhouse's leaf-nosed bat (*Macrotus waterhousii*) in the same subterranean locale.

Throughout the bat's range, nectar and pollen top the diet, with food provided by flowers of banana, columnar cacti, agave (*Agave*), balsa (*Ochroma pyramidale*), calabash (*Crescentia cujete*), and orchid tree (*Bauhinia ungulata*), among many others, as well as backyard feeders intended for hummingbirds. Pallas's long-tongued bat uses various senses to find a nectar-producing flower, including echolocation, olfaction, and even ultraviolet vision. The mammal hovers as it temporarily depletes the nectar with rapid flicks of the tongue; however, the flowers slowly refill, and the bat may return to the same flower ten or twenty times before dawn, relying primarily on spatial cues to relocate that specific flower. Nectar is very dilute, and a single 10 g bat may require from 15 to 20 g of the sugary liquid each night to meet the animal's needs.

Fruits and insects consistently supplement the nutrients and calories provided by nectar, but the amount of either item varies considerably depending on location and time of year. Sparse data from Jamaica indicate that this winged mammal eats soft fruits from the trumpet tree (*Cecropia schreberiana*) and various pepper shrubs (*Piper*). Although Pallas's long-tongued bat can digest pollen, the majority of its protein comes from insects, and this mammal actively pursues flies, beetles, and noctuid moths. However, unlike most aerial insectivorous species, Pallas's long-tongued bat does not process its prey while in continuous flight; instead, the bat must land before manipulating an insect with wrists, thumbs, and jaws.

This species has diverse foraging strategies. At some locations, it is active for just a few hours after sunset, but in other places, this bat forages primarily after dusk and before dawn; on Jamaica, in contrast, this animal seems active at any time of night. If food is particularly abundant in a small area, Pallas's long-tongued bat may defend a feeding territory. However, at other times, individuals "trapline," following a regular path and visiting several feeding sites in succession. At each stop along the line, the animal drinks from multiple flowers and occasionally assesses the status of new inflorescences, before moving to the next spot, from 150 to 250 m away; the bat follows the same route multiple times during the evening and on consecutive nights. When flying, Pallas's long-tongued bat emits echolocation calls that decline steeply from 83 to 45 kHz over 2 milliseconds.

Elsewhere in the range, the species typically has two birth cycles per year, one in the rainy season and one in the dry period, but little data on reproductive timing exist for Jamaica specifically. Biologists have captured pregnant individuals between 9 June and 9 August, although one female with a small embryo was caught on 29 January; the only dates for lactating bats are 8 and 10 June. Unlike many vespertilionids that roost head-up when giving birth, phyllostomid mothers generally hang in the usual head-down position. The single offspring is lightly haired at birth, and in captivity, a youngster begins clumsy flights at 20 to 25 days of age but continues to nurse for approximately 2 months. Akin to a human child crying, a dependent pup separated from its mother emits an ultrasonic "isolation call" that signals the offspring's distress and is individual specific, allowing the adult to recognize her baby by sound alone and come to the rescue. Before weaning, a mother shares regurgitated nectar with her pup; the young animal hangs in close contact and repeatedly dips its tongue in and out the corner of the adult's mouth to obtain the watery fare. A captive individual lived for 11 years.

Status and Conservation. Pallas's long-tongued bat is probably the most common and widespread nectar-feeding bat in the Neotropics, and therefore, the IUCN deems it a species of Least Concern. However, a recent study examining subtle differences in skull shape suggests that Pallas's long-tongued bat is actually a complex of five separate species, and according to this hypothesis, animals on Jamaica should be considered a distinct species and called *Glossophaga antillarum*. If further research supports this concept, then the conservation status of the Jamaican long-tongued bat may need changing to reflect that it inhabits only one island.

Selected References. Genoways et al. 2005; Harper et al. 2013; Clairmont et al. 2014; Rose et al. 2019; Calahorra-Oliart et al. 2021.

Macrotus waterhousii

Waterhouse's Leaf-Nosed Bat, Murciélago Orejón de Waterhouse, Macrotus de Waterhouse

Name. The generic name is formed by combining the Greek root *macros*, meaning "large," and *otos*, denoting "ear," and together they reference the conspicuously large pinnae of these bats. The specific epithet *waterhousii* honors the British natural historian George R. Waterhouse, who helped Charles Darwin catalogue the mammals that he had collected after the expedition of the H.M.S. *Beagle*.

Distribution. Waterhouse's leaf-nosed bat is one of only three species in the genus *Macrotus* and the only one present in the West Indies. It roams throughout the Bahamas, the Cayman Islands, Jamaica, Cuba, Hispaniola, and the Turks and Caicos Islands, al

though fossils have been unearthed far to the east—on Puerto Rico, as well as the Lesser Antillean islands of Anguilla, Barbuda, and Saint Martin—indicating a much broader distribution in the past. On the mainland, this species occurs from west-central Mexico to Guatemala.

Measurements. Total length: 70–108 mm; tail length: 25–42 mm; hindfoot length: 13–15 mm; ear height: 26–33 mm; forearm length: 45–58 mm; body mass: 12–19 g.

Description. Overall color varies from reddish-brown to greyish-brown, with the latter more common in the Caribbean. Individual hairs are whitish at the base and darker at the tips, especially on the back; belly hairs are lighter and may appear frosted. The wings are broad with grey membranes. In all other phyllostomids living on these particular islands, the uropatagium is a narrow band of skin mostly paralleling the legs, but in Waterhouse's leaf-nosed bat, this structure is broad and completely spans the space between the legs, all the way to the ankles. Moreover, the tail itself is unusually long

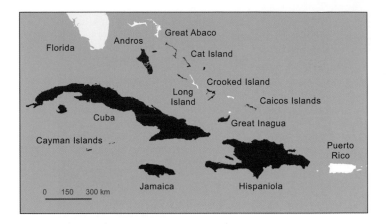

for a leaf-nosed bat, continuing back as far as the feet, and about 20% of the tail is free of the membrane. Other distinguishing characters include a simple lanceolate nose leaf, a triangular tragus that is taller than the nose leaf, and huge ears that are joined at their bases above the forehead.

Natural History. In the Bahamas and Turks and Caicos Islands, this species inhabits dry broadleaf or coppice forests and also coniferous woodlands dominated by Bahamian pine (*Pinus caribaea*) near sea level. However, on Cuba, Hispaniola, and Jamaica, it occurs in both xeric and humid forests at elevations up to 1400 m. This animal primarily roosts in cool, open chambers of caves, although it sometimes takes shelter in mineshafts or abandoned buildings. Within a site, resting areas vary from completely dark to fairly well lit, with building roosts generally brighter than subterranean locations.

These bats may be solitary, but they usually form colonies of about fifty animals and sometimes up to five hundred individuals. Although the bats commonly hang in groups, rather than spread throughout a room, individuals do not roost in contact with each other. Like natalids, Waterhouse's leaf-nosed bats often suspend themselves from the ceiling by just a single foot. Occasionally, this mammal is the only species within a cave, but Waterhouse's leaf-nosed bat frequently associates with other species that tolerate cool environments, such as the Jamaican fruit-eating bat (*Artibeus jamaicensis*), big brown bat (*Eptesicus fuscus*), Brazilian free-tailed bat (*Tadarida brasiliensis*), buffy flower bat (*Erophylla sezekorni*), and Bahamian lesser funnel-eared bat (*Chilonatalus tumidifrons*). Nightly activity begins from 30 to 65 minutes after sunset, and most individuals remain outside the home cave for the entire evening. A large influx of bats occurs during the 2 hours before dawn, with all individuals returning by about 20 to 50 minutes before sunrise.

The flight of Waterhouse's leaf-nosed bat is rather slow but highly maneuverable and typically takes place within 1 m of the ground. This mammal is capable of hovering. Diet varies with location but usually includes moths, and varying amounts of dragonflies, beetles, katydids, cockroaches, and crickets, including many pests of economic importance to humans, such as adult corn earworms (*Helicoverpa zea*). Unlike aerial insectivores that snatch insects out of the sky with a wing or tail membrane, Waterhouse's leaf-nosed bat is a gleaning species that uses its feet or mouth to pluck large prey, with wingspans from 25 to 80 mm, from the ground, leaves, or branches. The bat usually transports each morsel to a temporary feeding roost in a building, under an awning, or inside a cave, where the mammal culls membranous wings and other inedible parts before consuming the juicier abdomen and thorax. At some feeding sites that multiple bats use repeatedly over many weeks, a large number of insect parts can accumulate. At one feeding roost in a building, Gilberto Silva-Taboada noted "that the floor was virtually upholstered (2 cm thick) with scraps of insects . . . that surely represented tens of thousands of individuals." In addition to insects, a few old observations suggest that these bats incorporate fruit, such as berries of the fustic tree (*Maclura tinctoria*), into their diet, similar to the North American pallid bat (*Antrozous pallidus*), another gleaning species that occasionally feeds on nectar and fruit produced by cacti.

These bats emit low-intensity high-frequency calls as they fly through the interior of a forest. The calls are frequency modulated, with most energy in the second harmonic, which sweeps from an average of 84 to 56 kHz over 1.3 milliseconds; time between pulses is about 23 milliseconds. When foraging, these bats probably behave in a manner similar to the closely related California leaf-nosed bat (*Macrotus californicus*). That species supplements data gained through echolocation with information obtained by passively listening to sounds made by the bat's quarry, such as mating calls or rustling noises as the insect moves through leaves. Furthermore, the large eyes of animals in this genus apparently allow them to use vision under very low levels of light, and the bats often pinpoint the position of their prey without the aid of echolocation.

Timing of reproduction apparently varies across the range. On Cuba, the Turks and Caicos Islands, and Crooked Island in the Bahamas, wildlife biologists have captured pregnant females from February to April, and nursing mothers mostly between May and August. On Jamaica, though, both gravid and lactating individuals have been caught much earlier, in mid-December. Events on Hispaniola, in contrast, may occur slightly later, with records of five pregnant females on 16 May but also one as late as September. The long reproductive period in the region suggests that either some females conceive twice annually or that the reproductive period is highly asynchronous. Litter size is one. Ashy-faced (*Tyto glaucops*) and American barn owls (*T. furcata*) are common predators in the Dominican Republic.

Status and Conservation. Waterhouse's leaf-nosed bat is widespread across the Greater Antilles and the Bahamas, and the IUCN classifies it as a species of Least Concern.

Selected References. Silva-Taboada 1979; Bell 1985; Muscarella et al. 2011; Speer et al. 2015; Sánchez and Wilson 2016.

Micronycteris buriri

Saint Vincent Big-Eared Bat, Murciélago Peludo, Micronyctère de Saint Vincent

Name. In Greek, *nyckteris* indicates a "bat," and the word stems from *nyx*, which means "night." *Micro*, of course, denotes "small," so the generic name refers to the small size of bats in the genus. The specific epithet, in contrast, is derived from Garifuna, a language that originally developed on Saint Vincent in the 1600s, after the mixture of peoples from West Africa and indigenous Caribs. The Garifuna word *búriri* also means "bat" and comes from another Garifuna word, *buriga*, which stands for "dark."

Distribution. This species is endemic to Saint Vincent, in the southern Lesser Antilles, and is the sole member of the genus currently living on any of the islands north of Tobago.

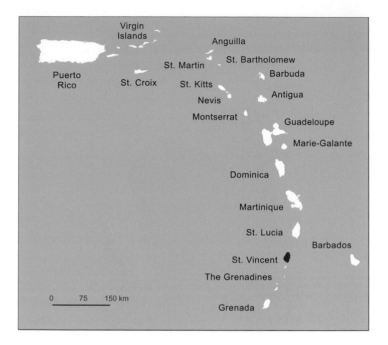

Measurements. Total length: 66–72 mm; tail length: 12–18 mm; hindfoot length: 10–12 mm; ear height: 19–25 mm; forearm length: 36–39 mm; body mass: 7.4–10 g.

Description. These mammals form a recently described species of *Micronycteris*, a genus that comprises small-to-medium-bodied bats with large ears relative to the size of their heads. Overall, Saint Vincent big-eared bats have brown pelage that varies little in color between the belly and the back. Individual hairs of adults are bi-colored, with white at the base; however, the white occurs only in a narrow band on ventral hairs but occupies from 10 to 50% of the shaft of each dorsal hair. Hairs from juveniles, in contrast, lack the white base entirely and are unicolored. A distinct vertical groove oc-curs in the middle of the lower lip. The ears are hairy along the front margin, rounded at the tip, connected across the top of the head by a narrow band of skin, and disproportionately large (≥50% as long as the forearm). Membranes are naked, but the thumbs and feet are hairy. A small body, tall ears, and a robust nose leaf are sufficient to separate this bat from any other on the island of Saint Vincent.

Natural History. Biologists have captured the Saint Vincent big-eared bat in localities ranging from sea level at Mount Wynne Bay to 646 m in elevation on the southeastern flank of La Soufrière Volcano. Most nocturnal captures occur in or near rain forest. On Trinidad and the tropical mainland, the closely related little big-eared bat (*Micronycteris megalotis*) utilizes various diurnal retreats, including animal burrows, logs, caves, buildings, and tree hollows. However, the only reported roosting sites for the Saint Vincent big-eared bat are Mount Wynne Cave and Dennis' Cave, both of which are shallow sea caves that also provide shelter for Schwartz's fruit-eating bat (*Artibeus schwartzi*).

No specific information exists concerning dietary or foraging habits, except one observation of Saint Vincent big-eared bats hunting beneath the canopy of a rain forest. Other big-eared bats, though, obtain food primarily by gleaning, which is a foraging technique that involves a bat grabbing a stationary insect from the ground, a leaf, or a tree trunk, rather than catching the prey while it is in flight. Typical food items for these other gleaning species include hefty insects such as cicadas, crickets, cockroaches, katydids, and dragonflies, as well as moths. After grabbing a large-bodied insect, the bat usually transports its food to a nightroost, where the mammal culls the indigestible wings and legs and eats the more nutritious body and head. Most big-eared bats also supplement their diet with soft-bodied fruits, such as banana, guava (*Psidium guajava*), and hog plum (*Spondias mombin*), especially during the dry season. Similar to other members of the genus, the Saint Vincent big-eared bat probably flies slowly, occasionally hovers, and emits low-intensity echolocation calls that generally include very high frequencies, sometimes over 100 kHz.

Little information is available concerning reproduction in the Saint Vincent big-eared bat. Ecologists captured three females in early August that were not pregnant or lactating, and similarly, sixteen males caught between late May and early August were not in breeding condition. One volant juvenile was caught on 3 August.

As in the little big-eared bat from Trinidad, though, the Saint Vincent big-eared bat probably gives birth to a single young at the start of the rainy season.

Status and Conservation. In 2011, taxonomists determined that the Saint Vincent big-eared bat was a distinct species, as opposed to a subspecies of the little big-eared bat, and no information is available concerning population size or whether it is increasing or decreasing. Consequently, the IUCN classifies the Saint Vincent big-eared bat as Data Deficient. Nonetheless, restriction to one small island, which is located in a hurricane-prone region, and the eruption of La Soufrière Volcano in 2021 make this insular endemic a focus for conservation concern.

Selected References. Vaughan and Hill 1996; Larsen et al. 2011; Kwiecinski et al. 2018; Morales-Martínez et al. 2021.

Monophyllus plethodon
Insular Single-Leaf Bat, Murciélago Lengüilargo de las Antillas Menores, Monophylle des Petites Antilles

Name. The name *Monophyllus* derives from the Greek words *mono*, indicating "single" or "alone," and phyllon, meaning "leaf." The name of the genus describes the simple structure of the distinctive nose leaf, which lacks deep grooves or lateral projections. Gerrit S. Miller coined the word *plethodon* in 1900, based on an animal from Barbados. In Greek, *plethore* indicates "fullness of" and *odon* refers to "teeth," and the name likely denotes the lack of space between the upper premolars of this bat compared with the other member of the genus, the Greater Antillean long-tongued bat (*M. redmani*).

Distribution. The insular single-leaf bat is endemic to the Caribbean region and occurs on all major islands of the Lesser Antilles, from Anguilla in the north to Saint Vincent and Barbados in the south. Fossilized skulls indicate that this species once lived on Puerto Rico as well.

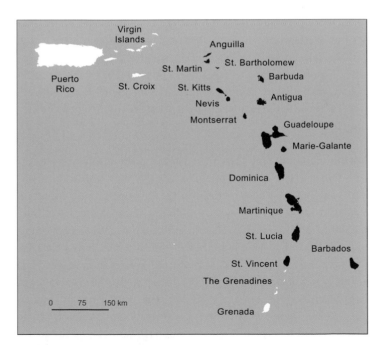

Measurements. Total length: 67–84 mm; tail length: 8–16 mm; hindfoot length: 12–13 mm; ear height: 14–15 mm; forearm length: 39–46 mm; body mass: 12–17 g. Males are about 12% heavier than females.

Description. The fur of the insular single-leaf bat is short and ranges in color from brown to grayish-brown on the head and back and light tan on the belly. The ears and flight membranes are dark brown to black. The insular single-leaf bat has an elongated and narrow snout, and the tongue is highly protrusible and tipped with multiple papillae that help the animal to lap viscous nectar from flowers. The short tail protrudes beyond the margin of the narrow uropatagium.

Within in its current range, this bat differs from most other species by having a small but well-formed and pointed nose leaf, along with delicate teeth and a slender rostrum. This widespread mammal is somewhat similar to Miller's long-tongued bat (*Glossophaga longirostris*), although the geographic ranges of the two species overlap only on Saint Vincent. On that southern island, Miller's long-tongued bat is generally smaller (forearm length usually <39 mm) and has a tail that does not extend beyond the uropatagium; in addition, the upper incisors noticeably point forward, whereas those of the insular single-leaf bat are more typical and are directed downward.

Natural History. Biologists have captured the insular single-leaf bat in mistnets set at sea level up to 950 m of elevation and in habitats ranging from cloud forest on Guadeloupe to rain forest on Saba and Saint Lucia to xeric woodlands on Barbuda, Saint Bartholomew, and Saint Kitts. Nevertheless, acoustic studies on Martinique indicate that these bats are much more active in moist forests than dry and more common at elevations greater than 400 m. This animal visits both secondary and primary forests but is also regularly encountered in small groves and large plantations of banana. The insular single-leaf bat varies considerably in abundance from island to island, possibly due to variations in rainfall and the corresponding abundance of flowers and fruits; for instance, these bats represent 9% of all bats captured on mesic Saint Lucia but only 2% on the drier Antigua. The only documented roosting sites are caves and shallow mines, called tarrish pits, although some biologists suspect that these mammals also hide by day in deep rock crevices.

Unlike the related Greater Antillean long-tongued bat that forms colonies containing tens of thousands of individuals, populations of the insular single-leaf bat are typically quite small, usually varying from a single animal to groups of less than one hundred bats. The largest colony ever reported is an estimated nine hundred of these mammals in Harrison's Cave on Barbados. The insular single-leaf bat occasionally roosts in a cave with other species, and the most frequent of these is the Antillean fruit-eating bat (*Brachyphylla cavernarum*).

These bats are active beginning about 30 minutes after sunset. When flying, the insular single-leaf bat emits multiharmonic echolocations calls that are very low in intensity (quiet). Maximum energy is in the fundamental or second harmonic. The fundamental steeply sweeps from 62 to 28 kHz over just 2.1 milliseconds, with a moderately long interpulse interval of 70 milliseconds, whereas the second harmonic begins at 115 kHz and ends near 58 kHz. Such calls are suitable for detecting food and obstacles in highly cluttered space under the forest canopy or among dense shrubs.

The insular single-leaf bat is capable of hovering while foraging. These bats apparently consume the fruits of pepper plants (*Piper*) and often visit banana plantations, presumably to drink the copious nectar provided by the flowers. However, no detailed account of the diet is available. On Puerto Rico, the closely related Greater Antillean long-tongued bat is primarily a nectar feeder, but it also depends on small fruits and tender-bodied insects, such as flies and moths; foods of the insular single-leaf bat are probably comparable. Many bats respond to a temporary lack of food by lowering their body temperature, that is, they turn down their internal thermostat to save energy. After being held overnight in a collecting bag, one insular single-leaf bat registered a body temperature of 30.5 °C but raised it to 35.2 °C in just 5 minutes when removed from confinement.

Females give birth to a single hairless pup at a time. The greatest amount of data on reproductive timing comes from Saint Lucia, where these bats apparently copulate in December and January. Young are born primarily at the beginning of the rainy season, from late April to early June, and become volant during July and August, at the peak of the wet period. Biologists have never reported lactating females carrying small embryos on any of the islands, which indicates that a postpartum estrus does not occur. Consequently, the presence of a few pregnant females in late June and early August on Saint Lucia, as well as 26 July on Dominica, suggests that the breeding season is unusually prolonged and asynchronous, compared with other phyllostomid bats in which all parturition occurs over a tighter period of 1 to 2 months.

The only known predator is the Lesser Antillean barn owl (*Tyto insularis*). Strong storms destroy the fruits and flowers that the insular single-leaf bat relies on for food, and after Hurricane Hugo, the number of these animals declined by 63% on Montserrat.

Status and Conservation. This nectar-feeding bat is common on many islands throughout its extensive range and is classified as Least Concern.

Selected References. Pedersen et al. 2005, 2018a; Genoways et al. 2007b; Barataud et al. 2017.

Monophyllus redmani

Greater Antillean Long-Tongued Bat, Murciélago Lengüilargo, Monophylle de Redman

Name. In Greek, *phyllon* means "leaf," and *mono* indicates "single" or "alone." The name of the genus refers to the simple structure of the well-formed nose leaf. William Elford Leach, who described the species in 1821, was comparing the Greater Antillean long-tongued bat to other bats, such as the lesser long-eared bat of Australia (*Nyctophilus geoffroyi*), which he considered to have two vertical nose leafs, and the Jamaican fruit-eating bat (*Artibeus jamaicensis*), which he portrayed as having both a vertical and horizontal nose

leaf. The word *redmani* refers to R. S. Redman, who obtained an individual of this species from Jamaica and presented it to Leach.

Distribution. This animal lives only in the West Indies, where it roams over Jamaica, Cuba (including Juventud), Hispaniola, and Puerto Rico, as well as Crooked and Acklin Islands in the southern Bahamas and Middle Caicos Island.

Measurements. Total length: 58–80 mm; tail length: 7–11 mm; hindfoot length: 10–12 mm; ear height: 9–14 mm; forearm length: 35–43 mm; body mass: 6–15 g. This species varies greatly in size, depending on island. The bats in Jamaica weigh twice as much as animals on Puerto Rico, whereas individuals living on Cuba and the Bahama Islands are intermediate in size.

Description. The body appears gray to brown on the head and back and noticeably lighter on the belly, where hairs often are tipped with white or silver. Many long-tongued bats also have faint, irregular white spots or blotches on top of the head and neck that become more distinct under ultraviolet light; the pattern of these

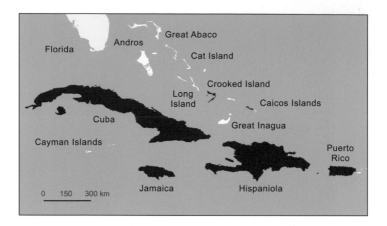

spots appears unique to each individual. Ears and flight membranes, in contrast, are dull black. Greater Antillean long-tongued bats feed on nectar, and consequently, they have a somewhat long and narrow snout, as well as a cylindrical, highy protrusible tongue that is studded with bristle-like papillae toward the tip. Teeth are small and delicate.

The lengthy snout and tongue distinguish the Greater Antillean long-tongued bat from other insular species, except nectar-feeding bats with overlapping distributions—the buffy flower bat (*Erophylla sezekorni*), Cuban flower bat (*Phyllonycteris poeyi*), Jamaican flower bat (*P. aphylla*), and Pallas's long-tongued bat (*Glossophaga soricina*). Nevertheless, the prominent, spear-shaped nose leaf of the Greater Antillean long-tongued bat readily differentiates it from the three flower bats in which the nose leaf is much smaller or even nonexistent. In Pallas's long-tongued bat on Jamaica, the narrow tail membrane encases all caudal vertebrae, but in the Greater Antillean long-tongued bat, a nubbin of a tail obviously protrudes from the trailing edge.

Natural History. The Greater Antillean long-tongued bat occurs in virtually all habitats, from secluded rain forests to desert-like environments to urban plazas and backyards. Biologists find this mammal at altitudes ranging from 43 m below sea level, at Cabritos

Island, within Enriquillo Lake, in the Dominican Republic, to 1515 m above sea level, in the Blue Mountains of Jamaica. This bat invariably spends the day underground, typically roosting in the warmest room (caldarium) of hot caves, where air temperature exceeds 28 °C. However, it also frequents cooler subterranean retreats and, overall, tolerates a wider range of temperature than any other cave-dwelling phyllostomid. In the caldarium, each animal usually hangs alone and exposed on the cave wall, but in less extreme settings, these bats are more likely to form small clusters or retreat into solution cavities. Up to five hundred thousand individuals of this species occupy a single hot cave, although the coolest sites shelter as few as fifteen to twenty animals. Almost complete segregation of the sexes occurs on Puerto Rico when births are occurring, and during this time, adult males often dominate the small populations that occupy cooler locales. At night, many of these bats briefly rest between foraging bouts in locations other than their diurnal home, sometimes using human-made structures, such as abandoned military bunkers.

These nocturnal mammals avoid hungry hawks and falcons and wait until total darkness to begin foraging, generally between 28 and 69 minutes after sunset. Greater Antillean long-tongued bats often share their hot cave with tens of thousands of mormoopid bats from multiple species, and onset of activity by this nectar-feeder often overlaps with that of the last group of insect-eating mormoopids. During these crowded times, the different species sometimes use separate exit routes, presumably to lessen congestion. At the cave entrance, activity of Greater Antillean long-tongued bats wanes and waxes throughout the night, without an obvious peak of returning animals near dawn; all activity, though, ceases by 28 to 71 minutes before sunrise. Detailed data on movements and home range are lacking, but the immense size of many colonies suggests that some of these tiny mammals disperse tens of kilometers over the countryside in their nightly search for food.

Naturalists long surmised that the Greater Antillean long-tongued bat consumed nectar, based on its long tongue, the capture of animals covered in pollen, and anecdotal observations of this bat hovering in front of large night-blooming flowers, such as those of banana. Furthermore, after dissecting some of these animals, the eminent Cuban ecologist Gilberto Silva-Taboada noted that the stomachs contained a translucent liquid that he found to be sweet to the taste. By examining the microscopic structure of pollen taken from the fur, biologists know that these bats typically visit two or three kinds of plants per night and rely on nine to eleven species over a period of a few months. Flowers visited include those of guava (*Psidium guajava*), maga (*Thespesia grandiflora*), silk cotton (*Ceiba pentandra*), wild tamarind (*Leucaena leucocephala*), various palms, columnar cacti, and many others, depending on island. Most Greater Antillean long-tongued bats also ingest pollen while feeding at flowers, and although the animals obtain some protein in this way, insects often are the most important source of that essential substance. The dainty teeth and weak jaw muscles of these mammals, though, are not well suited to crushing hard insects, such as beetles, so the bats consume mostly the softer flies and moths. Small, soft-bodied fruits, such as Panama berry (*Mutingia calabura*) and spiked pepper (*Piper aduncum*), contribute to the diet. Hence, this animal is actually an omnivore, relying on a combination of nectar, fruit, and insects for energy and nutrients. The basic diet is strikingly similar on Cuba and Puerto Rico, where, on any given night, 91 to 95% of the individuals lap nectar, 70 to 73% prey on insects, and 20 to 22% nibble fruit.

The Greater Antillean long-tongued bat always gives birth to just a single pink-skinned offspring, but details of the annual reproductive cycle differ slightly among islands. On Jamaica, the females appear tightly synchronized; pregnancies begin in November, births occur only from March through May, and lactation lasts into July. On Cuba and Hispaniola, though, a small proportion of the population may breed twice, with most pregnant individuals caught between January and June and a few appearing in autumn. On Puerto Rico, pregnancy historically occurred from February through July and then again in September and October with reduced frequency. However, 15 months after Hurricane Maria, 60% of the females were already lactating in January, although the reason for such a shift in timing of the annual cycle is unknown.

Predators include house cats, the endemic Puerto Rican boa

(*Chilabothrus inornatus*), and the American barn owl (*Tyto furcata*). Like many bats worldwide, some Greater Antillean long-tongued bats die after colliding with the fast-moving blades of wind turbines, although why these bats are flying so high and so far from any shrub or tree is not understood. Nothing is known concerning longevity.

Status and Conservation. The Greater Antillean long-tongued bat inhabits multiple large islands and is abundant throughout its distribution; hence, the IUCN considers it a species of Least Concern.

Selected References. Soto-Centeno and Kurta 2006; Mancina et al. 2007a; Clairmont et al. 2014; Soto-Centeno et al. 2014.

Phyllonycteris aphylla

Jamaican Flower Bat, Murciélago Caripelado, Phyllonyctère de Jamaïque

Name. *Phyllonycteris* comes from the Greek nouns *phyllon*, meaning "leaf," and *nykteris*, denoting "bat," and refers to a fleshy projection near the nostrils, the nose leaf, which is characteristic of

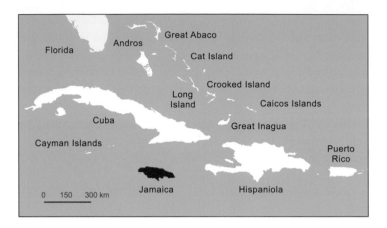

bats in the family Phyllostomidae. Conversely, the specific epithet *aphylla* stems from the Greek word *áphullos*, meaning "leafless," and indicates the rudimentary nose leaf of a Jamaican flower bat compared with those of other members of the family.

Distribution. This species is endemic to Jamaica, where most records of the species occur in the central mountains and along the north coast.

Measurements. Total length: 72–76 mm; tail length: 7–10 mm; hindfoot length: 16–18 mm; ear height: 15–18 mm; forearm length: 44–48 mm; body mass: 14–15 g. Males are slightly larger than females in several external measurements, including length of forearm.

Description. Dorsal hairs are short, about 6 mm long, and overall, the back appears blond or light gold in color; ventral hairs are shorter, and the belly looks almost white. The naked wing membranes are dark brown, nearly black in color, and attach to the body as far back as the distal part of the tibia. The sparsely haired muzzle, which is noticeably broad and deep, often exhibits dark spots or blotches and terminates in a small, disk-shaped nose leaf that gives the bat a piglike appearance. Ear tips are rounded and light brown. The tongue is long and pointed at the tip, which is covered in numerous stiff papillae; a midline groove in the lower lip allows the tongue to slip in an out with ease. The tail protrudes for about half its length beyond the uropatagium, which is extremely narrow (≤6 mm wide) and ends above the ankle; consequently, a calcar is absent. On Jamaica, this species could be confused with the buffy flower bat (*Erophylla sezekorni*); however, the tail membrane in that species extends to the ankle, where a small calcar helps support the trailing edge. In addition, the skulls of the Jamaican flower bat and the related Cuban flower bat (*Phyllonycteris poeyi*) lack a complete zygomatic arch, making them unique among West Indian species.

Natural History. Gerrit S. Miller, a taxonomist with the Smithsonian Institution, first described this species in 1898, based on a single specimen preserved in alcohol. However, for more than 50 years afterward, no living animals were caught, and conservationists presumed that the Jamaican flower bat was extinct. After its rediscovery in St. Clair Cave in 1957, mammalogists captured it again at Riverhead Cave in 1965 and then at multiple locations across central Jamaica between 1966 and 1985. Surveys conducted after 1985 failed to find the Jamaican flower bat at any of its known cave roosts, and once more, biologists believed that the species was extirpated. In 1997, though, a new population was discovered in Stony Hill Cave, and the Jamaican National Environment and Planning Agency is now monitoring this only known maternity roost as part of a long-term conservation strategy for the species. Although no other maternity sites have been located, a few individuals, apparently males, were photographed as they emerged from Green Grotto Cave, about 125 km from Stony Hill Cave, in December 2019.

This species apparently prefers the warmest part of hot caves, where air temperature exceeds 30 °C and humidity approaches saturation. A need for a constantly warm and moist environment may explain the absence of this species south of the Central Range and Blue Mountains, where long dry spells cause some variation in conditions inside the local caves. Unlike the Greater Antillean long-tongued bat (*Monophyllus redmani*), which forms immense colonies containing tens of thousands of individuals in hot caves, populations of the Jamaican flower bat are considerably smaller. Only seventy-five of these bats roosted in St. Clair Cave during the 1970s, and the largest colony ever discovered, the one at Stony Hill Cave, included just over five hundred individuals. This mammal shares its cave with as many as eight other kinds of bats, although the various species typically have differing thermal requirements and often segregate themselves accordingly into different chambers within the cavern. The buffy flower bat is an exception, and it occasionally intermixes with Jamaican flower bats in the same subterranean passage. This species does not emerge from its roost until almost complete darkness, perhaps as a predator-avoidance measure. Sporadic activity occurs at the cave entrance throughout the night, indicating that at least some bats return to their dayroost for periods of nightroosting, during which they rest, socialize, and digest their latest meal.

The Jamaican flower bat has a papillose and protrusible tongue, which implies a diet emphasizing pollen and nectar, although this animal likely shifts from a reliance on flowers in the dry season to eating more fruits in the wet season, as does the Cuban flower bat. Reports of specific dietary items are few but include the ripening fruits of the fustic tree (*Maclura tinctoria*). These bats are occasionally caught as they fly through plantations of banana, mango (*Mangifera indica*), and papaya (*Carica papaya*), which suggests that the Jamaican flower bat consumes the nectar or fruits of these plants as well. In captivity, though, one individual seemed unable to bite into firm fruit and instead fed on juice and pulp from overripe and damaged items.

Little data on reproduction are available. Biologists report single pregnant females in January and June, lone nursing mothers in January and July, and pinkish newborn pups clinging to the cave walls or attached to adults in late May. Although the closely related Cuban flower bat reproduces only once per year, this scant information concerning Jamaican flower bats suggests that births occur twice annually.

Status and Conservation. The Jamaican flower bat is protected under Jamaica's Wildlife Protection Act and is categorized as Critically Endangered on the IUCN Red List. Information on optimal roosting and foraging habitat is limited, but the species appears highly sensitive to changes in its environment. Although agricultural intensification and mining operations often are cited as reasons for the decline of the species, many historic roosts are not located in areas with intense land-use change. Instead, a combination of actions and disturbance events, such as extraction of guano, increasing utilization of caves for recreation, and predation by feral house cats, likely contributed to the decrease in numbers directly or perhaps indirectly by changing the temperature and humidity of the caves themselves. At Stony Hill Cave, for instance, house cats killed at least eighty-nine of these bats as they emerged during a single month, and guano harvesters entered the cave and removed material fourteen times over a month-long period. The National Environment and Planning Agency of Jamaica and Bat Conservation International acquired the cave and surrounding land in May 2022 and hope to install predator-proof fences to eliminate the threats.

Selected References. Henson and Novak 1966; Goodwin 1970; Genoways et al. 2005.

Phyllonycteris poeyi

Cuban Flower Bat, Murciélago de Poey, Phyllonyctère de Cuba

Name. The generic name originates from the Greek words meaning "leaf" (*phyllon*) and "bat" (*nykteris*) and presumably refers to the small nose leaf in these animals. The specific epithet honors Felipe Poey y Aloy, a prolific Cuban naturalist of the nineteenth century.

Distribution. This mammal lives on Hispaniola and Cuba, including Isla de la Juventud and some keys in the Sabana-Camagüey Archipelago, such as Salinas and Palmas. There are at least two records from the Florida Keys, but these animals likely represent accidental occurrences and not a resident population. Fossil remains from the Late Quaternary period exist on the Bahamas (Abaco and

New Providence) and Cayman Brac, suggesting that the geographic distribution was more extensive than today.

Measurements. Total length: 75–87 mm; tail length: 6–18 mm; hindfoot length: 12–20 mm; ear height: 12–16 mm; forearm length: 42–51 mm; body mass: 15–29 g. Males are, on average, larger than females.

Description. The pelage is pale cream or beige. Individual hairs are short (3–6 mm long) and bicolored with whitish bases. Wing and tail membranes are brownish-black. The snout is long and narrow, the nose leaf is rudimentary, and the lower lip has a median groove ridged with papillae. Ears are moderately large and well separated on the forehead, whereas the tail is short and extends just 5 mm beyond the back border of the uropatagium. The hind legs are long, with the tibia as long or longer than half the length of the forearm. A calcar is absent, and the narrow uropatagium extends from the tail to only the middle of the tibia.

Within its range, this bat is most similar in appearance and size to the two species of *Erophylla*, the buffy flower bat on Cuba and

the brown flower bat on Hispaniola. However, those species have a uropatagium that reaches the ankle, where the membrane is supported by a calcar, and they have a proportionately smaller tibia that is less than half as long as the forearm. In addition, the Cuban flower bat tends to have smaller ears and a shorter nose leaf than either *Erophylla*.

Natural History. Mammalogists have collected the Cuban flower bat in many habitats, including evergreen, semideciduous, gallery, and semiarid scrub forests, as well as suburban parks, plantations, and areas of secondary growth. This bat occurs at elevations ranging from sea level to about 1110 m at La Gran Piedra, in eastern Cuba, and 1700 m in southern Haiti. In evergreen forest of western Cuba, these bats seem equally active in lowland areas and sites of moderate elevation (>320 m).

This mammal is an obligate cave dweller and one of the most gregarious bats in the Greater Antilles, roosting in colonies consisting of tens of thousands or hundreds of thousands of animals. The Cuban flower bat usually inhabits the innermost parts of blind-ending galleries in hot caves. These rooms typically have a single, low, restricted entrance that limits ventilation and entraps the metabolic heat and moisture produced by the bats, thus resulting in ambient temperatures above 35 °C and relative humidity of 90% or greater. At least eighty-five caves supporting populations of this species occur throughout the Cuban archipelago, and biologists have reported this animal from nearly thirty localities on Hispaniola. The sooty mustached bat (*Pteronotus quadridens*), Parnell's mustached bat (*Pteronotus parnellii*), Cuban fruit-eating bat (*Brachyphylla nana*), and buffy flower bat (*Erophylla sezekorni*) often roost within the same subterranean site but segregate inside based on temperature preferences.

The Cuban flower bat emerges from its daytime retreat between 30 and 60 minutes after sunset and returns about 1 hour before sunrise. In evergreen forests, the highest capture rate occurs during the first 3 hours of the night as these bats search for food. Adult males generally begin foraging later than females, and pregnant and lactating females leave their roost earlier than do nonreproductive individuals. Unlike insectivorous species, such as the big brown bat (*Eptesicus fuscus*), the amount of time spent foraging is unaffected by air temperatures as low as 8 °C. However, the Cuban flower bat drastically decreases activity on bright moonlit nights, perhaps to avoid predation by owls.

When flying in open spaces, this species emits intense single-harmonic echolocation calls that drop in frequency from 46 to 34 kHz over 4.7 milliseconds. The structure of these calls is more akin to that of insectivorous bats, such as the Cuban evening bat (*Nycticeius cubanus*) or big brown bat, than to other fruit-eating species. However, in enclosed spaces, such as the interior of a forest, calls are shorter in duration (2.4 milliseconds) and often contain two harmonics that sweep from 51 to 31 and from 95 to 56 kHz; the shorter duration allows faster updates, and the higher frequencies allow perception of more detail and detection of smaller items.

The Cuban flower bat is a generalist nectarivore that also includes varying amounts of fruit, pollen, and insects in its diet. These bats feed on fruits produced by at least twelve species of plants and visit flowers from a minimum of twenty-one species. The fruits of understory shrubs, such as various peppers (*Piper*) and Panama berry (*Mutingia calabura*), are year-round dietary staples. During the dry months (November–April), however, when fruits are less available, the bats increase consumption of pollen and nectar, especially from flowers of royal palm (*Roystonea regia*) and blue mahoe (*Talipariti elatus*). Beetles, fig wasps, flies, moths, and thrips contribute to the insect fare. Although both sexes eat the same types of food, fruits dominate the diet of adult females, whereas males rely more on pollen and nectar. Short, broad wings permit maneuverable flight through the forest, but Cuban flower bats have little capacity for hovering; nevertheless, their long hind limbs make them somewhat agile as they feed along flowering or fruiting branches.

Females bear a single young once per year. Biologists typically find pregnant females between February and June, although three bats on Hispaniola carried small fetuses in mid-December; lactation occurs between June and September. Both sexes occupy the same cave throughout the year, but during the reproductive season, the females gather in maternity groups and roost separately from the males.

Several vertebrates feed on this species. Avian adversaries include the American barn owl (*Tyto furcata*), ashy-faced owl (*T. glaucops*), and stygian owl (*Asio stygius*), whereas the Cuban boa (*Chilabothrus angulifer*) and Cuban racer (*Cubophis cantherigerus*) are among the reptilian predators. The introduced house cat also catches and eats the Cuban flower bat.

Status and Conservation. This species is common and widespread on Cuba and the Bahamas, and the IUCN deems it a species of Least Concern.

Selected References. Silva-Taboada 1979; Mancina 2010; Sánchez and Mancina 2019.

Phyllops falcatus

Cuban Fig-Eating Bat, Murciélago Frutero Chico, Phyllops de Cuba

Name. *Phyllops* comes from the Greek nouns *phyllon*, meaning "leaf," and *ops*, signifying "face," and refers to the distinct nose leaf in these mammals. The specific name "*falcatus*" is a Latin

adjective that means "sickle-shaped" and denotes the highly curved index finger that runs along the front margin of the outer wing.

Distribution. This species is widely distributed on Hispaniola and Cuba, as well as on several keys off the northern Cuban coast, such as Cayo Coco, Lucas, Paredón Grande, and Sabinal. The species is apparently present but rare on the Cayman Islands, where most records consist of bones recovered from fresh owl pellets rather than actual captures of living animals. A lone Cuban fig-eating bat is known from Key West, in the Florida Keys, and a small population has become established on Cozumel Island, about 20 km from the Yucatán Peninsula of Mexico; in both instances, the bats likely reached their new island home after being blown from Cuba during a storm. Recent fossils of this species have also been uncovered in caves on Isla de la Juventud.

Measurements. Total length: 55–65 mm; tail length: 0 mm; hindfoot length: 9–12 mm; ear height: 11–13 mm; forearm length: 38–48 mm; body mass: 16–26 g. Females are much larger than males.

Description. This mammal is the sole living member of the genus *Phyllops*. Overall, the pelage is dense and silky, grayish-brown on top with paler underparts. The dorsal fur of animals from Cuba is considerably darker than that of bats from Hispaniola. Individual hairs are from 7 to 10 mm in length on the back and shorter on the belly, but all hairs are tricolored, with dark tips and bases and a pale median band. A small but prominent patch of white fur adorns each shoulder. The Cuban fig-eating bat is one of the four species of short-faced fruit bats that live in the West Indies. The snout is markedly short and broad, whereas the nose leaf is tall and wide at the base. The ears have rounded tips, and the tragus is thick and yellowish or pale pinkish. The thumbs are unusually long (9–12 mm). The wing is large and broad and blackish in color, except the anterior portion of the distal wing between the second and third fingers (the dactylopatagium minus), which is lighter in color and actually translucent. A tail is absent, the tail membrane is narrow and hairy, and the calcar is very short. The semitransparent dactylopatagium minus and the curved index finger distinguish this short-faced bat from all other species in the West Indies.

Natural History. Biologists catch the Cuban fig-eating bat in various wooded environments, including evergreen, gallery, pine, and semideciduous forest, and even in patches of secondary forest in suburban Havana. This species is more frequent at sites of low and moderate elevation, but captures have occurred at 680 m above sea level in Alejandro von Humboldt National Park in eastern Cuba, at 1220 m at Constanza in the Dominican Republic, and at an astounding 2150 m in the Haitian Massif de la Selle.

The Cuban fig-eating bat is a foliage-roosting species that hangs quietly in both microphyllous and broadleaf trees, such as Australian pine (*Casuarina equisetifolia*), binucao (*Garcinia binucao*), black olive (*Terminalia buceras*), and West Indian mahogany (*Swietenia mahagoni*). This animal roosts singly, in pairs, or in small compact groups of up to twelve individuals in a single tree. The bats may hang either within dense foliage, where the animals hide among the shadows, or in exposed sites, where they are readily observed and photographed.

Cuban fig-eating bats are active throughout the night, although biologists capture this species most frequently between 2 and 4 hours after sunset. These mammals fly slowly, and occasionally, two individuals, usually a male and female, appear to chase each other. This species is a frugivore. Although details of its diet are unknown, seeds of trumpet tree (*Cecropia schreberiana*) and various figs (*Ficus*) occur in some fecal samples, and one individual was captured while carrying a fruit of the rose apple (*Syzygium jambos*).

Echolocation calls during free flight in uncluttered space are brief (4.5 milliseconds), with a long interpulse interval (112 milliseconds). As in many bats, one pulse is emitted with every downstroke of the wings so that essentially the animal is harnessing the internal pressure generated by flight muscles to produce intense sounds at little additional energetic cost. Calls usually contain two harmonics, although the first harmonic (i.e., the fundamental) is the most intense and sweeps over a broad range of frequencies, from 74 to 24 kHz. Echolocation pulses produced by other Cuban phyllostomids under similar conditions involve a smaller range of frequencies.

The highest percentage of females that are pregnant or lactating occurs at the end of the dry season, in April and May, but mammalogists have captured reproductively active individuals throughout the year. Some Cuban fig-eating bats are simultaneously pregnant and nursing, indicating that they undergo a postpartum estrus and mate soon after giving birth. Litter size is one. At several localities on both Cuba and Hispaniola, the number of females caught in mistnets greatly exceeds the number of males. Although the reason for an unbalanced sex ratio is unknown, it might indicate that the sexes forage in different habitats, or perhaps the predominance of females is attributable to sexually related behavior, such as males guarding harems of females.

Bony remnants of the Cuban fig-eating bat frequently occur in pellets regurgitated by the American barn owl (*Tyto furcata*), ashy-faced owl (*T. glaucops*), and stygian owl (*Asio stygius*). Capture rates of these bats decline markedly following hurricanes or other severe storms, which knock the animals from their leafy roosts and destroy the fruits on which they feed.

Status and Conservation. The Cuban fig-eating bat is abundant and found throughout much of Cuba and Hispaniola, and the IUCN regards this animal as a species of Least Concern.

Selected References. Klingener et al. 1978; Silva-Taboada 1979; Mancina and García-Rivera 2000; Tavares and Mancina 2008; Rivas-Camo et al. 2020.

Stenoderma rufum

Red Fig-Eating Bat, Murciélago Frutero de Puerto Rico, Sténoderme Roux

Name. *Stenoderma* is from the Greek words *stenos* and *derma* meaning "narrow skin" and indicates the slender tail membrane of this bat. The specific name *rufum* is Latin for "red" and likely refers to the rusty-hued fur of the type specimen.

Distribution. Anselme Desmarest first described the red fig-eating bat as a species in 1818, based on a single animal from an unknown location. Although fossils of this bat were discovered on Puerto Rico 100 years later, no living individuals were reported until 1957,

when biologists captured three of these mammals on the island of Saint John. Today, this species inhabits Puerto Rico and the Virgin Islands of Saint Croix, Saint John, Saint Thomas, Tortola, and Vieques. All these islands, except Saint Croix, formed a single large land mass until sea levels rose at the end of the last glacial period. This species is the only bat living on the Virgin Islands that does not also inhabit the Lesser Antilles.

Measurements. Total length: 60–73 mm; tail length: 0 mm; hindfoot length: 12–15 mm; ear height: 16–19 mm; forearm length: 46–51 mm; body mass: 20–31 g. Females are somewhat larger than males, although the difference in size is typically only 4 to 6%.

Description. The color of the back varies from tan to a dark chocolate-brown in Puerto Rican specimens, whereas individuals from Saint John and Saint Croix are a lighter brown and have a reddish tint to the fur. The belly is always paler than the back. Length of the pelage varies from 8 mm on the back to 6 mm on the belly. About 40 to 50% of the forearm distal to the elbow is well haired, as are the adjacent parts of the wing. A thin crescent

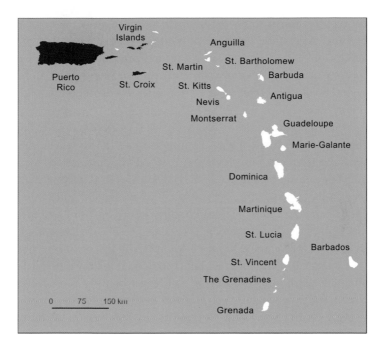

of white hairs begins below each ear and projects forward, and a noticeable white patch, about 4 mm in diameter, occurs on each shoulder where the wing attaches. Although white epaulettes occur on a few other leaf-nosed bats in the West Indies, such as the Cuban fig-eating bat (*Phyllops falcatus*), the red fig-eating bat is the only species living on Puerto Rico or the Virgin Islands with such markings. A tail is absent, and the interfemoral membrane is narrow and sparsely haired. Unfurred areas of the flight membranes are pale to dark brown, as are the hairless pinnae. Red fig-eating bats have a nose leaf that is simple, erect, and lanceolate in shape. The rostrum of this short-faced bat is small, relative to the head, and the upper teeth occur in a nearly semicircular arc, rather than a straight line from front to back.

Natural History. Unlike most other bats that form colonies and spend the day in caves, hollow trees, or buildings, the red fig-eating bat typically roosts singly within thick foliage, high in the canopy of forest trees. Radio-tracking of nonreproductive adults and juveniles on Puerto Rico indicates that these mammals change their roosting site every day or two. However, they do not travel far, and a typical home range containing all resting sites and foraging patches is only about 2.5 ha.

The red fig-eating bat is a frugivore that leaves its dayroost in the first hour after sunset. Within 10 minutes or so, the animal locates a suitable fruit, often in a forest gap or disturbed area; the bat then grabs the morsel with its teeth and quickly flies to a nearby tree to process the selected item. Typical foods are the fruit of bulletwood (*Manilkara bidentata*), Jamaican pepper (*Piper hispidum*), trumpet tree (*Cecropia schreberiana*), and sierra palm (*Prestoea montana*), but despite the common name of this animal, there is no evidence that it eats figs (*Ficus*) of any kind. Throughout the night, the bats alternate short bouts of foraging and longer bouts of roosting so that, overall, the animals spend only 27% of the nighttime actually in flight. Each bat uses multiple trees for roosting every night, moving about 60 m from one to the next, sometimes returning to the same tree but often going to a new one. These bats are active from dusk to dawn and do not alter their activity based on phases of the moon, as do many species on the mainland. Little data are available on echolocation, but calls appear to be multiharmonic; most energy is in the fundamental, which sweeps from 95 to 26 kHz over about 3.1 milliseconds, with 76 milliseconds between calls.

The greatest known density of red fig-eating bats occurs in tabonuco (*Dacryodes excelsa*) rain forest along the lower slopes (<650 m in elevation) of the Luquillo Mountains on Puerto Rico, where up to 25% of the bats caught have represented this species. However, this mammal is not restricted to stands of tabonuco, and biologists consistently encounter it, albeit in lesser numbers, in other types of forest on Puerto Rico and on different islands. On Vieques, for example, one capture site is located at sea level in subtropical moist forest containing mostly coconut (*Cocos nucifera*) and seagrape (*Coccoloba uvifera*), whereas the single successful netting locality on Saint Croix is in semievergreen rain forest dominated by silk cotton (*Ceiba pentandra*), sandbox (*Hura crepitans*), and mango

(*Mangifera indica*). On Saint John, in contrast, ecologists captured this bat in an area of dry woody vegetation at the mouth of a canyon opening onto lowlands grazed by livestock.

Reproductive information primarily comes from Puerto Rico, where biologists have caught pregnant females in January, March, June, July, and August. Each female produces a single young that weighs about one-third of the mother's mass. Dense, buffy gray hair covers the back of a newborn, and its distinctive white shoulder spot is already evident. Eyes are open at birth. Females that are simultaneously lactating and pregnant occur on a regular basis, indicating that many adult females experience a postpartum estrus and are able to produce two pups per year.

In September 1989, Hurricane Hugo hit the Luquillo Mountains resulting in great damage and defoliation. Although cave-dwelling bats ride out such storms in their underground retreats, the fierce winds likely blew many red fig-eating bats from their trees, and the survivors had to cope with the loss of the fruits on which they depended. Consequently, the population of red fig-eating bats plummeted by about 80% immediately after the storm. The remaining individuals responded by increasing the size of their home ranges by as much as fivefold in order to find sufficient food, but the regional population did not return to prehurricane levels until 5 years later. Similar precipitous declines likely occur after all major hurricanes.

Status and Conservation. The IUCN lists the red fig-eating bat as Near Threatened. Human-mediated deforestation and woodland fragmentation are potential threats, and climate change endangers the future of these bats, because of the resulting increased frequency and intensity of tropical storms, particularly hurricanes.

Selected References. Gannon and Willig 1994; Gannon et al. 2005; Kwiecinski and Coles 2007; Alexander and Geluso 2013; Calderón-Acevedo et al. 2021.

Sturnira angeli

Angel's Yellow-Shouldered Bat, Murciélago Esturnira Guadalupense, Sturnire d'Ángel

Name. *Sturnus* is the Latin name of a common European bird, the starling. The British ship H.M.S. *Starling* took part in a voyage of scientific discovery to South America from 1836 to 1842, and a little yellow-shouldered bat (*Sturnira lilium*) collected in Brazil during that expedition became the basis for the description of the genus *Sturnira*. Luis de la Torre first described Angel's yellow-shouldered bat in 1966, based on animals from Dominica, and he named that particular species in honor of his father, Angel de la Torre.

Distribution. Angel's yellow-shouldered bat lives in the central Lesser Antilles, on the islands of Martinique, Dominica, Guadeloupe, and Montserrat.

Measurements. Total length: 63–82 mm; tail length: 0 mm; hindfoot length: 12–17 mm; ear height: 14–19 mm; forearm length:

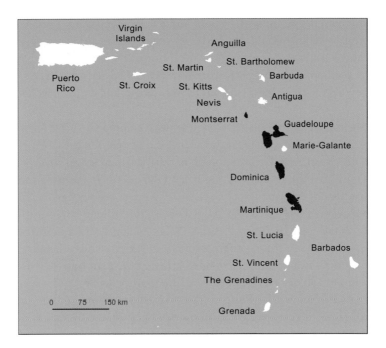

40–48 mm; body mass: 17–28 g. Males weigh more and have longer forearms than females on Martinique, although females have longer forearms than males on Dominica.

Description. The fur on the back appears dark grayish-brown, with each hair having four bands of color—pale at the base, then dark, followed by pale, and dark again at the tip. The belly, in contrast, is dull dark brown, with each hair having only three bands, with white near the root, dark brown in the middle, and grayish-brown at the outer end. Hairs at the shoulder are stained reddish-yellow by secretions from scent glands located below the hairs; the glands are better developed and the color is more intense in adult males than in females or juveniles. The ears and nose leaf are medium brown, whereas the wings are dark brown. A tail is absent. Although many species of bats have a broad membrane between the legs, often stretching from ankle to ankle, the "tail" membrane in all yellow-shouldered bats is an inconspicuous narrow strip of hairy skin paralleling the legs. Lack of an obvious membrane separates Angel's yellow-shouldered bat from other species inhabiting the same islands, and compared with Paulson's yellow-shouldered bat (*S. paulsoni*), which lives farther south in the Lesser Antilles, Angel's yellow-shouldered bat is larger, on average, and darker in color.

Natural History. This species appears least common on Montserrat and most abundant on Dominica and Martinique. On the latter island, one of every five bats caught in mistnets is an Angel's yellow-shouldered bat, and it is second in abundance only to the Jamaican fruit-eating bat (*Artibeus jamaicensis*). Angel's yellow-shouldered bat occurs from sea level to 740 m in elevation, although most captures take place between 151 and 450 m. Ecologists have captured or recorded this bat in banana groves and various natural habitats, including swamp forests, wooded stream corridors, rain forests, and wet and dry evergreen forests. The roosting habits of this species are unknown, but various species of yellow-shouldered bats on the mainland occupy cavities in trees or hang among the fronds of palms or in tangles of vines. On Martinique, this mammal remains active throughout the evening, on clear as well as cloudy and rainy nights, and it is equally abundant in the dry and wet seasons.

Other species of yellow-shouldered bat frequently forage in the forest understory, and the wing shape and echolocation calls of Angel's yellow-shouldered bat suggest that it behaves similarly. The wings are short and broad, which is necessary for maneuvering in complex environments, and the structure and timing of the echolocation calls allows detection of obstacles at close range. On Dominica, for example, these bats rely on echolocation calls that are frequency modulated and have a low duty cycle. The second harmonic usually has the most energy and typically has a starting frequency of 93 kHz and ends at 46 kHz; individual pulses are short, only 2.7 milliseconds in duration, and the time between calls is somewhat long, about 72 milliseconds. Calls from the bats on Guadeloupe are broadly similar to those from Dominica, although the sounds produced on Martinique apparently include higher frequencies, shorter durations, and smaller intervals between calls.

These bats are frugivores that pluck their food from the branches of trees, shrubs, vines, and even epiphytes. As the bats munch on the soft tissue of a fruit, any seeds inside, especially small ones, are swallowed whole and eventually appear in the feces. Defecated seeds are important for forest regeneration, but they also offer mammalogists a way of identifying what these bats eat. Seeds extracted from the fecal pellets of seven yellow-shouldered bats on Martinique reveal that they include the fruits of turkey berry (*Solanum torvum*) and trumpet tree (*Cecropia schreberiana*) in their diet. Similarly, material produced by thirty-one bats from Guadeloupe shows that these animals consume fruits from guava (*Psidium guajava*), monkey paws (*Marcgravia umbellata*), various peppers (*Piper*), and philodendrons (*Philodendron*), as well as the trumpet tree and turkey berry. Most of these yellow-shouldered bats (79%) have only one type of seed in their feces at a time, although some have two or three kinds.

In general, pregnant females occur from March to July, primarily during the dry season, and lactating females are evident from May to September. More specifically, reports of pregnant individuals exist for March and April on Martinique, July on Guadeloupe, and March, April, June, and July on Dominica. Some young bats become independent as early as June and July, just as fruits are becoming abundant in the surrounding forests. Longevity is unknown, and the only documented predator is the Lesser Antillean barn owl (*Tyto insularis*).

Status and Conservation. Until 2013, taxonomists believed that this animal and Paulson's yellow-shouldered bat were subspecies of the little yellow-shouldered bat, which was widespread on the mainland, from Mexico to Brazil; however, genetic studies indicated that both Angel's and Paulson's yellow-shouldered bats are distinct species endemic to the Lesser Antilles. The IUCN classifies Angel's yellow-shouldered bat as Near Threatened.

Selected References. de la Torre 1966; Genoways et al. 2001; R. J. Larsen et al. 2007; Barataud et al. 2015; Catzeflis et al. 2019.

Sturnira paulsoni

Paulson's Yellow-Shouldered Bat, Murciélago Esturnira de Paulson, Sturnire de Paulsen

Name. British biologists discovered the type specimen for the genus *Sturnira* during an expeditionary voyage to South America that included the H.M.S. *Starling*. That ship was named after the common European bird, which the Romans called *sturnus*. *Paulsoni* refers to Dennis Paulson, who provided the specimen used to describe this particular species of *Sturnira*.

Distribution. This mammal is restricted to the southern Lesser Antilles, where it occupies the islands of Saint Lucia, Saint Vincent, and Grenada.

Measurements. Total length: 60–67 mm; tail length: 0 mm; hindfoot length: 11–15 mm; ear height: 14–16 mm; forearm length: 40–46 mm; body mass: 14–25 g.

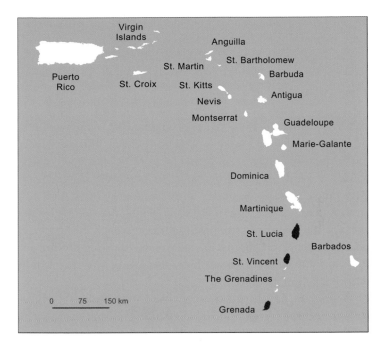

Description. Overall, the fur appears pale grayish-brown on the back but yellowish-brown on the belly. Each dorsal hair has four bands of color, alternating pale and dark, from the base to the tip of the shaft. Each ventral hair, though, is tricolored—white near the root, darker in the middle, and a light brown at the end. As in other members of the genus, the shoulders appear somewhat reddish-yellow, although this coloration is not caused by pigments in the hair. Instead, glands in the skin discharge an odiferous substance that stains the fur and provides the color; the glands and the resulting yellow color are best developed in adult males. The ears, nose leaf, and wing membranes are brown. There is no tail, and the membrane between the hind legs is just a narrow strip of skin, perhaps 5 mm wide at the knee, that parallels the legs. The feet, legs, and membrane are densely furred. The thin, hairy membrane along the inner legs is sufficient to separate this bat from all other species in the Caribbean, except Angel's yellow-shouldered bat (*S.*

angeli) and the Guadeloupean big-eyed bat (*Chiroderma improvisum*), both of which live only on islands north of Saint Lucia. In addition, Angel's yellow-shouldered bat is darker colored, whereas the big-eyed bat is considerably larger (forearm length >56 mm) than either yellow-shouldered bat.

Natural History. Paulson's yellow-shouldered bat is most common on Saint Vincent and least abundant on Grenada. This animal occurs primarily in interior montane rainforests, with open understories, above an elevation of 200 m. In addition, mammalogists have caught this species in dry forest, secondary forest, and cultivated areas, up to 650 m in elevation. Successful netting locales are typically over roads, streams, and trails through the forest and occasionally in locations that are more open. Roosting sites are unknown, but these bats probably hide within hollow trees or within a thick tangle of palm fronds.

Similar to Angel's yellow-shouldered bat, Paulson's yellow-shouldered bat has echolocation calls that sweep from 83 to 36 kHz at low duty cycle, which means that individual pulses are short in duration (1.4 milliseconds) and the interval between consecutive calls is comparatively long (42 milliseconds). Typical calls include four harmonics, with most energy in the second harmonic. Calls with these characteristics are useful for short-range detection of obstacles while flying in cluttered space, such as beneath the canopy of a forest. Similarly, the wings are short and broad (i.e., low aspect ratio), and the tips are moderately pointed; these traits enhance maneuverability when moving through complicated environments. This bat easily flits from tree to tree, but the animal is also a reasonably strong flyer, capable of traveling long distances across open areas to find new foraging patches. No specific information on diet is available, although biologists have snagged this mammal in nets set near pepper (*Piper*) plants, the fruits of which provide food for many species of bats.

Most females give birth to a single young by the end of the dry season in May and are lactating at the beginning of the wet season in June. During July and August, reproductive activity ceases in most individuals, but about 30% of the females are pregnant. This limited data suggest that two cycles of reproduction occur during the year but that not all females become pregnant the second time. As in other species of *Sturnira*, the shoulder glands of adult males are more active during the mating season, presumably because the odor of the secretions is attractive to potential mates. Nothing is known concerning longevity or mortality factors.

Status and Conservation. Taxonomists long considered this bat as a subspecies of the little yellow-shouldered bat (*S. lilium*), which roamed over the Lesser Antilles and from Mexico to Brazil; however, genetic data confirmed that Paulson's yellow-shouldered bat is a distinct species. The IUCN currently lists the species as Near Threatened, although that assessment does not consider the effects of the volcanic eruption on Saint Vincent in 2021.

Selected References. Genoways et al. 1998; Vaughan Jennings et al. 2004; Kwiecinski et al. 2018; Pedersen et al. 2018a.

Vesper Bats

The word *vesper* in Latin means "evening," and these bats are often simply called evening bats or vesper bats. The family has a nearly global distribution, occurring in temperate and tropical regions and on all continents except Antarctica. With 502 species arranged into 57 genera, the Vespertilionidae is the largest family of bats and the third-largest family of mammals, surpassed only by the rodent families Muridae and Cricetidae. Five genera and thirteen species of vesper bats occur in the West Indies, and eleven of the species are endemic to the islands.

A number of vesper bats, such as the tiny pipistrelle (*Pipistrellus nanulus*) and Himalayan whiskered myotis (*Myotis siligorensis*), weigh a mere 2–3 g and are among the smallest mammals in the world, whereas the largest vespertilionid is the giant house bat of Africa (*Scotophilus nigrita*), at about 90 g. Vespers typically have a short face and an ill-defined neck region and lack elaborate nasal or facial appendages. Ears are generally well separated and vary from short to very long; for instance, the translucent pinnae of a spotted bat (*Euderma maculatum*) from Colorado are almost as long as the body itself. The uropatagium is large, and it totally encases the tail vertebrae, which continue to the posterior margin of the V-shaped membrane or slightly beyond (see Fig. 0.2). Color ranges from brownish to blackish with gray undertones, but some species display white, creamy, yellowish, or reddish-to-orangish pelage; the latter pattern is particularly evident in the spectacular painted woolly bat (*Kerivoula picta*) of Indonesia. The face, ears, and membranes of a vesper are usually naked, although the dorsal surface of the tail membrane is totally or partly furred in some species, such as the northern hoary bat (*Lasiurus cinereus*) and silver-haired bat (*Lasionycteris noctivagans*), respectively.

The skull varies widely in structure, but most, including all those in the West Indies, have a distinct notch in the palate that clearly separates the right and left incisors. Dilambdodont dentition (W-shaped ridges on the cheek teeth) is typical of vespers

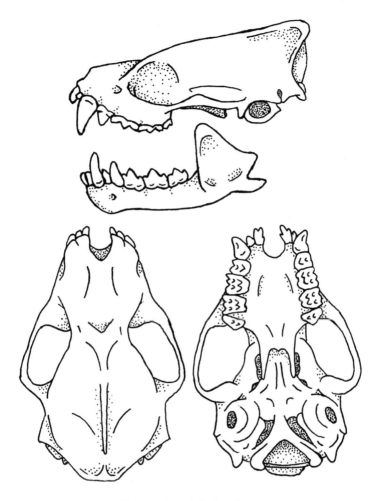

Big brown bat, *Eptesicus fuscus*

and other insect-eating bats, and these ridges are useful in slicing through the tough exoskeletons of their prey. The number of teeth varies from 28 to 38, depending on genus.

Vespertilionids occupy a myriad of habitats, from lowlands to highlands, and from deserts to rain forests. In addition, many species have adapted to human intrusion and frequent agricultural areas and urban environments. Although many vespers dwell in caves or cavelike environments, others roost in rock crevices, tree hollows, foliage, hollow joints of bamboo, tropical flowers, bird's nests, buildings, and bridges, as well as under rocky slabs or loose bark on tree trunks. Some species lead solitary lives, but most are social, forming colonies varying in size from less than ten individuals to hundreds of thousands of animals, depending on species.

Vesper bats feed mainly on insects, including many agricultural pests and insects of human-health concern. Noninsect prey includes spiders, scorpions, or fish, and a few of these mammals, such as the great evening bat (*Ia io*) of China, actually attack and consume migrating songbirds. Vespers forage preferentially in open areas, over land and water, and usually capture insects in flight using their wings or tail membrane; a few species are able to glean their prey from surfaces. These mammals are a major component of bat communities in temperate regions and are faced with the seasonal loss of their insect prey as winter approaches; consequently, many vespers move in autumn to a nearby mine or cave, where they hibernate for 6 to 9 months, whereas others migrate long distances to warmer climates. The common noctule (*Nyctalus noctula*) of Europe, for example, migrates up to 1500 km each spring and fall.

Species from tropical and subtropical regions sometimes mate year-round and experience two or more pregnancies, whereas bats from temperate regions give birth only once per year. Temperate vespers utilize a process called delayed fertilization; the animals mate in autumn, females store the sperm inside their uterus until spring, and ovulation occurs and pregnancy begins when the bats leave hibernation. At birth, the mothers hang head-up, curl their large tail membrane into a basket, and catch the newborn as it leaves the birth canal. Litter size is typically one, although many species consistently produce twins. Most female bats worldwide have two nipples, but vespers in the New World genus *Lasiurus* and the Old World *Otonycteris* have four; consequently, litters of three and four offspring are not unusual in those groups.

Selected Reference. Moratelli et al. 2019.

Antrozous koopmani

Koopman's Pallid Bat, Murciélago de Koopman, Antrozon Blond de Koopman

Name. *Antrozous* is derived from Greek nouns *ántro*, denoting a "cave," and *zóo*, meaning "animal." The species was named in honor of Karl F. Koopman, a remarkable twentieth-century taxonomist from the American Museum of Natural History.

Distribution. All records of Koopman's pallid bat come only from the main island of Cuba.

Measurements. Total length: >65 mm; tail length: 50–60 mm; hindfoot length: 13–15; ear height: 24–25 mm; forearm length: 49–62 mm; body mass unknown. Very few modern specimens have been examined.

Description. Koopman's pallid bat closely resembles the smaller North American pallid bat (*A. pallidus*), which is the animal in the

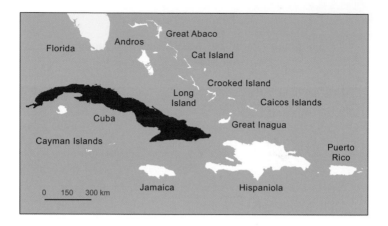

accompanying photograph. The fur of the Cuban species is slightly dense, with pale yellow hairs that are darker at the base. Dorsal hairs are longer (7–10 mm) than the ventral (5–8 mm). This magnificent animal sports large eyes, noticeably tall and broad ears that are well separated above the head, and an elongate tragus that extends more than half the height of the pinna. The muzzle is short, wide, and blunt, with a fleshy bulge on each side between the nostril and eye. Like other members of the family Vespertilionidae, this bat lacks a nose leaf, and the long tail protrudes at most 2–3 mm beyond the posterior margin of the V-shaped uropatagium. Wings and tail membrane are hairless. At first glance, this species might be confused with Waterhouse's leaf-nosed bat (*Macrotus waterhousii*), which also has large ears; however, that species has a short but well-defined nose leaf. The molars of Koopman's pallid bat have distinct ridges in the form of a W that are characteristic of insectivorous species.

Natural History. Koopman's pallid bat is the rarest of all Cuban bats. The description of the species is actually based on a single skull recovered from an owl pellet in Cueva del Hoya de García, Pinar del Río, in 1958, and biologists have captured only three living individuals—two in 1920–1921 and one in 1956. Charles Ramsden collected the first two specimens, both adult females, in Caney and Guantánamo, in southeastern Cuba, and they eventually became part of the collection at the Universidad de Oriente, in Santiago de Cuba. However, Ramsden, a herpetologist, apparently misidentified the mammals, and they were stored in a jar with more than forty Waterhouse's leaf-nosed bats; it was not until at least 40 years later that the mistake was discovered. Unfortunately, Ramsden recorded no other details concerning the captures, not even the specific date. Karl Koopman and Gilberto Silva-Taboada caught the third individual in a mistnet set in the foothills of Pan de Guajaibón, Pinar del Río, in October 1956. That animal became entangled at 2300 hours in the lowest part of the net so that the bat's weight caused the net to sag to the ground. Incredibly, the bat escaped soon after capture, and no further information was obtained. Since 1956, a few naturalists have reported capturing this species, but none has been verified through photos or actual specimens. The only other definitive records of this mammal come from bones hidden within owl pellets.

Obviously, little is known of the roosting habits, reproduction, or diet of this uncommon animal. The one other member of the genus *Antrozous*, the North American pallid bat, lives in arid to semiarid environments, from southern British Columbia in Canada to Jalisco in the central highlands of Mexico. This continental species typically shelters in caves and rock crevices and occasionally inside buildings or hollow trees, and Koopman's pallid bat may use equivalent sites on Cuba. This mammal, like its continental cousin, probably gives birth once per year, usually to one or two offspring, and one of the specimens collected by Ramsden carried a single tiny embryo. The stomach from one of Ramsden's animals contained unidentified parts of spiders and dictyopteran insects (roaches or mantids), but no other dietary information is available. Like the North American species, Koopman's pallid bat is most likely a gleaning insectivore that snatches large insects and maybe other arthropods, such as spiders and scorpions, from the vegetation or the ground, consistent with the bat's capture at the bottom of the net in 1956. American barn owls (*Tyto furcata*) are the only known predator.

Status and Conservation. Fossilized remains suggest that the species occurred throughout most of Cuba until the nineteenth and twentieth centuries, although it is extremely rare today and possibly extinct. Deforestation and fragmentation of mature forests may have played a role in its scarcity. For example, soon after the animal was mistnetted in 1956, the surrounding forest was felled. The IUCN believes that Koopman's pallid bat is a subspecies of the wide-ranging North American pallid bat and does not separately categorize the status of this insular population.

Selected References. Orr and Silva-Taboada 1960; Silva-Taboada 1979; Mancina 2012; Borroto-Páez and Mancina 2017.

Eptesicus fuscus

Big Brown Bat, Murciélago Alí-Oscuro, Sérotine Brun

Name. *Eptesicus* comes from Greek words meaning "house flyer"; these bats often form colonies in barns and houses in eastern North America, where the species was first described. *Fuscus* means "dusky" or "somber," and denotes the color of the animal's fur.

Distribution. The big brown bat is widely distributed from Canada in North America, through Central America, to Colombia, Venezuela, and Ecuador in South America. In the West Indies, this bat inhabits the Bahama Islands (including Acklins, Andros, Crooked, Great Abaco, Great Exuma, Little Exuma, Long, New Providence, and San Salvador), Cayman Islands, Cuba (including Isla de la Juventud), Hispaniola, and Puerto Rico. This species is likely absent from all islands between Puerto Rico and Trinidad. A big brown bat supposedly obtained in Barbados in 1878 is either incorrectly labeled or represents an accidental arrival, and although a report from Dominica in 1985 describes this species, biologists have not captured one since that time.

Measurements. Total length: 86–125 mm; tail length: 39–50 mm; hindfoot length: 8–14 mm; ear height: 12–20 mm; forearm length: 39–51 mm; body mass: 11–23 g. Females are marginally larger than males. Within the Greater Antilles, big brown bats are smallest on Juventud, largest on Puerto Rico, and intermediate in size on Cuba and Hispaniola.

Description. Dorsal fur varies from tan to dark chocolatey brown, and the long soft hairs have somewhat shiny tips; ventral fur is paler and varies from yellowish to olive-buff. The big brown bat has a wide muzzle and large broad head, with short rounded ears. Face, ears, tragus, and wings are dark brown to black and naked. The uropatagium is well developed, stretching from ankle to ankle, with the posterior margin forming a backward-pointing V; the top of the tail membrane is mostly naked with sparse hairs close to the body. The actual tail is shorter than the body alone, and only the terminal 1 to 3 mm protrude past the membrane.

The only bats in the region with a backward-V shape to the tail membrane are funnel-eared bats and other vespertilionids. Similar-sized funnel-eared bats, though, have a tail longer than the body and a fringe of hairs along the trailing edge of the tail membrane that is missing in big brown bats. Among the vesper bats that lack dense fur on the uropatagium, the Cuban evening bat (*Nycticeius cubanus*) is much smaller, and Koopman's pallid bat (*Antrozous koopmani*) has tremendously larger ears than a big brown bat. In addition, all other vespertilionids residing on the same West Indian islands have a single pair of upper incisors, whereas the big brown bat has two pairs.

Natural History. In the West Indies, the big brown bat is a habitat generalist, living in dry and moist woodlands, agricultural regions, and urban areas, although it seems most common in forested highlands up to 2100 m and less abundant in coastal sites. These animals most frequently spend the day inside cool, well-ventilated caves or cavelike structures, such as tunnels and culverts, often near the entrance, where light still penetrates, and occasionally deeper as long as the air temperature remains low. Colonies of up to three hundred bats may take shelter in these underground sites, and inside, the animals roost singly or in groups of ten to twenty individuals that tuck themselves into cracks or deep solution cavities in the ceiling; occasionally, a Brazilian free-tailed bat (*Tadarida brasiliensis*) joins these clusters. About 25% of the time, big brown bats day-roost within human-made structures, including houses, tobacco barns, churches, schools, and industrial buildings.

The big brown bat is an aerial insectivore that typically forages above the forest canopy, along edges, or in open areas, flying at speeds up to 33 km/hr. Individuals start leaving their shelters near sunset and typically mill about the entrance for several minutes before dispersing to their hunting grounds. Occasionally during the night, the bats roost alone or in small groups, for a few minutes to several hours, so that total nightly time in flight by nonreproductive individuals may be as little as 100 minutes. As in other insectivorous species, nightroosting occurs at sites that are distant from the dayroost and closer to the foraging grounds. Nocturnal roosts on Cuba and Puerto Rico occur in buildings and under highway bridges, but these bats likely use rock shelters or other natural sites as well. Mothers do not carry their youngsters on foraging flights and must periodically return to the dayroost to nurse, although most other members of the colony do not come home until dawn.

Big brown bats have a less maneuverable flight than many other species, although their echolocation is incredibly precise. During search and approach phases, echolocation calls of big brown bats from Cuba last less than 8 milliseconds and sweep downward from an initial frequency of about 47 kHz to a minimal and almost constant frequency of about 33 kHz. A big brown bat is capable of capturing one insect every 3 seconds, and the bat's body mass may increase by 24% during a single foraging bout. Mothers producing milk need large amounts of energy and can consume up to 100% of their own weight in insects in a single night. Adults prey mainly on flying beetles, especially June bugs (*Phyllophaga*), although juveniles may take more soft-bodied insects. Moths, flies, ants, lacewings, and flying cockroaches contribute to the diet.

In temperate regions, big brown bats hibernate during winter, lowering their body temperature to 2 to 10 °C and subsisting for months on fat stored in autumn. Hibernation does not occur in the sultry West Indies, but in Cuba, these mammals still become fat by October and then lose 32% of their body weight by March, suggesting that insects preyed on by the bats are less available during the dry season. Although the Cuban bats do not hibernate, they are able to lower their body temperature (i.e., become torpid) for a few hours or even a few days, to save energy during cool weather; active body temperature in this species is 37 °C, the same as humans, but it can fall to 26 °C and probably lower. If air temperature at sunset is less than 20 °C, these tropical bats do not leave their roost and instead remain torpid and inactive throughout the night.

Females give birth once per year, and 85% of the mothers produce twins. Biologists report finding pregnant females from March through early June on various islands, with parturition occurring as early as 5 May and probably continuing until mid-June. On the North American mainland, juveniles start flying between 18 and 35 days after birth, and lactation lasts from 32 to 40 days.

The Cuban treefrog (*Osteopilus septentrionalis*) is a documented predator, and remains of the big brown bat frequently appear in pellets regurgitated by the ashy-faced owl (*Tyto glaucops*) and American barn owl (*T. furcata*) on Cuba. The spinning blades of a wind turbine killed one of these animals at a wind farm on Puerto Rico. In North America, big brown bats live up to 26 years in captivity and 20 years in the wild; however, longevity in bats correlates with use of hibernation, and the maximum lifespan of free-living individuals inhabiting the West Indies is presumably shorter.

Status and Conservation. The big brown bat is widespread on the mainland. Furthermore, this species is common on Cuba, locally abundant on Hispaniola, and occurs throughout Puerto Rico and the Bahamas; consequently, it is a species of Least Concern to the IUCN.

Selected References. Silva-Taboada 1979; Timm and Genoways 2003; Speer et al. 2015; Cláudio 2019a; Yi and Latch 2022.

Eptesicus guadeloupensis

Guadeloupean Big Brown Bat, Murciélago Marrón Guadalupense, Sérotine de la Guadeloupe

Name. *Eptesicus* combines Greek words that mean "house flyer." The naturalist Constantine Samuel Rafinesque first applied this name in 1820 to the big brown bat (*E. fuscus*) of North America that frequently lives in houses and barns. The specific epithet indicates that this particular mammal is native to Guadeloupe.

Distribution. To date, biologists have been able to find the Guade-

loupean big brown bat only on Basse-Terre, the largest island of the Guadeloupean Archipelago.

Measurements. Total length: 129–132 mm; tail length: 54–60 mm; hindfoot length: 11–14 mm; ear height: 22–24 mm; forearm length: 49–54 mm; body mass: 19–22 g.

Description. Individual hairs are black at the base, on both the back and belly; however, the tips are dark chocolate-brown dorsally but dark buff to almost white in some areas of the underside. Ears, wings, and tail membrane are black. The wings are mostly naked but lightly haired near the body. As in other vespertilionids, the interfemoral membrane is broad, extending from ankle to ankle, and its trailing edge points backward. The tail vertebrae extend the length of the membrane and protrude slightly beyond it.

This species is most closely related to the widespread big brown bat, which occurs on some of the Greater Antilles and Bahamas. Nevertheless, the ears of a Guadeloupean big brown bat are taller than are those of the big brown bat, and the length of the tibia (24–

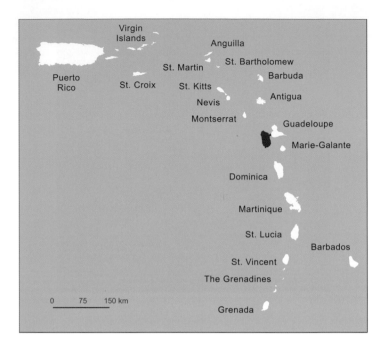

26 mm) is greater than that of any North American member of the genus *Eptesicus*. Twelve additional species of bats occur on Guadeloupe, but the Guadeloupean big brown bat could be confused on that island only with the Dominican myotis (*Myotis dominicensis*) or possibly the Lesser Antillean funnel-eared bat (*Natalus stramineus*), which also have V-shaped tail membranes. However, both those species are much smaller than a Guadeloupean big brown bat, with forearm lengths that are less than 37 and 42 mm, respectively.

Natural History. Our knowledge of the natural history of the Guadeloupean big brown bat is quite limited. Since its discovery in 1974, mammalogists have caught or acoustically detected this animal at fewer than ten locations, all of which are along the eastern slopes of the north-south-trending mountains of Basse-Terre, below 300 m in elevation. No Guadeloupean big brown bats are yet known from the sister island of Grande-Terre, which is more dominated by urban zones and agriculture than Basse-Terre, where most of the Guadeloupean forest remains.

Similar to the Dominican myotis, the Guadeloupean big brown bat appears restricted to woodlands. The original capture location in Baie-Mahault, in northeastern Basse-Terre, is at the edge of a swamp forest, with trees up to 15 m tall forming a closed canopy and little understory present. Other habitats where ecologists have found this bat are variously described as mesophilic forest, rain forest, coastal forest, semideciduous forest, and a woodland adjoining a banana plantation. The Guadeloupean big brown bat typically flies over the canopy or along the forest edge. Some biologists speculate that this bat spends the day in trees, although no roosts have been located yet.

All members of the genus are insectivores, and limited information indicates that the Guadeloupean big brown bat includes winged termites and beetles in its diet. When cruising along a forest edge, this bat emits echolocation pulses that start near 43 kHz and fall to 25 kHz over 9 milliseconds; the frequency with maximum energy is 28 kHz. When flying high above the treetops, though, the bat uses calls that are almost constant in frequency, varying only from 28 to 24 kHz. The two unique features of this bat's anatomy—large pinnae and long legs—are partly involved in gathering prey; the large outer ears help focus returning echoes, and the extended tibia supports a large interfemoral membrane, which is used to scoop flying prey out of the air. Whether or not these traits allow the Guadeloupean big brown bat to seek insects in a different manner or to consume different types of food than related species is unknown.

Field biologists have never captured a pregnant or lactating individual, but both subadult and postlactating females are known from February and July. Such widely spaced dates for postlactating animals may indicate that these bats breed more than once during the year. Two males taken in April appeared to be sexually active, with testes in the scrotal position.

No members of the genus *Eptesicus* live on any adjacent islands, so how did the Guadeloupean big brown bat get to Guadeloupe? One possibility is that the founders came from the north, from the

Greater Antilles, where the nearest known population of the big brown bat is located on Puerto Rico, 500 km northwest of Guadeloupe. The second hypothesis is that immigration came from South America, about 600 km to the southwest in Venezuela, and big brown bats from the latter area more closely resemble the Guadeloupean big brown bat in size than do individuals from Puerto Rico. In either case, though, the ancestral bats probably moved in a steplike fashion, from island to island, up or down the Lesser Antilles, until they occupied Guadeloupe. Why the bats survived on Guadeloupe yet disappeared from all intervening islands is a mystery.

Status and Conservation. The IUCN lists the Guadeloupean big brown bat as Endangered. Threats to the species include deforestation and fragmentation of wooded habitats to make way for human habitations and expanded croplands; for example, the swamp forest in Baie-Mahault where the bat was first discovered amounted to 800 ha in 1974 but is now less than 200 ha. Intensive use of agricultural pesticides, especially organochlorines that accumulate in the food chain, may affect this bat as well.

Selected References. Genoways and Baker 1975; Baker et al. 1978; Barataud and Giosa 2013; Barataud et al. 2015.

Eptesicus lynni

Jamaican Brown Bat, Murciélago Marrón Jamaiquino, Sérotine de la Jamaique

Name. The famous American-French naturalist Constantine Samuel Rafinesque coined the word *Eptesicus* in 1820 as the generic name for the big brown bat (*E. fuscus*) in North America. The last line of his description of that animal indicated that "it comes often in the house at night," and the word *Eptesicus* actually combined Greek words meaning "house flyer." H. Harold Shamel named the Jamaican brown bat, a beautiful little bat, in 1945, in honor of the collector of the original specimens—professor William Gardner

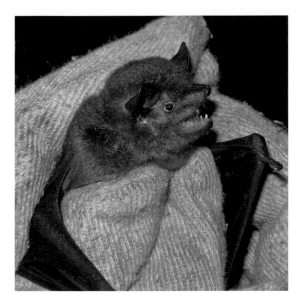

Lynn, a member of the faculty at Johns Hopkins University and later Catholic University, who cowrote *The Herpetology of Jamaica* (1940).

Distribution. This bat inhabits only Jamaica.

Measurements. Total length: 91–102 mm; tail length: 35–42 mm; hindfoot length: 8–12 mm; ear height: 12–16 mm; forearm length: 43–49 mm; body mass: 20–25 g.

Description. Hairs of the back are unicolored to the base. The overall color is brown but sometimes reddish or somewhat paler, with the underside always lighter than the back. Ears and membranes are dark brown to black. The hairless tail membrane points backward, and the tail vertebrae travel through the membrane to the tip of the V but not noticeably beyond. Length of the tail is about the same as that of the body or less. On Jamaica, this species could be mistaken for the Jamaican funnel-eared bat (*Natalus jamaicensis*), which has a similar-shaped uropatagium, but that species has

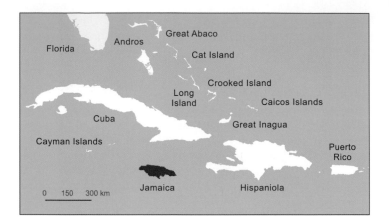

a fringe of hairs along the posterior margin of the membrane, a tail that is obviously longer than the body alone, larger ears rounded into a funnel shape, and lighter-colored fur.

Natural History. Biologists have recorded the Jamaican brown bat at ten or so sites throughout the island, but the species is abundant at only two locales—Green Grotto Caves near Discovery Bay and an unnamed cave near Montego Bay. Both of these north-shore sites are at or near sea level in areas characterized by native vegetation heavily influenced by human activity. In addition, biologists have collected the Jamaican brown bat at several other lowland locations, including arid areas of Portland Point on the central southern coast, as well as from spots high on the southern slope of the Blue Mountains, at elevations approaching 1500 m.

This small bat is probably an obligate cave-roosting species, although places such as deep crevices and rock ledges may provide shelter on occasion. The largest colonies occur in flank-margin limestone caves along the north coast. Like other species in the genus, the Jamaican brown bat likely preys almost exclusively on flying insects. Reproductively inactive females are known from July and November, but no other breeding data are available. Mammalogists identified the remains of at least four individuals of this mammal from regurgitated food pellets of the American barn owl

(*Tyto furcata*) that were found in caves near Jackson's Bay on the south coast of the island; the Jamaican brown bat accounted for 4.8% of the nonrodent remains in the pellets.

Status and Conservation. Populations living on oceanic islands have long intrigued biologists because of the possibility that such isolated groups might slowly change over time, become different from their mainland ancestors, and ultimately become different species. The Jamaican brown bat illustrates some of the issues, though, involved in identifying a species, which is a group of interbreeding individuals that is genetically isolated from other such groups. Of course, biologists do not experimentally breed animals from different populations (islands), and the decision as to whether two groups represent different species is often based on their overall morphologic or genetic differences. H. Harold Shamel believed that these animals were most closely related to a group of bats typified by the Brazilian brown bat (*E. brasiliensis*) in Central and South America. Jamaica lies not far to the east of Central America, and several other species on the island have relatives on this part of the mainland, making the supposition plausible. However, based on additional structural details, other scientists disagreed and placed the Jamaican brown bat as a subspecies of the big brown bat that resided in North America and on nearby islands in the Greater Antilles, such as Cuba and Hispaniola.

A genetic analysis offered further insights into the relationships among these populations. That study determined that the Jamaican brown bat and big brown bat were somewhat similar, but overall, about 40% of their genes differed from the species in Central and South America, thus making it highly unlikely that the Jamaican big brown bat evolved from the Brazilian cohort and more likely that it was related to the big brown bat. Nevertheless, subpopulations of the same species generally differ at an average of 15% or less in that type of genetic study, yet the Jamaican brown bat differed by 20% from the big brown bat, thus supporting the idea that the Jamaican population is on its own genetic path and probably a distinct species.

So, is the Jamaican brown bat a "true" species? The IUCN says no, and calls it a subspecies of the big brown bat, and consequently, it has no individual conservation ranking by that organization. However, the National Environment and Planning Agency of Jamaica recognizes it as a true species that is endemic to their island. If the Jamaican brown bat is a separate species, then an appropriate classification would be Vulnerable, because these animals occur on a single island, their cave roosting sites are few in number and prone to human disturbance, and habitat loss related to human development along the north coast of Jamaica is ongoing.

Selected References. Shamel 1945; Arnold et al. 1980; McFarlane and Garrett 1989; Genoways et al. 2005.

Lasiurus cinereus

Northern Hoary Bat, Murciélago Canoso, Lasiure Cendré

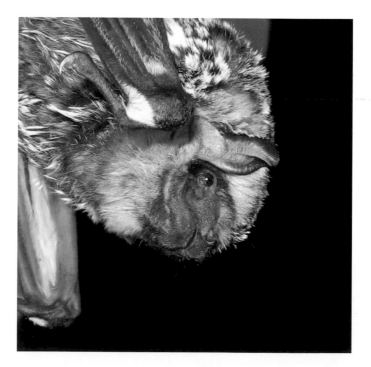

Name. *Lasiurus* derives from the Greek words *lasios* and *oura*, meaning "hairy tail" and describes the furred tail membrane observed in members of this genus. In Latin, *cinereus* is an adjective describing something that is "ashen" or "ash gray" in appearance, and the word denotes the typical grayish-white look of this species.

Distribution. The northern hoary bat is one of the most widely distributed mammals in the New World, occurring from northern Canada to southern Mexico. In the Caribbean, though, this bat is known from a single animal preserved in alcohol and multiple skeletal fragments from Hispaniola and Gonave.

Measurements. Total length: 127–140 mm; tail length: 47–61 mm; hindfoot length: 8–13 mm; ear height: 13–19 mm; forearm length: 50–57 mm; body mass: 20–35 g. Measurements are from animals dwelling on the mainland, and the size of fossilized bones from Hispaniola suggests that individuals from the island are somewhat smaller.

Description. This mammal is arguably one of the most visually appealing and unique bats in the region. The fur of a northern hoary bat is long and full and covers the entire body. Hairs on the back are mostly dark brown or brownish-gray along the shaft, although

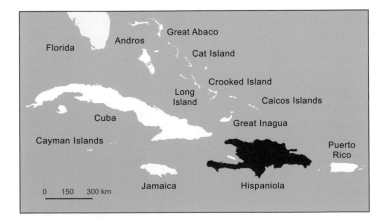

the one fluid-preserved individual from Hispaniola appears some-what reddish-brown. Many hairs are tipped with white, giving the animal a frosted appearance, but the amount of this frosting varies among individuals. Small patches of white or blond hairs are also present on the wrists and along the forearms, and their color contrasts sharply with the dark brown of the wing membrane. Hairs tipped with light brown coat the wing membranes near the body and along the upper arm toward the elbow, and the top of the tail membrane is completely furred. Similar to other lasiurines, this bat has a short rostrum and very compact snout, and its face is nearly naked except a patch of short hairs behind the nostrils. The skin of the face, wing, and tail membranes is dark brown. Ears are rounded and edged in dark brown or black, and the pinna is lightly furred on the inside. On Hispaniola, the only other bat with a compact snout, rounded ears, frosted fur, and a hairy tail membrane is the minor red bat (*L. minor*). However, that species has distinctly reddish-to-orange fur, as well as an orange tint to the skin of its face and ears, and it lacks the dark border on the pinnae.

Natural History. The only whole specimen of a northern hoary bat from Hispaniola is a male preserved in the collection of the United States National Museum. The bat was collected by L. Peters on 22 March 1900, at an unspecified location in the Dominican Republic, but no other information accompanied the specimen. However, biologists have unearthed numerous bones and skulls on Gonave and Hispaniola. For instance, the bones of at least twenty-two individuals were recovered at Trouing Jean Paul, a limestone cave in southeastern Haiti. The bats presumably were captured by ashy-faced (*Tyto glaucops*) or barn owls (*T. furcata*), which frequently roost just inside the entrance to a cave but hunt widely in the surrounding countryside. After capturing prey, owls often return to the home roost for internal processing of their meal. These birds prevent bones, teeth, and claws from entering their intestines, and the indigestible material is instead regurgitated and expelled from the mouth in the form of an ovoid pellet, which explains why the remains of this tree-dwelling bat were discovered in a cave. Radiocarbon dating indicates that some of these bones were only about 700 years old.

On the mainland, northern hoary bats inhabit temperate deciduous forests, pine-oak woodlands, and coniferous forests, in uplands and lowlands, in dry and wet areas, with either dense or open vegetation. This species also occupies agricultural zones if some trees are present, and it permeates urban environments, where this bat often resides in wooded city parks. On Hispaniola, the skeletal material suggests that this species occurs at elevations from 100 m above sea level on Gonave to over 1200 m in the Massif de la Selle. Today, Gonave includes forested, agricultural, shrubby, and barren habitats that are unprotected and rapidly changing. Massif de la Selle, in contrast, is designated a biosphere reserve by UNESCO (the United Nations Educational, Scientific, and Cultural Organization) and has a diversity of protected habitats, such as rain forest, mountain pine forest, deciduous forest, and high-altitude dry forest. No additional natural history information is available for this species on Hispaniola.

Elsewhere, the northern hoary bat is solitary. Its cryptic coloration provides excellent camouflage, and this animal primarily finds cover among the branches and leaves of trees, typically from 10 to 15 m above the ground; it rarely rests inside tree cavities, buildings, or caves. These bats are remarkably strong fliers, and they are capable of seasonal migrations over 1000 km, although the animals on Hispaniola are probably year-round residents. Like molossids, northern hoary bats have long, slender wings that allow them to fly swiftly, at speeds varying from 21 to 39 km/hr, but such a wing shape also restricts maneuverability. Consequently, this mammal typically flies high over the treetops while commuting or foraging.

These bats usually leave their tree late in the evening, up to 1 hour after sunset, and peak activity may occur even later into the night. Northern hoary bats are insectivorous, with moths and beetles dominating the diet, although dragonflies, flies, grasshoppers, and other prey complete the menu. These aerial predators concentrate on large insects, with body lengths ranging from 12 to 29 mm, and often remove the mostly inedible head, wings, and legs, while still in flight, before chewing and swallowing the more nutritious abdomen and thorax.

Echolocation calls vary depending on environment. When flying close to the ground, pulses start at 30–50 kHz and end at 20–30 kHz, but the initial and terminal frequencies vary unpredictably, up and down, from one pulse to the next. Such inconsistency is characteristic of bats in this genus. At greater heights or far from obstacles, the calls become much more consistent and lower in frequency, sweeping from perhaps 20 to 16 kHz. During the mating season, though, male hoary bats appear to emit very low-intensity calls or even to fly silently, perhaps as a way of avoiding detection by other aggressive males.

The breeding period of northern hoary bats on Hispaniola is unknown. On the continent, this mammal produces litters containing one to four pups and gives birth once per year. Neonates have a sparse coating of hair; their ears become erect after 3 days, and eyes open after 12 days. Mothers leave their offspring behind, resting among the branches while she forages, until the youngsters begin to fly at about 1 month of age.

Status and Conservation. The IUCN considers this wide-ranging mammal as a species of Least Concern. The northern hoary bat, though, is the species that most commonly dies at commercial wind farms across North America, and the long-term effects of such consistent killing is unknown. The status of this species on Hispaniola is very uncertain. Although it may be extinct, the lack of recent records could simply reflect the difficulty of catching these high-flying bats in ground-based mistnets. A survey using acoustic techniques may have a better chance of detecting this elusive creature within the rugged mountains of Hispaniola.

Selected References. Soto-Centeno et al. 2017; Corcoran and Weller 2018; Friedenberg and Frick 2021; Soto-Centeno and Simmons *in review.*

Lasiurus degelidus

Jamaican Red Bat, Murciélago Rabi-Peludo Jamaiquino, Lasiure de Jamaïque

Name. *Lasiurus* is derived from the Greek words *lasios*, meaning "hairy," and *oura*, signifying "tail," and refers to this animal's densely furred uropatagium, which is typical of bats in this genus. In the eastern red bat (*L. borealis*) of North America, adults are sexually dimorphic. Females are a subdued yellowish-red, and their dorsal hairs are tipped with white, giving them a frosted appearance; males in contrast are bright red, and many of their back hairs lack the white ends. When describing the Jamaican red bat for the first time in 1931, Gerrit S. Miller, a taxonomist with the Smithsonian Institution, relied on two female specimens. He noted that the Jamaican females were as red as male eastern red bats and similarly lacked the white tips. In Latin, *gelidus* means "frost," and *de* often indicates the "reverse" of something.

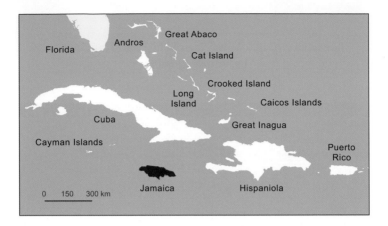

Distribution. The Jamaican red bat is endemic to Jamaica. Biologists have captured this species at only six widely separated locations across the island, all of which are below 400 m in elevation.

Measurements. Total length: 106–110 mm; tail length: 53 mm; hindfoot length: 7.6–8 mm; ear height: 13 mm; forearm length: 41–47 mm. Although data are limited, females appear larger than males in linear measurements; body mass, though, has never been determined for either sex.

Description. Dorsal hairs are bright red and lack white tips, although ventral hairs may have some frosting. Distinct white patches of fur are present on each shoulder where the wing membrane reaches the body. The wide uropatagium is densely clad with reddish hairs. A red color combined with small ears and eyes, a short snout, and a hairy tail membrane readily separates this species from any other bat on Jamaica. Although no photographs of this rare mammal exist, it appears similar to the male eastern red bat, which appears in the photo.

Natural History. Similar to other bats in this genus, the Jamaican red bat probably spends the day by itself, hidden among dead palm fronds or within clumps of green leaves on a tall tree. In the 1970s, mammalogists mistnetted four of these bats over two artificial ponds (earthen tanks), located in disturbed areas of St. Ann Parish, presumably as the animals came to drink. They also reported finding another dead in the driveway of a home in Kingston.

The only other natural history observations come from the book by Philip Henry Gosse in 1851. He described how an assistant used an insect net to catch two male red bats flying a little before sunset, underneath an avocado (*Persea americana*) tree near a small stream. After capture, one individual, "in its impatience of restraint, was perpetually clawing with the hind feet at anything within reach, by which it tore several holes in the membrane of its own wings," and "for the same reason, it strove to bite, seizing its hind feet or tail with its jaws" by accident. During the struggle, the frantic mammal "emitted a harsh hissing sound, with the mouth open, and occasionally a little peculiar 'click.'" Such hyperactivity and distinctive vocalizations in response to handling by humans are also typical of eastern red bats and northern hoary bats (*L. cinereus*) in the United States and Canada and probably most other members of the genus.

Lasiurines are unusual because they have four, rather than two, mammary glands. Although litter size of the Jamaican red bat is unknown, other species in the genus produce litters that typically contain two or three, occasionally four, and exceptionally five offspring. Two adult male and one female Jamaican red bat captured in July were not reproductively active, and there are no reports of pregnant or lactating individuals from the island. Scant data from related species on Cuba and Puerto Rico suggest that births occur once per year, probably in May or June. No information on predators is available, but two Jamaican red bats caught in 1851 were infested with parasitic wingless flies (Nycteribiidae), "which ran in and out of the fur with much agility."

Status and Conservation. The Jamaican red bat is one of the rarest and most poorly understood species of bats in the West Indies, and wildlife biologists have not caught one in over 40 years. The species is protected under Jamaica's Wildlife Protection Act and is categorized as Vulnerable on the IUCN Red List, primarily because ongoing agricultural intensification and forest clearing presumably are negatively affecting the habitat of this mammal.

Selected References. Gosse 1851; Miller 1931; Genoways et al. 2005.

Lasiurus insularis

Cuban Yellow Bat, Murciélago Rabi-Peludo Grande, Lasiure de Cuba

Name. *Lasiurus* derives from the Greek *lasios*, signifying "hairy," and *oura*, meaning "tail." Although most bats have naked tail membranes, members of this genus always have hairs covering 50 to 100% of the membrane's upper surface. *Insularis* in Latin means "of or pertaining to an island" and refers to the limited distribution of this species compared with other yellow bats on the mainland.

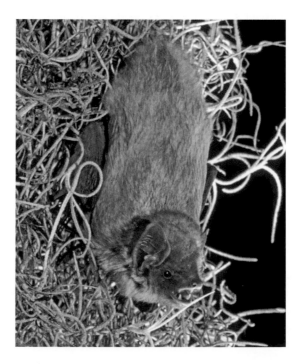

Distribution. The current range includes just Cuba and the adjacent Isla de la Juventud. However, Cuba and Hispaniola at one time were part of the same ancient land mass, and fossils of the Cuban yellow bat have been located on both islands.

Measurements. Total length: 150–164 mm; tail length: 65–81 mm; hindfoot length: 12–13 mm; ear height: 13–15 mm; forearm length: 57–64 mm; body mass: 20–30 g.

Description. The fur is long and dense. Individual hairs are 13 to 17 mm long on the back but shorter (8–12 mm) on the belly. Each dorsal and ventral hair is golden yellow at the tip but dark at the base. As in other members of the genus, hairs also occur along the forearm and extend from the wrist onto the fifth finger, and the proximal half of the tail membrane is furred. Ears are short and rounded. The tail is almost as long as the body and totally encased in the well-formed tail membrane. No other species on Cuba has hairs along the fifth finger and a long tail wrapped in a furry membrane, except Pfeiffer's red bat (*L. pfeifferi*), which belongs to the same genus. Nevertheless, Pfeiffer's red bat, with a forearm length of only 40 to 50 mm, is 30% smaller than a Cuban yellow bat and has a definite reddish cast to its fur. No photograph of the Cuban yellow bat is available; however, this uncommon animal originally

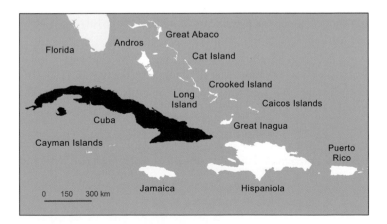

was described as a subspecies of the smaller northern yellow bat (*L. intermedius*), which is pictured in this account.

Natural History. Information concerning the life of this rare mammal is quite sparse. The Cuban yellow bat is known from fewer than a dozen locations throughout the island, and most records consist of bones recovered from pellets of the American barn owl (*Tyto furcata*), with the last discovered in 2013–2014. As do other *Lasiurus*, the Cuban yellow bat probably spends the day hanging alone among the leaves of a tree, rather than joining colonies within caves or buildings. A biologist discovered one dead individual on the ground, under trees in the Yumurí River Gorge, near the city of Matanzas, suggesting a tree-roosting habit for the species, but only two roosts have actually been located. In each case, a single bat rested among the fronds of a *Thrinax* palm, one of which was guano de costa palm (*T. radiata*); the trees were located in the botanical gardens of Havana and Cienfuegos, rather than in forested areas.

The Cuban yellow bat is an aerial insectivore—locating an insect with echolocation, using a wing or tail membrane to grab the invertebrate out of the air, and then transferring the prey to the waiting mouth—all while the mammal is in continuous flight. Bats in the genus *Lasiurus* give birth once per year, usually to two, three, or even four offspring at a time. The only pregnant female examined on Cuba supported three small embryos; she was caught in May, which suggests a possible birth in late June or July, with lactation perhaps lasting into August.

Status and Conservation. The IUCN regards the Cuban yellow bat as Vulnerable, but biologists do not know the size of the current population or whether the number of these insular endemics is growing or declining.

Selected References. Silva-Taboada 1976, 1979; García-Rivera and Mancina 2011.

Lasiurus minor

Minor Red Bat, Murciélago Rabi-Peludo, Lasiure Rouge Mineur

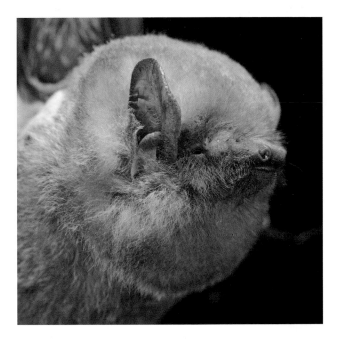

Name. *Lasiurus* derives from the Greek words *lasios* and *oura*, meaning "hairy tail," and the word indicates the furred tail membrane in members of this genus. The specific epithet *minor* originates from a Latin word for "smaller" and denotes the size of this species relative to Pfeiffer's red bat (*L. pfeifferi*) and the Jamaican red bat (*L. degelidus*).

Distribution. The minor red bat occurs on Hispaniola and Puerto Rico and throughout the Bahamas, except Andros Island.

Measurements. Total length: 79–91 mm; tail length: 30–47 mm; hindfoot length: 7–8 mm; ear height: 7–14 mm; forearm length: 37–43 mm; body mass: 6–12 g.

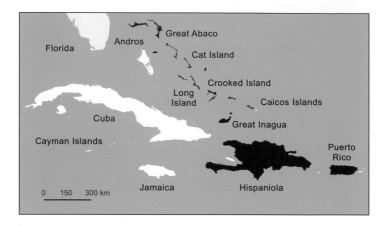

Description. Overall, this handsome mammal appears orange-red to yellowish, becoming darker toward the rump. Many hairs, depending on individual, are tipped with white, giving a frosted appearance, and off-white epaulettes are noticeable in most individuals. The minor red bat has a distinctively compact snout; the ears are short with a rounded tip, and the eyes are small and jet black. Naked skin on the face and ears has a slight orange tinge—an unusual color for a mammal. The skin of the wings and uropatagium is generally blackish, although areas around the forearm and fingers display light orange dots and splotches. As in other vespertilionids, tail vertebrae continue to the posterior end of a broad membrane that, in this species, is fur-covered over at least 75% of its surface. The combination of a short muzzle, reddish or orangish color, and a hairy uropatagium separates this mammal from most Caribbean bats. Within its range, the single species that could be confused with the minor red bat is the northern hoary bat (*L. cinereus*), both of which occur on Hispaniola; however, the latter is easily distinguished by its larger size, dark face, dark-rimmed ears, brownish fur, and more frosted appearance of the back hairs.

Natural History. Minor red bats use habitats ranging from subtropical dry forests to wet forests, at elevations from sea level up to 1300 m in the Sierra del Baoruco of the Dominican Republic. At low elevations, as in the Bahamas, these bats frequently fly high in open space above standing fresh or salt water, and in high-elevation forests on Hispaniola and Puerto Rico, they forage along forest edges and wide corridors through the trees or over the woodland canopy. This bat occupies both rural and urban sites; for example, local citizens captured one on the campus of the University of Puerto Rico at Río Piedras and in a public plaza in Toa Baja, Puerto Rico.

Similar to other red bats, this mammal is probably solitary, except mothers with dependent young, and hides during the day within foliage and not sheltering in tree cavities, rock crevices, caves, or buildings. The bat from Toa Baja was roosting in an unidentified tree when local children knocked the animal to the ground, and another individual was photographed hanging in a pigeon pea (*Cajanus cajan*) shrub in Dorado, Puerto Rico. No other roosting sites have been reported.

Onset of foraging by minor red bats on Great Inagua, in the Bahama Islands, occurred close to sunset, when mammalogists watched at least five individuals flying in an open area over a salt marsh, a foraging space that they shared with Brazilian free-tailed bats (*Tadarida brasiliensis*). Reports from Hispaniola suggested peak activity from 2 to 3 hours after sundown at mountainous sites, where some red bats were captured in mistnets. On Puerto Rico, field biologists acoustically documented and visually observed this mammal as it chased insects attracted to a streetlight, in a parking lot at the El Verde Field Station, in the El Yunque National Forest.

This insectivorous species has elongate wings that permit swift flight but limit maneuverability. Detailed accounts of this bat's diet are lacking, but two individuals collected on Puerto Rico consumed moths, winged termites, and flying ants. When foraging, minor red bats produce echolocation calls with a duration varying from 2 to 6 milliseconds and a downward frequency-modulated sweep from as high as 60 to as low as 32 kHz, although starting and ending frequencies vary from pulse to pulse, as is typical of members of the genus. When held in the hand after capture, these bats tend to close their eyes and make audible sounds, often with the mouth remaining open.

Minor red bats, like other members of the genus *Lasiurus*, consistently have litters of two or three pups. In North America, birds, such as the blue jay (*Cyanocitta cristata*), frequently attack eastern red bats (*L. borealis*) for unknown reasons, often causing mothers with multiple large babies to become grounded. The animal from Río Piedras was caught under similar circumstances—a bird attacked the mother, and burdened with the weight of three young, she crashed into the side of a building while attempting to escape. Eastern red bats are one of the two most commonly killed species of bats at wind farms in North America, and biologists have documented mortality of the minor red bat at such facilities in Puerto Rico. The first record of this bat on Puerto Rico actually resulted from a collision with the windshield of a fast-moving automobile near Moca.

Status and Conservation. The minor red bat is uncommon throughout its range, but conservationists have no knowledge as to whether its population is increasing or decreasing. The IUCN classifies the species as Vulnerable. These bats are particularly susceptible to the increased severity of tropical storms brought on by climate change, which hurl the bats into the air and damage their leafy roosts.

Selected References. Miller 1931; Koopman et al. 1957; Rodríguez-Durán 1999; Speer et al. 2015; Soto-Centeno and Calderón-Acevedo 2022.

Lasiurus pfeifferi

Pfeiffer's Red Bat, Murciélago Rabi-Peludo Rojo, Lasiure de Pfeiffer

Name. The generic name of this bat with a heavily furred uropatagium comes from the Greek words *lasios,* meaning "shaggy" or "hairy," and *oura,* meaning "tail." Juan Cristobal Gundlach first described this species in 1862 and chose the specific epithet to honor Ludwig Pfeiffer, a renowned European botanist, conchologist, and physician of the nineteenth century, who lived on Cuba for 2 years and studied many organisms throughout the West Indies.

Distribution. This species is endemic to Cuba. It is widespread on the main island and also inhabits several keys off the northern coast, such as Romano, Sabinal, and Las Brujas.

Measurements. Total length: 96–126 mm; tail length: 42–71 mm; hindfoot length: 7–8 mm; ear height: 7–11 mm; forearm length: 40–50 mm; body mass: 8–14 g. Females are generally larger than males.

Description. The pelage is dense and long. The color is brick-red, with underparts slightly paler and a distinct white patch at the shoulders. Dorsal hairs are about 8 mm long and tricolored, with black bases, pale middle sections, and reddish tips. Ventral hairs, in contrast, are about 6 mm in length, and bicolored, with dark bases and reddish ends. The dorsal and ventral fur extends slightly onto the wing membranes, and dense hairs cover the top surface of the uropatagium. The snout is short and slightly concave between the nostrils. Eyes are small and beady, and the ears are low and rounded; the tragus is somewhat triangular and reaches half the height of the pinna or less. The wings are long and pointed, and the

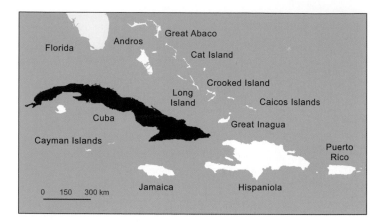

black membranes are marbled with orange. The long tail extends to the rear edge of the uropatagium and is completely enclosed within that membrane. The calcar is thin but supports nearly half the length of the free edge of the uropatagium. Red fur and a broad and hairy tail membrane distinguish this bat from all others on Cuba.

Natural History. Conservationists long considered Pfeiffer's red bat as a rare species; however, since 2000, ecologists have captured this beautiful mammal in several localities throughout the Cuban archipelago. This animal lives in diverse wooded environments, including evergreen, pine, and semideciduous forests, and it even occupies patches of secondary forest in urban and suburban localities. This species is more frequent at low and moderate elevations, but captures have occurred at 1000 m above sea level on Pico Cuba, in the Sierra Maestra, on the eastern end of the island.

This bat generally seeks shelter among the leaves of various trees, such as Australian pine (*Casuarina equisetifolia*), orchid tree (*Bauhinia*), blue mahoe (*Talipariti elatus*), crabwood (*Carapa guianensis*), guacimilla (*Trichospermum lessertianum*), mango (*Mangifera indica*), West Indian elm (*Gauzuma ulmifolia*), and various figs (*Ficus*). There is also one old observation of Pfeiffer's red bat apparently roosting in a cave, but such behavior certainly is not common. Although this mammal is usually solitary, from two to four adult animals hang together on rare occasions. Captures with mistnets and harp traps indicate that these bats are active at any time of night as they search for flying insects, and analysis of a few fecal pellets and stomach contents suggests that moths and beetles are common prey.

Female *Lasiurus* differ from other bats in having four mammary glands, and most laisurines produce litters of two or three offspring. In Pfeiffer's red bat, biologists also report finding two or three fetuses in the few pregnant individuals that have been examined. The Cuban species probably reproduces once per year. Females with well-developed embryos have been captured between February and May, and the earliest date of parturition is 22 May. The American barn owl (*Tyto furcata*) is the lone reported predator.

Status and Conservation. Pfeiffer's red bat is uncommon, but it is not a rare mammal. Increasing frequency and severity of tropical storms that destroy the roosting sites of these bats and drive them from their trees is a potential concern, as is expansion of wind power and deforestation for human developments. The IUCN considers this species Near Threatened.

Selected References. Silva-Taboada 1979; García-Rivera and Mancina 2011; Sánchez-Lozada et al. 2018; Orihuela et al. 2020a.

Myotis dominicensis

Dominican Myotis, Murciélago Rajiero de Dominica, Murin de la Dominique

Name. *Myotis* is from the Greek words *mys* and *otus*, meaning "mouse" and "ear," respectively, and indicates the mouselike shape of the pinnae. The specific epithet *dominicensis* is a latinized term meaning "of Dominica" and describes where the type specimen originated.

Distribution. This species live only in the Lesser Antilles, where it inhabits Dominica, as well as the island of Basse-Terre in the nearby Guadeloupean Archipelago.

Measurements. Total length: 63–78 mm; tail length: 26–34 mm; hindfoot length: 4–8 mm; ear height: 8–11 mm; forearm length: 32–36 mm; body mass: 4–6 g.

Description. The Dominican myotis is one of the smallest bats in the Lesser Antilles. It is uniformly dark brown on the back and slightly burnt brown below. Individual hairs on the back are black at the base, but those on the underparts are salty black through the proximal half. The ears are dull brown and the tragus is tall and pointed; the wing membranes are black. The tail passes to the posterior margin of the large tail membrane, which is furred on its inner half. The Lesser Antillean funnel-eared bat (*Natalus stramineus*), which dwells on the same islands, is similar in size; however, its body is lighter in color and the uropatagium is hairless, except a short fringe along the rear margin. Schwartz's myotis (*M. martiniquensis*), on neighboring Martinique, has a slightly larger body, darker pelage, and lighter wing membranes than the Dominican myotis.

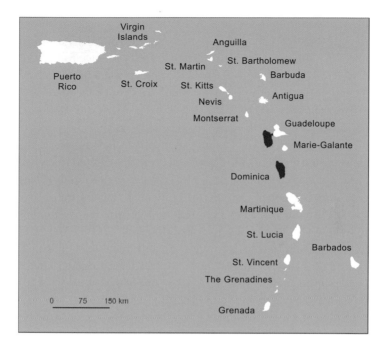

Natural History. On Dominica, at the Clark Hall Estate, mammal-ogists captured this bat in the same net as the Jamaican fruit-eating bat (*Artibeus jamaicensis*), on a trail adjacent to banana groves and open vegetable plots, and caught another in an insect sweep net as the animal flew through the rooms of the main house. About thirty Dominican myotis were netted above the Layou River, a small stream with cleared cattle pastures and coconut groves on either side.

On Basse-Terre, mistnetting efforts yielded two males near the Grand Etang, an 8 ha pond surrounded by tropical forest. One of these bats was snared over a forest path bordered by dense un-dergrowth, whereas the second male became entangled along the edge of the pond; both bats were flying about 1 m above the ground, shortly after sunset, when caught. A third Dominican my-otis touched the surface of the pond, perhaps drinking or plucking an insect from the surface. Acoustic studies from Guadaloupe showed that this mammal was most active in small clearings in low-elevation moist forests on Basse-Terre. This species was not detected in northeastern Basse-Terre nor on the adjacent island of Grande-Terre, both of which contain more open farmland and urbanized zones where Pallas's mastiff bat (*Molossus molossus*) dominates.

Roosting sites for the Dominican myotis included caves, rocky crevices, and buildings. On Dominica, these bats roosted in the bell tower of a church in Portsmouth; two clusters of one hundred bats each were positioned 2 m apart on a slanted wooden wall above the bell. An estimated two hundred to three hundred individuals roosted in Glo Manioc Cave, located near Bells, during March; of fifty animals examined, all were male and many seemed torpid. Five months later, in August, only two bats were present, suggesting seasonal variation in choice of roost; one of the bats was wedged into a narrow crevice in the cavern wall.

Unlike the local free-tailed bats that limit their activity to periods near sunset and sunrise, the Dominican myotis forages sporadically throughout the night. Most of these small animals search for insects around the crowns of trees and along forest paths and restrict their feeding activity to approximately 2 ha of woodland. Biologists tagged one individual with a luminescent capsule and watched for 50 minutes as the glowing mammal hunted in the upper canopy, from 15 to 20 m above the ground. Despite the preference for wooded sites, one individual pursued insects around a street lamp at the edge of a village, although the light was only about 30 m from the forest edge. The only specific dietary information comes from the fecal pellets of two Dominican myotis on Guadeloupe that contained numerous moth scales and parts of beetles and nematoceran flies, which are a group of flies with aquatic larval stages, such as mosquitoes and midges. While searching for prey or commuting, the bat produces concave echolocation calls that sweep from 95 to 43 kHz; maximum energy in the fundamental occurs between 48 and 61 kHz. The animal emits a pulse, on average, every 66 milliseconds, and each call lasts just 4.2 milliseconds.

Reproductive information is limited to observations of pregnant females during early April, nursing individuals in July, and males with well-developed testes during the same months. The Barbadian myotis (*Myotis nyctor*) on Barbados and the black myotis (*Myotis nigricans*) from mainland South America are reproductively active during the dry season months of April, May, and June, as well as the wet season months of September and October. The Dominican myotis may follow a similar pattern.

Status and Conservation. The IUCN classifies the Dominican myotis as Vulnerable, due to its restricted distribution and threats that include bioamplifcation of pesticides, severe storms, and habitat loss caused by the growth of recreational activities and tourism.

Selected References. Masson and Breuil 1992; Genoways et al. 2001; Ibéné et al. 2007; Larsen et al. 2012; Barataud et al. 2015.

Myotis martiniquensis

Schwartz's Myotis, Murciélago Rajiero de Martinica, Murin de la Martinique

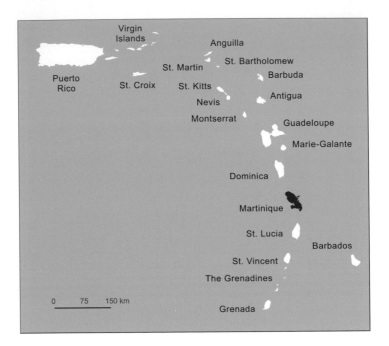

Name. The generic name is a combination of the Greek nouns for "mouse" (*mys*) and "ear" (*otus*) and refers to the well-formed, mouselike external ears of these bats. The specific name *martiniquensis* is a Latinized term meaning "of Martinique" and indicates the origin of the type specimen. The common name, Schwartz's myotis, refers to zoologist Albert Schwartz, who studied terrestrial vertebrates throughout the West Indies during the middle of the twentieth century.

Distribution. This species is endemic to the Lesser Antillean island of Martinique. Mammalogists originally believed that Schwartz's myotis also existed on nearby Barbados; however, genetic analyses published in 2012 indicated that that population represented a distinct species, called the Barbadian myotis (*M. nyctor*).

Measurements. Total length: 75–84 mm; tail length: 34–38 mm;

hindfoot length: 7–8 mm; ear height: 12–13 mm; forearm length: 35–39 mm; body mass: 5–7 g.

Description. Schwartz's myotis is a small West Indian bat. The fur appears woolly, and the overall color varies from dark gray to brown to orangish-brown but generally looks lighter on the belly. Hairs on the back are from 6 to 9 mm long and are brownish throughout, whereas those on the underside are shorter and distinctly bicolored, with dark bases and buffy brown tips. The pinnae and wing and tail membranes are dark brown. Hairs coat the top surface of the well-developed uropatagium, from the body to past the knees. The uropatagium totally encloses the tail vertebrae, which continue to the back edge of the pointed membrane.

On Martinique, Schwartz's myotis is comparable in size to the Lesser Antillean funnel-eared bat (*Natalus stramineus*), which also has a broad interfemoral membrane and a tail that reaches to the rear margin. However, the funnel-eared bat is lighter in color and has very broad ears, dark-rimmed pinnae, a twisted tragus, and a naked uropatagium. Only two other species of *Myotis* occur in the entire West Indies, one on Barbados and the other on Dominica and Guadeloupe, and those bats also are similar in size to Schwartz's myotis. The Barbadian myotis, though, has a smaller head, slightly shorter forearm, and fur that appears more silky than woolly, whereas the Dominican myotis has a smaller forearm, lighter-colored fur, and darker membranes.

Natural History. Little information is available on the natural history and ecology of Schwartz's myotis. However, it is a common species on Martinique, representing about 10% of all bats mistnetted on the island, and this species is equally abundant in the dry (January–June) and wet (July–December) seasons. About 60% of the captured individuals are female. Schwartz's myotis frequents seasonal evergreen forests, wooded wetlands, and forest clearings, especially near rivers and streams. Although maximum height of the mountains on Martinique approaches 1400 m, mammalogists catch most Schwartz's myotis at sites ranging from 150 to 300 m above sea level, with an occasional capture at greater elevations.

Biologists often record the echolocation sounds of this animal near bodies of water, and a dearth of streams and ponds on the steep-sided mountains of Martinique may explain the decreased abundance of this mammal at higher elevations.

Confirmed roosting sites include a rocky crevice, a tunnel, and a narrow crack in a concrete bridge. Nevertheless, one individual roosted under the large umbrella-like leaf of a grandleaf seagrape (*Coccoloba pubescens*), about 5 m above the ground. The tree was on a steep slope in a dry tropical forest located near Les Trois-Îlets, in the interior of the southwestern peninsula of Martinique. Roosting among leaves is unusual for bats in the genus *Myotis*, but given the paucity of caves on Martinique, it may be a necessary behavior.

Schwartz's myotis is likely insectivorous, although no details of its diet are available. Data from radio-tracking show high site fidelity to hunting areas and peak activity in the first 2 hours after sunset. Some insectivorous bats, such as the Pallas's mastiff bat (*Molossus molossus*) or Davy's naked-backed bat (*Pteronotus davyi*), are drawn to street lamps and other artificial sources of light, presumably because the lights attract and concentrate insect prey for these mammals. Acoustic surveys, however, indicate that Schwartz's myotis avoids lit areas, which is consistent with the bat's apparent preference for the interior of forests. This mammal hunts primarily near the ground when little undergrowth is present, but above the shrubs and small understory trees in more dense situations. The echolocation calls of Schwartz's myotis are very similar to those of the Dominican myotis (*M. dominicensis*). While flying in open situations, these bats produce concave, frequency-modulated echolocation calls that sweep from 93 to 44 kHz; maximum energy occurs between 46 and 66 kHz. Average pulse duration is 4.1 milliseconds, and the time between calls is 82 milliseconds.

No reproductive information exists for Schwartz's myotis. Litter size is probably one, as it is in almost every species of *Myotis*. The morphologically similar Barbadian myotis is reproductively active from April to early October, and the same may be true for Schwartz's myotis.

Based on the geographic proximity of the Lesser Antilles and South America and structural similarities among closely related bats from those regions, evolutionary biologists believe that the common ancestor of Schwartz's myotis, as well as the Dominican myotis and Barbadian myotis, likely dispersed to the Caribbean from northern South America. Schwartz's myotis evolved over 4 million years ago in the Pliocene epoch, making it older than the Barbadian myotis and Dominican myotis, which arose 2.5 and 3.5 million years ago, respectively.

Status and Conservation. Schwartz's myotis is listed as Near Threatened. Its primary conservation concerns on Martinique include severe hurricanes, as well as continued deforestation and habitat fragmentation to make way for human developments.

Selected References. Larsen et al. 2012; Barataud et al. 2017; Dinets 2017; Moratelli et al. 2017; Catzeflis et al. 2019.

Myotis nyctor

Barbadian Myotis, Murciélago Rajiero Concubio, Murin de la Barbade

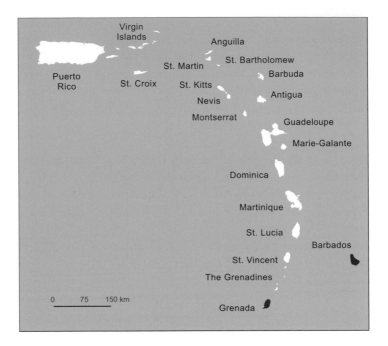

Name. The generic name *Myotis* comes from Greek words meaning "mouse ear." The authors of the name *nyctor* do not indicate its meaning, but it is probably from the Greek *nyct* referring to "night" or "darkness," perhaps reflecting that the original specimens were captured 2 or more hours after sunset. Biologists originally thought this animal was a subspecies of Schwartz's myotis (*M. martiniquensis*), which occurs on Martinique, but in 2012, genetic studies showed that the Barbadian and Schwartz's myotis were only distantly related.

Distribution. The Barbadian myotis inhabits just the islands of Barbados and Grenada. On Barbados, these bats are somewhat common and occur in the greatest concentrations in the central highlands and in the northeastern portion of the island. Records of this bat from Grenada, in contrast, include just two old specimens from unspecified localities and a third individual from near Beaton, St. David Parish, in the southeastern part of the island. The animals from Grenada may represent a distinct subpopulation of the Barbadian myotis.

Measurements. Total length: 78–85 mm; tail length: 32–38 mm; hindfoot length: 6–8 mm; ear height: 13–15 mm; forearm length: 34–38 mm; body mass: 4–7 g (based on individuals from Barbados).

Description. The Barbadian myotis is small for a West Indian bat and is the smallest species on Barbados. Dorsal hairs are approximately 6 mm long and have a silky appearance. Hairs on the back are medium brown, with little contrast between the tips and bases, whereas hairs on the underparts are buffy at the tips and dark brown at the bases. The top of the tail membrane is partly furred, with hairs occurring from the body to well past the knees, although these hairs do not continue laterally onto the tibias. The tail makes up about 45% of the bat's overall length and is totally enclosed in the uropatagium. Compared with Schwartz's myotis from Martinique, the Barbadian myotis has smaller wings and silky rather than woolly hair.

Natural History. On Barbados, the Barbadian myotis is most commonly associated with caves, rocky outcrops, and "gullies," in combination with stands of large trees. A gully is a winding ravine about 20 m deep, which runs from elevated inland areas to the coast and is often associated with a cave. Barbados is composed of coral limestone, and rainwater persistently seeps through the rock, leading to the formation of a large number of caves with underground streams. Over time, the roofs of many caves collapse, transforming the subterranean passage into a surface gully. Much of the original fauna and flora of Barbados has disappeared after centuries of agriculture aimed at producing tobacco, cotton, and sugar, but remnants of the original biodiversity persist in the moist environments of these gullies.

The Barbadian myotis was first discovered at the mouth of Cole's Cave in the central highlands of the island, and these bats were abundant in the associated Jack-in-the-Box Gully, where they roosted in cracks, crevices, and other small caves in the walls of the ravine. These mammals also were present in Harrison's Cave and the adjoining Welchman Hall Gully, which the Barbados National Trust now operates as a nature preserve and education center. Some of the dominant native trees in the gully are bearded fig (*Ficus citrifolia*), cabbage palm (*Roystonea oleracea*), fiddlewood (*Citharexylum spinosum*), sandbox (*Hura crepitans*), silk cotton (*Ceiba pentandra*), trumpet tree (*Cecropia schreberiana*), West Indian mahogany (*Swietenia mahagoni*), and whitewood (*Tabebuia heterophylla*). In addition, one of these bats was found in the attic of a building on the campus of Codrington College in St. John Parish. This record may be an accidental occurrence, but if the Barbadian myotis is capable of using human-made structures for daytime shelter, as do many temperate members of the genus, this species will have a much broader range of available roosts than currently believed.

At Welchman Hall Gully, the Barbadian myotis was active in the 30 minutes before sunset, fluttering in and out of the trees as they foraged. Similarly, biologists have mistnetted these bats before sundown along drainage culverts associated with Friendship Terrace Gully. Near the northern end of Hackleton's Cliff and along a nearby intermittent stream, Joe's River, these tiny bats were drinking and foraging on small insects, possibly flies and moths, as they flew over stagnant pools in the riverbed; the surroundings were densely covered in thorny trees, dominated by several species of acacia, including redwood (*Acacia glauca*). Other individuals on Barbados were caught while flying along the edge of mahogany forest and in a canopied flyway under a stand of bamboo. On Grenada, the sole animal from Beaton was captured in an area of semideciduous thorny bushes and trees, as the bat flew over a small stream that coursed through a poorly developed gallery forest.

The distribution of the Barbadian myotis does not seem affected by elevation. Captures on Barbados occur at locations that are as low as 54 m above sea level and as great as 290 m; the highest point on the island is only 336 m. The lone capture site known from Grenada is at an elevation of 115 m.

The overall reproductive pattern is not understood. However, reproductively active females have been captured in April, June,

July, late September, and early October; hence, some reproductive activity occurs both during the dry season (April) and the wet season (September and October). Females produce a single pup at each birth.

Status and Conservation. The IUCN lists the Barbadian myotis as Vulnerable. Threats to the species are a small geographic range, increasingly severe hurricanes and declining forest habitat, due to an increasing human population and the growth of tourism.

Selected References. Genoways et al. 1998, 2011; Grindal 2004; Larsen et al. 2012; Moratelli et al. 2017.

Nycticeius cubanus

Cuban Evening Bat, Murciélago Crepuscular, Nycticée de Cuba

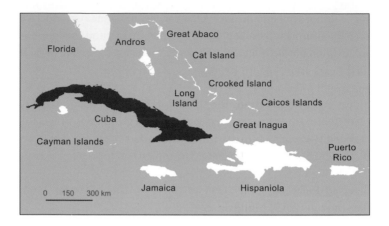

Name. The generic name *Nycticeius* is of Greek and Latin origin and means "belonging to the night." *Cubanus* refers to Cuba, which is the type locality of the species.

Distribution. The Cuban evening bat inhabits only Isla de la Juventud and the western part of the main island of Cuba, where the species is locally common in the province of Pinar del Río and the city of Havana.

Measurements. Total length: 76–85 mm; tail length: 24–36 mm; hindfoot length: 5–7 mm; ear height: 8–10 mm; forearm length: 29–32 mm; body mass: 4–7 g.

Description. The back of a Cuban evening bat is dark brown, but the belly is slightly lighter and more olive-brown. Both dorsal and ventral hairs are bicolored with dark bases and lighter tips. The ears, when folded forward, reach or surpass the point of the muzzle, and the height of the curved tragus is about half that of the ear. Wing and tail membranes are dark and naked. The rear margin of the uropatagium points backward and is supported for half its length by the calcar. Only the last 1 to 3 mm of the tail are free of the membrane. The Cuban evening bat is the smallest member of the Vespertilionidae on Cuba and could be confused only with the Caribbean lesser funnel-eared bat (*Chilonatalus micropus*), Gervais's funnel-eared bat (*Nyctiellus lepidus*), or the little goblin bat (*Mormopterus minutus*). However, the funnel-eared bats have larger ears and a bent or twisted tragus that is much less than half the height of the ear, and in the little goblin bat, at least 50% of the long tail projects beyond the back edge of the membrane.

Natural History. Capture localities range from near sea level to an elevation of just 250 m. The few known roosting sites are associated with humans; naturalists have found these mammals resting inside cracks above the window of a house, under roofing tiles, and within crevices in a wooden utility pole. Cuban evening bats spend the day alone or in small colonies consisting of up to twenty-six animals.

These insectivorous bats begin hunting between 20 minutes before and 5 minutes after sunset and remain active throughout the night, with individuals occasionally reentering the roost and then leaving again. Cuban evening bats forage from ground level to more than 6 m in the air and in different situations, ranging from open sites away from all obstacles to environments cluttered with vegetation and buildings. Flight is slow and maneuverable. During initial foraging bouts before it is dark, these bats continually loop back and forth over a localized area, rather than follow linear corridors. Biologists have watched these mammals pursue winged termites above streetlights and have detected the chewed remains of beetles, moths, and cockroaches in fecal samples.

Search-phase echolocation calls begin with a steep frequency-modulated component and end with a period of quasi-constant frequency. When foraging in the open, these bats emit pulses that typically decrease from 59 to 41 kHz and last about 8.4 milliseconds. Calls made in more cluttered environments have longer durations and higher initial frequencies, although the terminal frequency varies little from site to site. Individual bats are able to recognize each other based on slight differences in their echolocation calls.

The North American evening bat (*N. humeralis*) typically produces twins once per year in spring, and the same is probably true for the Cuban evening bat. However, information concerning reproduction on Cuba is limited to a single record of a breeding female captured in May while carrying two embryos.

Status and Conservation. The Cuban evening bat is endemic to the islands, uncommon, and in need of further study to document its natural history and determine its conservation needs. The IUCN classifies this animal as Near Threatened.

Selected References. Silva-Taboada 1979; Mora et al. 2005; Cláudio 2019b.

APPENDIXES

Summary of Body and Skull Measurements

Table A.1 Standard body and skull measurements[a]

Family/Species	Total length (mm)	Tail length (mm)	Hindfoot length (mm)
MOLOSSIDAE			
Eumops auripendulus	80–85	43–54	12–18
Eumops ferox	80–85	40–54	10–15
Molossus milleri	97–113	32–42	7–12
Molossus molossus	84–124	30–46	7–11
Molossus verrilli	85–126	30–46	7–11
Mormopterus minutus	68–82	22–34	5–6
Nyctinomops laticaudatus	90–116	30–45	8–9
Nyctinomops macrotis	73–89	40–63	9–11
Tadarida brasiliensis	52–98	27–41	7–11
MORMOOPIDAE			
Mormoops blainvillei	78–87	21–31	6–10
Pteronotus davyi	73–84	15–24	10–12
Pteronotus fuscus	90–99	20–27	12–14
Pteronotus macleayii	72–78	19–25	9–10
Pteronotus parnellii	76–86	18–22	10–13
Pteronotus portoricensis	78–83	13–22	8–13
Pteronotus pusillus	73–82	18–22	11–13
Pteronotus quadridens	59–80	13–21	6–11

Ear height (mm)	Forearm length (mm)	Weight (g)	Greatest length of skull (mm)	Zygomatic breadth of skull (mm)
19–25	55–68	25–35	24.1–25.5	13.7–14.5
17–23	57–64	30–45	22.9–25.0	13.8–15.4
11–14	37–40	13–19	14.6–17.5	9.3–11.4
11–15	36–43	9–17	15.1–17.9	9.3–11.1
11–15	37–41	13–20	15.7–17.9	10.0–10.7
8–11	28–33	4–8	12.6–13.9	7.6–8.5
15–17	40–45	8–13	17.1–18.4	9.5–10.2
24–32	55–65	17–34	15.3–23.2	9.0–12.4
8–15	36–43	6–13	15.0–16.6	8.7–10.4
11–18	45–49	8–11	12.7–14.8	8.0–9.0
14–17	43–48	7–9	15.3–16.5	8.8–9.3
20–25	58–63	14–20	21.5–22.8	12.2–12.8
16–18	39–44	4–7.5	15.2–17.0	7.3–8.5
19–22	50–55	10–16	19.9–20.9	10.7–11.5
16–21	50–53	11–16	20.1–20.8	10.6–11.4
19–21	48–51	8–13	18.9–19.4	10.3–10.8
13–19	36–39	4–7	13.8–15.3	7.1–8.1

Family/Species	Total length (mm)	Tail length (mm)	Hindfoot length (mm)
NATALIDAE			
Chilonatalus macer	85–92	46–56	7–8
Chilonatalus micropus	80–89	45–48	7–8
Chilonatalus tumidifrons	85–93	47–48	7–8
Natalus jamaicensis	105–112	57–60	?
Natalus major	100–112	50–59	8–11
Natalus primus	?	?	?
Natalus stramineus	98–105	49–57	7–10
Nyctiellus lepidus	63–70	25–35	?
NOCTILIONIDAE			
Noctilio leporinus	76–130	23–55	25–34
PHYLLOSTOMIDAE			
Ardops nichollsi	58–73	0	12–18
Ariteus flavescens	50–69	0	11–14
Artibeus jamaicensis	78–89	0	16–18
Artibeus lituratus	91–107	0	15–21
Artibeus schwartzi	90–107	0	17–21
Brachyphylla cavernarum	80–103	0	18–23
Brachyphylla nana	69–102	0	13–20
Chiroderma improvisum	70–87	0	15–19
Erophylla bombifrons	74–77	11–14	13–15
Erophylla sezekorni	69–84	11–16	13–17
Glossophaga longirostris	57–80	4–17	5–14
Glossophaga soricina	58–79	5–10	10–17
Macrotus waterhousii	70–108	25–42	13–15
Micronycteris buriri	66–72	12–18	10–12

Ear height (mm)	Forearm length (mm)	Weight (g)	Greatest length of skull (mm)	Zygomatic breadth of skull (mm)
13–16	32–34	2.5–4	14.2–14.9	6.5–6.7
13–16	31–35	2.6–5	13.5–14.7	6.2–6.8
15–17	32–36	3–3.5	15.0–16.0	7.1–7.4
18	44–46	5.9–7.3	17.2–18.2	8.6–9.5
13–19	41–45	5.5–10	17.0–18.1	9.3–9.6
20–22	46–52	6–13	18.1–19.9	9.1–10.0
12–16	37–42	4.3–6.5	15.7–17.8	7.9–8.9
8–14	26–32	2–3	12.5–14.0	6.2–6.7
16–31	81–92	48–87	24.7–29.6	18.1–20.7
12–18	41–56	13–30	20.2–24.6	12.2–16.0
13–18	36–44	9–16	18.4–21.3	12.7–14.7
20–27	52–67	36–48	21.5–30.2	15.1–18.2
20–27	66–76	61–73	30.3–32.4	17.9–19.8
19–23	56–69	56–73	28.4–32.9	16.9–20.6
18–24	59–71	29–54	28.4–32.7	14.7–18.1
12–22	52–63	23–65	26.8–29.8	13.9–16.0
21–22	57–60	35–60	28.7–29.9	18.5–19.0
15–19	42–49	15–19	23.8–25.4	10.9–12.0
17–22	43–51	13–21	23.6–25.7	10.7–11.8
7–17	33–43	6–15	21.7–23.4	9.4–10.2
10–16	36–38	10–14	21.3–22.8	9.0–9.9
26–33	45–58	12–19	22.6–27.6	10.5–12.9
19–25	36–39	7.4–10	20.7–21.8	9.6–10.2

Family/Species	Total length (mm)	Tail length (mm)	Hindfoot length (mm)
Monophyllus plethodon	67–84	8–16	12–13
Monophyllus redmani	58–80	7–11	10–12
Phyllonycteris aphylla	72–76	7–10	16–18
Phyllonycteris poeyi	75–87	6–18	12–20
Phyllops falcatus	55–65	0	9–12
Stenoderma rufum	60–73	0	12–15
Sturnira angeli	63–82	0	12–17
Sturnira paulsoni	60–67	0	11–15
VESPERTILIONIDAE			
Antrozous koopmani	>65	50–60	13–15
Eptesicus fuscus	86–125	39–50	8–14
Eptesicus guadeloupensis	129–132	54–60	11–14
Eptesicus lynni	91–102	35–42	8–12
Lasiurus cinereus	127–140	47–61	8–13
Lasiurus degelidus	106–110	53	7.6–8
Lasiurus insularis	150–164	65–81	12–13
Lasiurus minor	79–91	30–47	7–8
Lasiurus pfeifferi	96–126	42–71	7–8
Myotis dominicensis	63–78	26–34	4–8
Myotis martiniquensis	75–84	34–38	7–8
Myotis nyctor	78–85	32–38	6–8
Nycticeius cubanus	76–85	24–36	5–7

[a] Skull measurements were extracted from various sources (Miller 1931; de la Torre 1966; Buden 1975, 1976, 1977; Jones and Baker 1978; Klingener et al. 1978; Genoways et al. 2001, 2005, 2007a, 2007b, 2010, 2011; Timm and Genoways 2003; Gannon et al. 2005; Pedersen et al. 2005, 2006, 2018a, 2018b; Tejedor 2011; Larsen et al. 2012; Kwiecinski et al. 2018; Loureiro et al. 2018, 2020; and J. A. Soto-Centeno pers. comm., 6 December 2021).

Ear height (mm)	Forearm length (mm)	Weight (g)	Greatest length of skull (mm)	Zygomatic breadth of skull (mm)
14–15	39–46	12–17	22.1–23.7	8.9–10.7
9–14	35–43	6–15	19.4–23.6	8.1–10.2
15–18	44–48	14–15	24.0–26.1	–
12–16	42–51	15–29	23.6–26.3	–
11–13	38–48	16–26	19.0–21.4	13.1–14.2
16–19	46–51	20–31	21.4–23.6	14.4–15.8
14–19	40–48	17–28	22.5–25.1	11.9–13.7
14–16	40–46	14–25	21.6–23.8	11.5–13.7
24–25	49–62	?	21.5–24.3	12.9–14.1
12–20	39–51	11–23	16.3–20.2	12.4–13.8
22–24	49–54	19–22	22.5–23.1	13.7–13.9
12–16	43–49	20–25	15.9–17.7	10.1–12.0
13–19	50–57	20–35	?	?
13	41–47	?	12.9–14.2	9.2–10.1
13–15	57–64	20–30	21.6–23.0	14.1–15.6
7–14	37–43	6–12	11.4–13.2	8.4–9.5
7–11	40–50	8–14	13.6–14.7	9.4–10.6
8–11	32–36	4–6	12.6–13.7	7.6–8.1
12–13	35–39	5–7	14.1–14.8	8.8–9.0
13–15	34–38	4–7	14.0–14.9	7.5–8.9
8–10	29–32	4–7	12.2–13.0	9.3–9.4

Dental Formulas

Knowing the number and kinds of teeth sometimes is helpful in identifying a bat in the flesh or an isolated skull found in a cave or building. For example, the Barbadian myotis (*Myotis nyctor*) is the only mammal with a total of thirty-eight teeth living on Barbados. The Jamaican fig-eating bat (*Ariteus flavescens*) and big free-tailed bat (*Nyctinomops macrotis*) both occur on Jamaica and both have thirty teeth overall; however, the Jamaican fig-eating bat has two incisors on each side of the upper jaw, whereas the big free-tailed bat possesses only one. Occasionally, it is not even necessary to count the total number of teeth to distinguish among species; Pallas's mastiff (*Molossus molossus*), for instance, is the lone bat inhabiting Puerto Rico that has just a single pair of incisors in the top and bottom jaws.

How does one differentiate among incisors, canines, premolars, and molars? Technically, the upper incisors are the only teeth in the premaxillary bone, whereas the canine is the most anterior tooth in the maxillary bone; lower incisors and canines are simply the teeth that occlude (come together) with those upper teeth. Nevertheless, canines in bats of the West Indies are easy to recognize without having to identify individual bones; canines are the tallest teeth, located near the front, and sharply pointed. The incisors are the only teeth in front of the canines, whereas the premolars and molars are located behind the canines. Although premolars always are in front of molars, the two types of teeth often look similar, and the real distinction is how they develop. Just as in humans, bats have two sets of incisors, canines, and premolars, and early in the animal's life, the first set (the "baby" or deciduous teeth) is replaced by the adult dentition. In contrast, mammals have only a single complement of molars in their lifetime, so molars are just cheek teeth (i.e., teeth behind the canine) that are never replaced. Fortunately, it is not necessary to distinguish between premolars and molars to identify a bat; it is usually sufficient to know the total number of cheek teeth.

Rather than memorize the number and kinds of teeth for each type of bat, field biologists rely on a table of dental formulas, like the one below. Consider the first entry for *Chilonatalus* and three other genera. Uppercase letters in the formula refer to the different kinds of teeth—I (incisors), C (canines), P (premolars), and M (molars), and numerals above and below each line indicate the number of each kind of tooth in the upper and lower jaws, respectively. The formula indicates, for example, that the members of this genus have two upper incisors but three lower incisors on each side. After the first equal sign is the total number of teeth in the upper jaw (18) and lower jaw (20), and the final numeral indicates the total complement of teeth in the skull and lower jaw combined (38). Following each formula is an alphabetical list of genera displaying that specific dental pattern. All formulas, of course, apply only to adults.

Although number of teeth is usually quite constant within a genus and species, variation consistently occurs in some. For example, 13% of Schwartz's fruit-eating bats (*Artibeus schwartzi*) sport a third upper molar, and almost half the little goblin bats (*Mormopterus minutus*) have an additional, tiny, upper premolar. In addition, most Angel's yellow-shouldered bats (*Sturnira angeli*) have three molars on either side of the mandible, but some individuals have only two.

$\text{I} \frac{2-2}{3-3}, \text{ C} \frac{1-1}{1-1}, \text{ P} \frac{3-3}{3-3}, \text{ M} \frac{3-3}{3-3} = \frac{18}{20} = 38$ *Chilonatalus, Myotis, Natalus, Nyctiellus*

$\text{I} \frac{2-2}{2-2}, \text{ C} \frac{1-1}{1-1}, \text{ P} \frac{2-2}{3-3}, \text{ M} \frac{3-3}{3-3} = \frac{16}{18} = 34$ *Glossophaga, Macrotus, Micronycteris, Monophyllus, Mormoops, Pteronotus*

$\text{I} \frac{2-2}{2-2}, \text{ C} \frac{1-1}{1-1}, \text{ P} \frac{2-2}{2-2}, \text{ M} \frac{3-3}{3-3} = \frac{16}{16} = 32$ *Ardops, Brachyphylla, Erophylla, Phyllonycteris, Phyllops, Stenoderma, Sturnira*

$\text{I} \frac{2-2}{2-2}, \text{ C} \frac{1-1}{1-1}, \text{ P} \frac{2-2}{3-3}, \text{ M} \frac{2-2}{3-3} = \frac{14}{18} = 32$ *Artibeus*

$\text{I} \frac{2-2}{3-3}, \text{ C} \frac{1-1}{1-1}, \text{ P} \frac{1-1}{2-2}, \text{ M} \frac{3-3}{3-3} = \frac{14}{18} = 32$ *Eptesicus*

$I\frac{1-1}{3-3}$, $C\frac{1-1}{1-1}$, $P\frac{2-2}{2-2}$, $M\frac{3-3}{3-3} = \frac{14}{18} = 32$ *Lasiurus, Tadarida*

$I\frac{2-2}{2-2}$, $C\frac{1-1}{1-1}$, $P\frac{2-2}{2-2}$, $M\frac{2-2}{3-3} = \frac{14}{16} = 30$ *Ariteus*

$I\frac{1-1}{2-2}$, $C\frac{1-1}{1-1}$, $P\frac{2-2}{2-2}$, $M\frac{3-3}{3-3} = \frac{14}{16} = 30$ *Eumops, Nyctinomops*

$I\frac{1-1}{3-3}$, $C\frac{1-1}{1-1}$, $P\frac{1-1}{2-2}$, $M\frac{3-3}{3-3} = \frac{12}{18} = 30$ *Nycticeius*

$I\frac{1-1}{2-2}$, $C\frac{1-1}{1-1}$, $P\frac{1-1}{2-2}$, $M\frac{3-3}{3-3} = \frac{12}{16} = 28$ *Antrozous, Mormopterus*

$I\frac{2-2}{2-2}$, $C\frac{1-1}{1-1}$, $P\frac{2-2}{2-2}$, $M\frac{2-2}{2-2} = \frac{14}{14} = 28$ *Chiroderma*

$I\frac{2-2}{1-1}$, $C\frac{1-1}{1-1}$, $P\frac{1-1}{2-2}$, $M\frac{3-3}{3-3} = \frac{14}{14} = 28$ *Noctilio*

$I\frac{1-1}{1-1}$, $C\frac{1-1}{1-1}$, $P\frac{1-1}{2-2}$, $M\frac{3-3}{3-3} = \frac{12}{14} = 26$ *Molossus*

Summary of Information on
Echolocation Calls

Beginning and ending frequency, duration, and interpulse interval are stated as the mean ± 1 standard deviation or the mean followed by the range in parentheses, depending on source. All values are for the strongest harmonic, except *Erophylla sezekorni*, for which the fundamental and second harmonic contained the most energy with about equal frequency. CF = constant frequency; QCF = quasi-constant frequency; and FM = frequency modulated. All FM components decrease in frequency during a call unless otherwise noted. The specific protocol used by a researcher greatly affects the timing and frequency of recorded echolocation calls, and these parameters may differ among islands; consequently, we also provide a brief comment on the recording situation and location, as well as the original reference.

Table A.2 Characteristics of echolocation calls of 46 West Indian species of bats for which data are available

Family/Species	Description	Number of calls analyzed	Beginning frequency (kHz)
MOLOSSIDAE			
Eumops auripendulus	Paired pulses; shallow FM then QCF	41, 27[a]	32 ± 4, 36 ± 4
Eumops ferox	QCF	269	23 ± 4
Molossus milleri	Paired pulses; shallow convex	157	35 ± 1, 41 ± 2
	Paired or triple pulses; QCF	156	44 (34–55)
		10	41 ± 3
Molossus molossus	Triple pulses; QCF	136, 94, 56[a]	36 ± 1, 39 ± 1, 43 ± 1
	Paired or triple pulses; QCF	149	42 (33–54)
	QCF	12	50 (37–55)
Molossus verrilli	Paired or triple pulses; QCF	183	41 (31–51)
Mormopterus minutus	Highly variable	24	56 ± 15
Nyctinomops laticaudatus	Triple pulses; QCF	63, 36, 9[a]	27 ± 1, 29 ± 1, 32 ± 1
Nyctinomops macrotis	QCF	320	29 ± 5

Ending frequency (kHz)	Duration (milliseconds)	Interpulse interval (milliseconds)	Comment	Reference
19 ± 0.4, 20 ± 0.3	20 ± 7, 19 ± 4	269 ± 69, 216 ± 61	Hand release; mainland	Jung et al. 2014
15 ± 2	14 ± 4	294 ± 206	Free-flying; Cuba	Mora and Torres 2008
32 ± 1, 37 ± 1	11 ± 1		Free-flying; Cuba	Loureiro et al. 2020
35 (23–45)	9 (4–10)		Hand release; Cayman Islands	Loureiro et al. 2020
34 ± 4	6 ± 2		Free-flying; Jamaica	Emrich et al. 2014
34 ± 1, 37 ± 1, 40 ± 1	10 ± 1, 10 ± 1, 10 ± 2	143 ± 25, 109 ± 45, 83 ± 12	Hand release; mainland	Jung et al. 2014
33 (26–43)	8 (3–14)		Hand release; Nevis	Loureiro et al. 2020
23 (21–30)	5 ± (2–6)	41 (18–65)	Hand release; Trinidad	Pio et al. 2010
33 (24–42)	9 (5–15)		Hand release; Dominican Republic	Loureiro et al. 2020
40 ± 3	8.0 ± 2		Foraging; Cuba	Mora et al. 2011
24 ± 1, 24 ± 1, 25 ± 1	12 ± 1, 12 ± 1, 13 ± 3	394 ± 117, 283 ± 82, 214 ± 60	Hand release; mainland	Jung et al. 2014
16 ± 2	14 ± 3	328 ± 187	Open space; Cuba	Mora and Torres 2008

Family/Species	Description	Number of calls analyzed	Beginning frequency (kHz)
Tadarida brasiliensis	QCF	48	28 ± 3
	QCF	97	29 ± 2.7
	FM	20[b]	58 ± 1
		10	40 ± 3
MORMOOPIDAE			
Mormoops blainvillei	Steep FM	2[b]	68 (61–76)
	Steep FM; slope slightly convex	38	64 ± 0.3
		10	67 ± 2
Pteronotus davyi	CF then FM	5[b]	70 (69–72)
Pteronotus fuscus	Long CF component	28	62 ± 0.3
Pteronotus macleayii	Steep FM, then short QCF	512	70 ± 0.1
		10	71 ± 2
Pteronotus parnellii	Short upward FM, long CF, then downward FM	67	61 ± 0.1
		10	61 ± 1

Ending frequency (kHz)	Duration (milliseconds)	Interpulse interval (milliseconds)	Comment	Reference
24 ± 1	14 ± 2	273 ± 56	Hand release; mainland	Jung et al. 2014
26 ± 2	12 ± 2	268 ± 117	Free-flying; Guadeloupe	Barataud et al. 2015
21 ± 1	4.1 ± 0.2	55 ± 6	Hand release; Bahamas	Murray et al. 2009
33 ± 3	9 ± 1		Free-flying; Jamaica	Emrich et al. 2014
40 (39–40)	2.9 (2–3.8)	27 (26–27)	Hand release; Puerto Rico	Vaughan Jennings et al. 2004
55 ± 0.4	2.1 ± 0.1		Hand release; Cuba	Mancina et al. 2012
44 ± 4	3 ± 1		Free-flying; Jamaica	Emrich et al. 2014
51 (46–56)	4.6 (2.7–6.1)	41 (13–85)	Hand release; Dominica	Vaughan Jennings et al. 2004
54 ± 2	22 ± 4		Zip line; Venezuela	Martino et al. 2019
55 ± 0.1	4.3 ± 0.03		Hand release; Cuba	Mancina et al. 2012
55 ± 1	5 ± 1		Free-flying; Jamaica	Emrich et al. 2014
58 ± 0.4	15 ± 1		Hand release; Cuba	Mancina et al. 2012
49 ± 3	29 ± 4		Free-flying; Jamaica	Emrich et al. 2014

Family/Species	Description	Number of calls analyzed	Beginning frequency (kHz)
Pteronotus portoricensis	Long CF, then brief FM	4[b]	56 (52–63)
Pteronotus quadridens	CF, then FM	3[b]	82 (80–84)
	QCF, then FM	10	84 ± 0.5
		10	80 ± 1
NATALIDAE			
Natalus primus	FM	200	114 ± 6
Natalus stramineus	Steep FM	7	103 ± 14
Nyctiellus lepidus	Concave FM	28[b]	114 ± 1
NOCTILIONIDAE			
Noctilio leporinus	Short upward FM, moderate CF, then downward FM		57
PHYLLOSTOMIDAE			
Ardops nichollsi	Slightly convex FM	70	112 ± 17
Artibeus jamaicensis	Steep FM	14[b]	73 (64–83)
	Steep FM	20	89 (74–106)

Ending frequency (kHz)	Duration (milliseconds)	Interpulse interval (milliseconds)	Comment	Reference
47 (44–49)	22 (15–32)	56 (30–99)	Hand release; Puerto Rico	Vaughan Jennings et al. 2004
59 (57–62)	5.0 (4.5–5.7)	82 (61–103)	Hand release; Puerto Rico	Vaughan Jennings et al. 2004
68 ± 0.6	4.2 ± 0.1	53 ± 34	Free-flying; Cuba	Macías and Mora 2003
61 ± 2	4 ± 1		Free-flying; Jamaica	Emrich et al. 2014
64 ± 8	1.7 ± 0.5		Inside cave; Cuba	Sanchez et al. 2017
54 ± 1	3.1 ± 1.5	202 ± 58	Open forest; Guadeloupe, Martinique	Barataud et al. 2015
71 ± 0.4	2.7 ± 0.1	32 ± 1	Interior forest; Bahamas	Murray et al. 2009
30	5.6	23	Low foraging flight; Costa Rica	Schnitzler et al. 1994
32 ± 8	4.5 ± 1.3	100 ± 40	Flying in undergrowth; Guadeloupe, Martinique	Baratuad et al. 2015
40 (36–45)	2.6 (1.0–3.5)	73 (33–133)	Hand release; Dominica, Puerto Rico	Vaughan Jennings et al. 2004
55 (40–77)	2.5 (1–4)	58 (23–92)	Hand release; Trinidad	Pio et al. 2010

Family/Species	Description	Number of calls analyzed	Beginning frequency (kHz)
Artibeus lituratus	Steep FM		80 (43–94)
Artibeus schwartzi	Steep FM	3[b]	74 (71–78)
Brachyphylla cavernarum	Steep FM	4[b]	67 (64–72)
Brachyphylla nana	Steep FM	13	89 ± 3
Chiroderma improvisum	FM	24	95 ± 2
Erophylla bombifrons	Steep FM	5[b]	54 (50–59)
Erophylla sezekorni	Fundamental; FM	13[b]	60 ± 1
	2nd harmonic; FM	11	90 ± 5
Glossophaga longirostris	Steep FM	8[b]	119 (97–148)
Glossophaga soricina	Steep FM	27	119 (66–102)
Macrotus waterhousii	FM	16[b]	84 ± 1
		10	74 ± 7
Monophyllus plethodon	Steep FM	15[b]	62 (55–72)
	Steep FM	96	66 ± 4

Ending frequency (kHz)	Duration (milliseconds)	Interpulse interval (milliseconds)	Comment	Reference
51 (30–60)	2.3 (1.5–3.6)	68 (29–140)	Hand release; Trinidad	Pio et al. 2010
37 (31–40)	2.7 (1.7–3.7)	97 (73–116)	Hand release; Saint Vincent	Vaughan Jennings et al. 2004
38 (36–40)	2.6 (2.2–3.2)	76 (46–108)	Hand release; Puerto Rico	Vaughan Jennings et al. 2004
34 ± 6	2.4 ± 0.4	97 ± 16	Flying in room; Cuba	Macías et al. 2006b
66 ± 2	3.0 ± 1.0	74 ± 50	Undergrowth and forest corridor; Guadeloupe	Barataud et al. 2015
27 (25–30)	4.7 (3.3–5.8)	107 (67–197)	Hand release; Puerto Rico	Vaughan Jennings et al. 2004
32 ⊥ 1	2.3 ± 0.2	40 ± 4	Interior forest; Bahamas	Murray et al. 2009
52 ± 2	2.3 ± 0.2	42 ± 4	Interior forest; Bahamas	Murray et al. 2009
72 (64–82)	1.6 (1.2–2.3)	48 (22–85)	Hand release; Saint Vincent	Vaughan Jennings et al. 2004
72 (21–105)	2 (1–4)	37 (9–117)	Hand release; Trinidad	Pio et al. 2010
56 ± 1	1.3 ± 0.1	36 ± 3	Interior forest; Bahamas	Murray et al. 2009
46 ± 3	2 ± 1		Zip line; Jamaica	Emrich et al. 2014
28 (22–39)	2.1 (1–4.6)	70 (18–164)	Hand release; Dominica, Saint Vincent	Vaughan Jennings et al. 2004
33 ± 5	4.0 ± 1.4	90 ± 58	Open space; Guadeloupe, Martinique	Barataud et al. 2015

Family/Species	Description	Number of calls analyzed	Beginning frequency (kHz)
Phyllonycteris poeyi	Steep FM	88	46 ± 2
Phyllops falcatus	Steep FM	64	74 ± 4
Stenoderma rufum	Steep FM	1[b]	95
Sturnira angeli	Steep FM	5[b]	93 (86–108)
	FM	28	73 ± 6
	Steeper FM than on Guadeloupe	50	91 ± 7
Sturnira paulsoni	Steep FM	2[b]	83 (75–91)
VESPERTILIONIDAE			
Eptesicus fuscus	Concave FM, then QCF	138	47 ± 3
Eptesicus guadeloupensis	Concave FM, then QCF	72	43 ± 7
Lasiurus cinereus	Concave FM		20–50
Lasiurus minor	Concave FM	7[c]	42 ± 2
Myotis dominicensis	Concave FM	75	95 ± 11
Myotis martiniquensis	Concave FM	170	93 ± 14

Ending frequency (kHz)	Duration (milliseconds)	Interpulse interval (milliseconds)	Comment	Reference
34 ± 2	4.7 ± 1	11 ± 2	Foraging in open; Cuba	Mora and Macías 2007
24 ± 4	4.2 ± 0.6	112 ± 56	Free-flying; Cuba	Macías et al. 2005 <2009?>
26	3.1	76	Hand release; Puerto Rico	Vaughan Jennings et al. 2004
46 (36–59)	2.7 (1.3–3.3)	72 (41–92)	Hand release; Dominica	Vaughan Jennings et al. 2004
42 ± 3	6.0 ± 1.1	133 ± 91	Undergrowth; Guadeloupe	Barataud et al. 2015
48 ± 6	3.7 ± 1.3	91 ± 54	Undergrowth; Martinique	Barataud et al. 2015
36	1.4 (1.1–1.7)	42 (41–42)	Hand release; Saint Vincent	Vaughan Jennings et al. 2004
33 ± 1	4.6 ± 1		Free-flying; Cuba	Rodríguez and Mora 2006
25 ± 2	9 ± 2	193 ± 73	Free-flying; Guadeloupe	Baratuad et al. 2015
15–32			Free-flying; mainland	O'Farrell et al. 2000
31 ± 0.3	10 ± 0.4	290 ± 24	Foraging at light; Bahamas	Murray et al. 2009
43 ± 4	4.2 ± 0.9	66 ± 14	Slightly cluttered habitat; Guadeloupe	Barataud et al. 2015
44 ± 2	4.1 ± 0.9	82 ± 23	Martinique	Barataud et al. 2015

Family/Species	Description	Number of calls analyzed	Beginning frequency (kHz)
Nycticeius cubanus	Concave FM, then QCF	285	59 ± 7

[a] Pulses occur in groups of two or three, with the second and third higher in frequency than the preceding sound.

[b] Number of individual bats.

[c] Number of sequences.

Ending frequency (kHz)	Duration (milliseconds)	Interpulse interval (milliseconds)	Comment	Reference
41 ± 1	8.4 ± 1		Free-flying; open space; Cuba	Mora et al. 2005

Major Islands or Island Groups
Occupied by Each Species

Table A.3 Bat species occurring on the Greater Antilles, Bahama Islands, Turks and Caicos Islands, and Virgin Islands

Islands are listed in order of the approximate distance from the island or island group to mainland North America, from close to far. E = species endemic to the West Indies. Records from specific islands in the Bahamas, Turks and Caicos Islands, Cayman Islands, and Virgin Islands, as well as a number of satellite islands associated with the Greater Antilles, are mentioned in the text and detailed in Hoffman et al. 2019.

Family/Species	Bahama Islands	Turks and Caicos Islands	Jamaica	Cayman Islands	Cuba	Hispaniola	Puerto Rico	Virgin Islands	Saint Croix
NOCTILIONIDAE									
Noctilio leporinus	X		X		X	X	X	X	X
MORMOOPIDAE									
Mormoops blainvillei (E)			X		X	X	X		
Pteronotus macleayii (E)			X		X				
Pteronotus parnellii			X		X				
Pteronotus portoricensis (E)							X		
Pteronotus pusillus (E)						X			
Pteronotus quadridens (E)			X		X	X	X		

Family/Species	Bahama Islands	Turks and Caicos Islands	Jamaica	Cayman Islands	Cuba	Hispaniola	Puerto Rico	Virgin Islands	Saint Croix
PHYLLOSTOMIDAE									
Ariteus flavescens (E)			X						
Artibeus jamaicensis	X	X	X	X	X	X	X	X	X
Brachyphylla cavernarum (E)							X	X	X
Brachyphylla nana (E)		X		X	X	X			
Erophylla bombifrons (E)						X	X		
Erophylla sezekorni (E)	X	X	X	X	X				
Glossophaga soricina			X						
Macrotus waterhousii	X	X	X	X	X	X			
Monophyllus redmani (E)	X	X	X			X	X	X	
Phyllonycteris aphylla (E)			X						
Pyllonycteris poeyi (E)					X	X			
Phyllops falcatus (E)				X	X	X			
Stenoderma rufum (E)							X	X	X
NATALIDAE									
Chilonatalus macer (E)						X			
Chilonatalus micropus (E)			X			X			
Chilonatalus tumidifrons (E)	X								
Natalus jamaicensis (E)			X						
Natalus major (E)							X		
Natalus primus (E)					X				
Nyctiellus lepidus (E)	X				X				

Family/Species	Bahama Islands	Turks and Caicos Islands	Jamaica	Cayman Islands	Cuba	Hispaniola	Puerto Rico	Virgin Islands	Saint Croix
MOLOSSIDAE									
Eumops auripendulus			X						
Eumops ferox			X		X				
Molossus milleri (E)			X	X	X				
Molossus molossus							X	X	X
Molossus verrilli (E)						X			
Mormopterus minutus (E)					X				
Nyctinomops laticaudatus					X				
Nyctinomops macrotis			X		X	X			
Tadarida brasiliensis	X	X	X	X	X	X	X	X	
VESPERTILIONIDAE									
Antrozous koopmani (E)					X				
Eptesicus fuscus	X			X	X	X	X		
Eptesicus lynni (E)			X						
Lasiurus cinereus						?			
Lasiurus degelidus (E)			X						
Lasiurus insularis (E)					X				
Lasiurus minor (E)	X	X				X	X		
Lasiurus pfeifferi (E)					X				
Nycticeius cubanus (E)					X				

Table A.4 Bat species occurring on major islands of the Lesser Antilles

Islands are listed in order of the approximate distance from the island to mainland South America, from close to far. E = species endemic to the West Indies. Records from a number of smaller islands are listed in Hoffman et al. 2019.

Family/Species	Saint Vincent	Barbados	Saint Lucia	Martinique	Dominica	Guadeloupe	Montserrat	Saint Kitts	Nevis	Antigua	Barbuda	Anguilla
NOCTILIONIDAE												
Noctilio leporinus	X	X	X	X	X	X	X	X	X	X	X	
MORMOOPIDAE												
Pteronotus davyi			X	X	X	X						
Pteronotus fuscus	X											
PHYLLOSTOMIDAE												
Ardops nichollsi (E)	X		X	X	X	X	X	X	X	X		
Artibeus jamaicensis		X	?	X	X	X	X	X	X	X	X	X
Artibeus lituratus	X											
Artibeus schwartzi (E)	X		?									
Artibeus schwartzi x *jamaicensis*			X									
Brachyphylla cavernarum (E)	X	X	X	X	X	X	X	X	X	X	X	X
Glossophaga longirostris	X											
Monophyllus plethodon (E)	X	X	X	X	X	X	X	X	X	X	X	X
Chiroderma improvisum (E)						X	X	X	X			

Family/Species	Saint Vincent	Barbados	Saint Lucia	Martinique	Dominica	Guadeloupe	Montserrat	Saint Kitts	Nevis	Antigua	Barbuda	Anguilla
Micronycteris buriri (E)	X											
Sturnira angeli (E)				X	X	X	X					
Sturnira paulsoni (E)	X		X									
NATALIDAE												
Natalus stramineus (E)				X	X	X	X		X	X	X	X
MOLOSSIDAE												
Molossus molossus	X	X	X	X	X	X	X	X	X	X	X	X
Tadarida brasiliensis	X		X	X	X	X	X	X	X	X	X	X
VESPERTILIONIDAE												
Eptesicus guadeloupensis (E)						X						
Myotis dominicensis (E)					X	X						
Myotis martiniquensis (E)				X								
Myotis nyctor (E)		X										

Key References to the Ecology and Natural History of Bats by Island or Island Group

Table A.5 Key references to the ecology and natural history of bats in the West Indies by island or island group.

Entries for the Bahamas, Cuba, and Puerto Rico do not include older references listed in the comprehensive reports of Speer et al. (2015), Silva-Taboada (1979), and Gannon et al. (2005), respectively.

Island or island group	References
BAHAMA ISLANDS	Speer et al. 2015, 2017, 2019; Hoffman et al. 2019; Mathis and Reed 2021
CAYMAN ISLANDS	Morgan 1994; Blumenthal 2007, 2011
GREATER ANTILLES	Griffiths and Klingener 1988; Timm and Genoways 2003; Muscarella et al. 2011; Hoffman et al. 2019
Cuba	Silva-Taboada 1979; Mancina and García Rivera 2000; Macías and Mora 2003; Tejedor et al. 2004, 2005a; Mora et al. 2005, 2011; Macías et al. 2006a, 2006b, 2009; Rodríguez and Mora 2006; Mancina et al. 2007a, 2007b, 2012; Mora and Macías 2007; Silva-Taboada et al. 2007; Mancina 2008, 2010, 2012; Mora and Torres 2008; Tavares and Mancina 2008; Borroto-Páez and Mancina 2011, 2017; Mancina and Castro-Arellano 2013; Clairmont et al. 2014; Sanchez et al. 2017; Sánchez-Lozada et al. 2018; De la Cruz Mora and García Padrón 2019; Vela Rodríguez et al. 2019; Sánchez-Losada and Mancina 2020; Vela Rodríguez and Mancina 2020; Orihuela et al. 2020a; De la Cruz Mora 2021; Fundora Caballero et al. 2021

Island or island group	References
Jamaica	Gosse 1851; Koopman and Williams 1951; Henson and Novick 1966; Goodwin 1970; McNab 1976; McFarlane 1985, 1986; McFarlane and Garrett 1989; Vareschi and Janetzky 1998; Dávalos and Eriksson 2003; Genoways et al. 2005; National Environment and Planning Agency 2011; Lundberg and McFarlane 2009; Hayward 2013; Emrich et al. 2014; Newman et al. 2016; Gallant et al. 2019, 2021
Hispaniola	Klingener et al. 1978; Tejedor et al. 2005a; Wiley 2010; Núñez-Novas and León 2011; Rodríguez-Durán and Christenson 2012; Velazco et al. 2013; Núñez-Novas et al. 2014, 2016, 2019, 2021; Soto-Centeno et al. 2017
Puerto Rico	Gannon et al. 2005; Soto-Centeno and Kurta 2006; Kurta et al. 2007; Schaetz et al. 2009; Rodríguez-Durán and Padilla-Rodríguez 2010; Rodríguez-Durán et al. 2010, 2020; Rodríguez-Durán and Otero 2011; Rolf and Kurta 2012; Alexander and Geluso 2013; Rolfe et al. 2014; Soto-Centeno et al. 2014, 2017; Rodríguez-Durán and Feliciano-Robles 2015, 2016; Hirsbrunner et al. 2020; Rodríguez-Durán and Rosa 2020; Calderón-Acevedo et al. 2021; Presley et al. 2021
LESSER ANTILLES	Koopman 1968; Timm and Genoways 2003; Hoffman et al. 2019
Anguilla	Genoways et al. 2007b
Antigua	Pedersen et al. 2006; Lindsay et al. 2010
Barbados	Grindal 2004; Genoways et al. 2011
Barbuda	Pedersen et al. 2007
Dominica	Hill and Evans 1985; Genoways et al. 2001; Vaughan Jennings et al. 2004; Angin 2014; Stoetzel et al. 2021

Island or island group	References
Grenada	Jones 1951; Genoways et al. 1998; Larsen et al. 2010
Grenadines	Genoways et al. 2010; Larsen et al. 2010
Guadeloupe	de la Torre and Schwartz 1966; Genoways and Baker 1975; Genoways and Jones 1975; Baker and Genoways 1976; Baker and Genoways 1978; Masson et al. 1990; Masson and Breuil 1992; Masson et al. 1994; Ibéné et al. 2007; Barataud and Giosa 2013, 2014; Lenoble et al. 2014a; Rolfe et al. 2014; Barataud et al. 2015
Marie-Galante	Masson et al. 1990; McCarthy and Henderson 1992; Lenoble et al. 2014b, 2019; Stoetzel et al. 2016; Pelletier et al. 2017
Martinique	Picard and Catzeflis 2013; Barataud et al. 2015, 2017; Dinets 2017; Catzeflis et al. 2019
Montserrat	Jones et al. 1978; Jones and Baker 1979; Pierson et al. 1986; Pedersen et al. 1996, 2009, 2012; R. J. Larsen et al. 2007
Nevis	Pedersen et al. 2003
Saba	Genoways et al. 2009
Saint Bartholomew	Larsen et al. 2006
Saint Eustatius	Pedersen et al. 2018b
Saint Kitts	Pedersen et al. 2005; Beck et al. 2016; Reeves et al. 2016
Saint Lucia	Arendt and Anthony 1986; Larsen et al. 2010; Pedersen et al. 2018a
Saint Martin	Genoways et al. 2007a
Saint Vincent	de la Torre and Schwartz 1966; Jones 1978; Vaughan 1995; Vaughan and Hill 1996; Vaughan Jennings et al. 2004; Larsen et al. 2010, 2011; Kwiecinski et al. 2018; Morales et al. 2021; Pietsch and Marx 2021

Island or island group	References
TURKS AND CAICOS	Buden 1976, 1977
VIRGIN ISLANDS	Bond and Seaman 1958; Koopman 1959, 1975; Nellis and Ehle 1977; Bacle et al. 2007; Kwiecinski and Coles 2007; Lindsay et al. 2009; Kwiecinski et al. 2010

GLOSSARY

Anthropocene epoch. Proposed geological time unit, referring to the most recent past, after human activities began to impact earth's climate and ecosystems.

antitragus. Fleshy projection at the base of the outer (posterior) edge of the ear (see Fig. 0.6).

aspect ratio. Length of wings divided by their width, sometimes calculated as wingspan (measured from tip of left wing to tip of right wing) divided by the wing chord (average distance from front to back of wing). Long narrow wings have a high aspect ratio (see Fig. 0.3).

assemblage. Group of populations from related species that occur together in the same geographic area.

bat house. Artificial bat roost; a human-made structure specifically built to provide a roosting environment for bats.

beetle. Hard-bodied insect in the order Coleoptera.

bicolored. Having two colors.

bifid. Divided by a deep cleft into two parts.

braincase. Portion of skull that protects most of the brain (see Fig. 13.1).

breech birth. Birth during which the rear portion of a neonate leaves the vagina first.

broadband. Including a wide range of frequencies (e.g., see Fig. 0.4d and e).

bug, true. Insect with piercing/sucking mouthparts in the order Hemiptera.

bulldog bat. Bat in the family Noctilionidae.

bush cricket. Another name for a katydid.

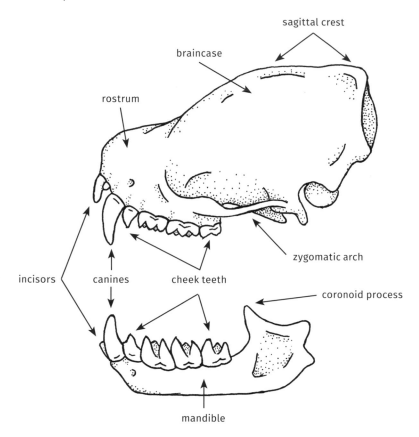

Figure 13.1. Basic parts of a bat skull.

calcar. Cartilaginous rod protruding from the ankle and supporting the rear edge of the tail membrane (see Fig. 0.1).

caldarium. Warmest room in a hot cave, where air temperature is
   generally between 28 and 40 °C.

call. Sound emitted by an echolocating bat; a pulse.

canine. First tooth in the maxillary bone of a bat (see Fig. 13.1);
   used for piercing food.

caudal. Referring to the tail.

cheek tooth. Any tooth located behind the canines; the premolars
   and molars (see Fig. 13.1).

chigger. Larval stage of an ectoparasitic mite in the family Trombiculidae.

cicada. Orthopteran insect in the family Cicadellidae.

climax community. Group of plant species that over time have reached a steady state in terms of composition.

cockroach. See roach.

columnar cacti. Upright, cylindrical cacti, often with white, night-opening flowers, as in the genera *Cereus* and *Pilosocereus*.

community. All species living and interacting in the same area.

conspecific. An individual of the same species.

constant frequency. Descriptive term for an echolocation call or portion of a call that varies little in frequency (e.g., see most of Fig. 0.4a).

coppice. Dry evergreen and broadleaf forest typical of the Bahama Islands.

coronoid process. Upward extension from the rear of the mandible for attachment of a major jaw-closing muscle, the temporalis (see Fig. 13.1).

cranium. Part of the skull encasing the brain and supporting the face; braincase (see Fig. 13.1).

crepuscular. Active primarily near sundown and sunrise.

Cretaceous period. Unit of geologic time that is part of the Mesozoic era, from 66 to 145 million years ago.

cricket. Orthopteran insect in the family Gryllidae.

cusp. Point or bump on the occlusal (chewing) surface of a tooth.

dayroost. To rest during the day; the site where a bat rests during daytime.

delayed fertilization. Reproductive process in some bats, particularly vespertilionids in temperate areas. After mating, sperm are stored in the uterus, but fertilization (fusing of egg and sperm) does not happen until weeks or months later.

dictyopteran. Roach or mantis; an insect in the superorder Dictyoptera.

distal. Occurring away from the base toward the end of a structure, such as a hair or a leg; opposite of proximal.

dorsal. Referring to the back of an animal; opposite of ventral.

dorsoventrally. From back (dorsum) to belly (venter).

dorsum. Top or backside of a bat; opposite of venter.

duty cycle. Duration of an echolocation pulse divided by the time between the beginning of successive sounds. Low duty cycle provides quiet time between pulses to hear the incoming echoes.

earwig. Flattened insect in the order Dermaptera, with a large pincer-like structure on its posterior end.

ectoparasite. Parasite living on the outside surface of an animal.

endemic. Restricted in distribution, usually to a small area, such as an island or group of islands.

Eocene epoch. Unit of geologic time, from approximately 35 to 55 million years ago.

exoskeleton. Outer covering of insects and other invertebrates.

fig wasp. Hymenopteran insect in the family Agaonidae that pollinates fig trees (*Ficus*).

flank-margin cave. Phreatic cave formed at the boundary (margin) of subterranean freshwater and seawater at the edge (flank) of a landmass.

fly. Insect with one pair of functional wings in the order Diptera.

forearm length. Distance from wrist to elbow of a bat, as measured on a folded wing.

frequency modulated. Descriptive term for an echolocation call or portion of a call that changes continually in frequency (e.g., see Fig. 0.4d).

free-tailed bat. A bat in the family Molossidae.

frigidarium. Coolest room in a tropical cave, where air temperature does not surpass 25 °C.

froghopper. Hemipteran insect in the superfamily Cercopoidea whose larvae hide within patches of foam produced while feeding on plants; a spittlebug.

frugivore. An animal that eats primarily fruit.

gallery forest. Woodland formed along a watercourse and extending into otherwise treeless areas.

gestation. Period between fertilization of the egg and birth in a mammal.

greatest length of skull. Distance from anteriormost bone to posteriormost projection of skull.

gular gland. External gland located in the throat region of many bats in the family Molossidae.

harem. Group of females that a lone male has access to for mating purposes.

harmonic. Different frequencies in a sound that are integral multiples of the base or fundamental frequency (e.g., see Fig. 0.4b).

harp trap. Apparatus used to capture flying bats, consisting of two or more sets of vertical wires, from 8 to 10 cm apart; bats fly through the first series, strike the second, and fall harmlessly into a cloth bag suspended below the trap.

hibernation. Protracted state of lowered metabolic rate and body temperature during the cold period of the year.

hindfoot length. Distance from heel to tip of claw on the longest toe.

Holocene epoch. Most recent unit of geologic time that followed the last ice age (i.e., the Pleistocene epoch).

home range. Area in which an animal performs its everyday activities, including roosting and foraging.

homopteran. Insect belonging to the suborder Homoptera, within the order Hemiptera; leafhoppers, cicadas, aphids, and scale insects are a few types.

horseshoe. Portion of the nose leaf of a phyllostomid bat lateral to and below the nostrils.

hot cave. Subterranean site in which warm air becomes trapped, resulting in a temperature that is higher than that of the outside air; see tepidarium and caldarium.

hot spot. Location with high species richness.

hutia. Burly rodents in the family Capromyidae that are vaguely similar in external appearance to guinea pigs, but with longer tails.

incisor. Front tooth of a bat; any tooth anterior to a canine (see Fig. 13.1).

induced drag. Force opposing the forward movement of a flying animal and created by differences in air pressure between the top and bottom of a wing.

interfemoral membrane. Uropatagium or tail membrane of a bat.

insectivore. Any animal that eats insects.

IUCN. International Union for the Conservation of Nature.

karst. Landscape typically derived from limestone, gypsum, or dolomite and characterized by caves, sinkholes, and underground streams.

katydid. Large, usually green, orthopteran insect with long antennae in the family Tettigoniidae.

kHz (kiloHertz). Unit of frequency, used with sound waves; 1 kHz = 1000 cycles/second.

labionasal. Compound adjective referring to a structure that combines the lips (labia) and the nostrils (nasal area).

lacewing. Soft-bodied insect with an aquatic larval stage in the order Neuroptera.

lanceolate. Having the appearance of a spear tip, often used to describe nose leafs (see Fig. 0.5).

lancet. The upright portion of the nose leaf located above the nostrils.

larynx. Voice box; cartilaginous box that contains the vocal cords.

lasiurine. Any bat in the genus *Lasiurus.*

leafhopper. Hemipteran insect in the family Cicadellidae that feeds on plant juices with piercing/sucking mouthparts.

leaf-nosed bat. Any bat in the family Phyllostomidae.

mandible. Lower jaw of a mammal.

metacarpals. Bones of the hand between the wrist and the fingers (phalanges) (see Fig. 0.1).

metabolic rate. Quantity of energy used by an animal per unit time.

mesic. Moderately moist.

microphyllous. Having small leaves.

millisecond. One one-thousandth of a second.

Miocene epoch. Unit of geologic time, from approximately 5.5 to 23 million years ago; the Miocene follows the Oligocene.

mistnet. Net made of fine strands of nylon or polypropylene and used to catch flying bats or birds; to capture a bat using such a net.

molar. Posterior cheek tooth, usually with multiple roots. A mammal has only one set of molars in its life; there are no molars among the deciduous (milk) teeth.

molossid. Any bat in the family Molossidae; a free-tailed bat.

mormoopid. Any bat in the family Mormoopidae; a mustached or ghost-faced bat.

moth. Nocturnal insect in the order Lepidoptera, having large, broad wings and body parts covered by tiny scales.

narrowband. Including a small range of frequencies.

natalid. Bat in the family Natalidae; a funnel-eared bat.

natalid organ. An obvious swelling, usually hairless, located near the junction of the rostrum and the cranium in adult males in the family Natalidae (see Fig. 6.1c); its specific function is unknown.

nematoceran. Group of flies typically with aquatic larval stages, like craneflies (Tipulidae), mosquitoes (Culicidae), and midges (Chironomidae).

nematode. Roundworm in the phylum Nematoda.

neonate. Newborn mammal.

Neotropics. Tropical regions of the New World.

nightroost. To rest between nocturnal foraging bouts or the actual site where a bat rests between foraging bouts.

noctilionid. Any bat in the family Noctilionidae; a bulldog bat.

nocturnal. Active between sunset and sunrise.

nose leaf. Leaf- or spear-shaped appendage near the nostrils in leaf-nosed bats (see Fig. 0.5).

nycteribiid. Ectoparasitic flies in the family Nycteribiidae, lacking wings and having a spiderlike appearance.

Oligocene epoch. Unit of geologic time, from approximately 23 to 34 million years ago; the Oligocene follows the Eocene and precedes the Miocene.

omnivore. Animal that eats diverse kinds of foods, including plant and animal products.

orthopteran. Insect in the order Orthoptera, which includes crickets, grasshoppers, and katydids.

palate. Roof or top of the mouth cavity.

papilla. Small, rounded bump on some part of the body, such as the skin or tongue (plural = papillae).

papillose. Covered with projections or small bumps.

parturition. Act of giving birth.

pelage. Broad term referring to the furry coat of a mammal.

phyllostomid. Any bat in the family Phyllostomidae; a leaf-nosed bat.

phreatic. Below the water table; used as an adjective to describe a cave formed in rocks saturated with water, as opposed to flowing water at or above the water table.

pinna. External ear (plural = pinnae).

Pleistocene epoch. Unit of geologic time, from approximately 12,000 years to 2.5 million years ago; the time of the most recent ice age.

Pliocene epoch. Unit of geologic time, from approximately 2.5 to 5.5 million years ago.

postpartum estrus. Period of heat or sexual receptivity that occurs soon after giving birth.

premaxillary bone. Anterior bone in the rostrum of a bat; the incisors are rooted in the premaxillary bone.

premolar. Tooth between the molars and canine.

promiscuous. Having more than one sexual partner in the same mating season.

protrusible. Capable of being forced outward or protruded.

proximal. Referring to a part close to the base of a structure, such as a hair or a leg; opposite of distal.

pulse. Single sound emitted by an echolocating bat; a call.

quasi-constant frequency. Echolocation call or portion of a call that varies only slightly, perhaps from 5 to 10 kHz, over time.

Quaternary period. Unit of geological time that combines the Pleistocene and the Holocene.

raptor. Predatory birds, such as owls, hawks, eagles, and falcons.

riparian. Associated with streams.

roach. A scavenging insect with a flattened, leathery body and chewing mouthparts in the order Blattodea, superorder Dictyoptera. Small roaches flying over fields and forests are eaten by some bats, whereas large species of roach feed on guano and bat carcasses inside a cave.

rostrum. Part of the skull that is in front of the eyes and supports the snout (see Fig. 13.1).

sagittal crest. Vertical ridge of bone along midline of braincase for attachment of the temporalis muscle.

search phase. A sequence of echolocation calls emitted by a bat when flying freely, without obstacles, before the mammal focuses on an insect or fruit.

short-faced bat. A leaf-nosed bat in the subtribe Stenodermatina characterized by a seemingly undersized rostrum; four species of short-faced bats live in the West Indies.

solenodon. Long-snouted mammals in the family Solenodontidae that look like very large shrews; solenodons are some of the few venomous mammals.

species richness. The number of species in an area.

stinkbug. Hemipteran insect in the family Pentatomidae that has a shieldlike outline and releases a potentially pungent defensive spray.

stochastic. Random.

succulent. Type of plant from arid areas, with thick fleshy leaves or stems that contain large amounts of water; cacti and agaves are succulents.

sweep. Noun or verb referring to the changing frequencies in an echolocation call.

systematist. Biologist studying the evolutionary relationships among various taxa.

tail length. Distance from the base of the tail to the end of the last vertebra.

tail membrane. Band of skin stretching from one leg to the other and enclosing some or all tail vertebrae (if present); also called the uropatagium or interfemoral membrane (see Fig. 0.1).

taxon. Group of animals classified together; species, genus, family, and order are each a taxon (plural = taxa).

taxonomist. Biologist who classifies organisms into different groups (taxa).

temporalis. Major jaw-closing muscle that extends from the side of the skull and sagittal crest (if present) to the coronoid process of the mandible.

tepidarium. Room in a hot cave, where air temperature is generally over 25 but under 28 °C; the tepidarium is cooler than the caldarium.

termite. Eusocial insect historically placed in the order Isoptera; it consumes plant products, including cellulose, which is broken down by microorganisms in the insect's gut.

thrip. Tiny, weak-flying insects in the order Thysanoptera that feed on plants.

tibia. Major bone of lower leg, between ankle and knee; the shin in humans.

tip index. Measure of the shape of the distal part of the wing; a high value indicates a rounded wing that is more suited for slow flight, and possibly hovering, compared with a wing with a low tip index.

torpor. Reversible state of lowered body temperature and metabolic rate.

total length. Distance from the nose to the tip of the tail.

tragus. A fleshy projection associated with the inner (anterior) part of the ear (see Figs. 0.1 and 0.6).

tricolored. Having three colors, as in individual hairs of some bats.

trilobed. Having three parts.

troglobitic. Referring to an animal that spends its entire life in caves, never coming to the surface.

type. Single specimen on which the name of a species or other taxon is based.

ultrasonic. Referring to sounds having frequencies exceeding 20 kHz, which are typically above the upper limit of human hearing.

uropatagium. Tail membrane or interfemoral membrane.

vampire. One of three New World species of bats in the family Phyllostomidae that consume blood.

vagility. Ability to move easily, especially over long distances.

ventral. Referring to the underside or lower part of an animal; opposite of dorsal.

vertebra. One of the bones forming the backbone, from the skull to the tip of the tail (plural = vertebrae).

vesper bat. Member of the family Vespertilionidae; a vespertilionid.

vespertilionid. Member of the family of vesper bats (Vespertilionidae).

volant. Capable of flight.

weevil. Plant-eating beetle with an elongated snout in the family Curculionidae.

wing loading. Body mass divided by wing area.

xeric. Dry.

zygomatic arch. Narrow strip of bone that begins at the anterior edge of the eye socket and extends back to the braincase, anterior to the ear opening (see Fig. 13.1).

zygomatic breadth. Greatest distance between the outer portions of the zygomatic arches.

REFERENCES

Abbad y Lasierra, I. 1788. Historia geográfica, civil y política de la isla de San Juan Bautista de Puerto Rico [Geographic, civil, and political history of the island of San Juan Bautista of Puerto Rico]. Madrid: Antonio Espinosa.

Abelleira Martínez, O.J. 2008. Observations on the fauna that visit African tulip tree (*Spathodea campanulata* Beauv.) forests in Puerto Rico. Acta Científica 22:37–43.

Abelleira Martínez, O.J., E.J. Meléndez Ackerman, D. García Montiel, and J.A. Parrotta. 2015. Seed dispersal turns an experimental plantation on degraded land into a novel forest in urban northern Puerto Rico. Forest Ecology and Management 357:68–75.

Acevedo Rodríguez, P., and M.T. Strong. 2008. Floristic richness and affinities in the West Indies. Botanical Review 74:5–36.

Aguiar, L.M.S., I.D. Bueno-Rocha, G. Oliveira, E.S. Pires, S. Vasconcelos, G.L. Nunes, M.R. Frizzas, and P.H.B. Togni. 2021 Going out for dinner—The consumption of agriculture pests by bats in urban areas. PLoS One 16(10): e0258066. https://doi.org/10.1371/journal.pone.0258066.

Agudelo, M.S., T.J. Mabee, R. Palmer, and R. Anderson. 2021. Post-construction bird and bat fatality monitoring studies at wind energy projects in Latin America: A summary and review. Heliyon 7(6): e07251. https://doi.org/10.1016/j.heliyon.2021.e07251.

Aide, T.M., J.K. Zimmerman, M. Rosario, and H. Marcano. 1996. Forest recovery in abandoned cattle pastures along an elevational gradient in northeastern Puerto Rico. Biotropica 28:537–548.

Aide, T.M., J.K. Zimmerman, L. Herrera, M. Rosario, and M. Serrano. 1995. Forest recovery in abandoned tropical pastures in Puerto Rico. Forest Ecology and Management 77:77–86.

Aizpurua, O., and A. Alberdi. 2018. Ecology and evolutionary biology of fishing bats. Mammal Review 48:284–297.

Alexander, D.E. 2015. On the wing: Insects, pterosaurs, birds, bats and the evolution of animal flight. New York: Oxford University Press.

Alexander, I.R., and K. Geluso. 2013. Bats of Vieques National Wildlife Refuge, Puerto Rico, Greater Antilles. Check List 9:294–297.

Allen, G.M. 1911. Mammals of the West Indies. Bulletin of the Museum of Comparative Zoology 54:175–263.

Álvarez-Castañeda, S.T., and T. Álvarez Solórzano. 1996. Etimologías de los géneros de mamíferos mexicanos. Ciencia 47:39–49.

Álvarez Ruiz, M., and A.E. Lugo. 2012. Landscape effects on structure and species composition of tabonuco forests in Puerto Rico: Implications for conservation. Forest Ecology and Management 266:138–147.

Anadón-Irizarry, V., D.C. Wege, A. Upgren, R. Young, B. Boom, Y.M. León, Y. Arias, K. Koenig, A.L. Morales, W. Burke, A. Pérez-Leroux, C. Levy, S. Koenig, L. Gape, and P. Moore. 2012. Sites for priority biodiversity conservation in the Caribbean Islands Biodiversity Hotspot. Journal of Threatened Taxa 4:2806–2844.

Angin, B. 2014. Bat predation by the Dominica boa (*Boa nebulosa*). Caribbean Herpetology 51:1–2.

Arendt, W.J., and D. Anthony. 1986. Bat predation by the St. Lucia boa (*Boa constrictor orophias*). Caribbean Journal of Science 22:219–220.

Arita, H.T. 1993. Conservation biology of the cave bats of Mexico. Journal of Mammalogy 74:693–702.

Arnett, E.B., E.F. Baerwald, F. Mathews, L. Rodrigues, A. Rodríguez-Durán, J. Rydell, R. Villegas-Patraca, and C.C. Voigt. 2016. Impacts of wind energy development on bats: A global perspective. *In* C.C. Voigt and T. Kingston, eds., Bats in the Anthropocene: Conservation of bats in a changing world, pp. 295–324. Cham, Switzerland: SpringerOpen.

Arnold, M.L., R.J. Baker, and H.H. Genoways. 1980. Evolutionary origin of *Eptesicus lynni*. Journal of Mammalogy 61:319–322.

Ávila-Flores, R., and R.A. Medellín. 2004. Ecological, taxonomic, and physiological correlates of cave use by Mexican bats. Journal of Mammalogy 85:675–687.

Ávila-Flores, R., J.J. Flores-Martínez, and J. Ortega. 2002. *Nyctino-mops laticaudatus*. Mammalian Species 697:1–6.

Bacle, J.-P., K.C. Lindsay, and G.G. Kwiecinski. 2007. Bats of St. Thomas and St. John, U.S. Virgin Islands: Priority conservation measures for species of greatest concern. Occasional Paper, Island Resources Foundation 60:1–13.

Baker, R.J., and H.H. Genoways. 1976. A new species of *Chiroderma* from Guadeloupe, West Indies (Chiroptera: Phyllostomidae). Occasional Papers of the Museum, Texas Tech University 39:1–9.

Baker, R.J., and H.H. Genoways. 1978. Zoogeography of Antillean bats. Academy of Natural Sciences of Philadelphia, Special Publication 13:53–97.

Baker, R.J., H.H. Genoways, and J.C. Patton. 1978. Bats of Guadeloupe. Occasional Papers of the Museum, Texas Tech University 50:1–16.

Barataud, M., and S. Giosa. 2013. *Eptesicus guadeloupensis*: Une espèce insulaire endémique en danger? [*Eptesicus guadeloupensis*: An endangered endemic insular species?]. Rhinolophe 19:177–187.

Barataud, M., and S. Giosa. 2014. Etude acoustique des chiroptères de Guadeloupe: Activité nocturne et utilisation de l'habitat [Acoustic study of bats from Guadeloupe: Nightly activity and habitat utilization]. Vespère 4:241–252.

Barataud, M., S. Giosa, F. LeBlanc, P.P. Favre, and J.-F. Desmet. 2015. Identification et écologie acoustique des chiroptères de la Guadeloupe et de la Martinique (Antilles Françaises) [Identification and acoustic ecology of bats from Guadeloupe and Martinique (French Antilles)]. Vespère 5:297–332.

Barataud, M., S. Giosa, S. Issartel, J. Jemin, M. Lesty, and J.-P. Fiard. 2017. Forêts tropicales insulaires et chiroptères: Le cas de la Martinique (Petites Antilles–France) [Insular tropical forests and bats: the case of Martinique (French Lesser Antilles)]. Vespère 7:411–457.

Barclay, R.M.R., and A. Kurta. 2007. Ecology and behavior of bats roosting in tree cavities and under bark. *In* M.J. Lacki, J.P. Hayes, and A. Kurta, eds., Bats in forests: Conservation and management, pp. 17–59. Baltimore: Johns Hopkins University Press.

Barreto, E., T.F. Rangel, L. Pellissier, and C.H. Graham. 2021. Area, isolation and climate explain the diversity of mammals on islands

worldwide. Proceedings of the Royal Society B 288:20211879.
https://doi.org/10.1098/rspb.2021.1879.

Beard, J.S. 1949. The natural vegetation of the Windward and Leeward islands. Oxford Forestry Memoirs 21. Oxford, United Kingdom: Clarendon Press.

Beck, J.D., A.D. Loftis, J.L. Daly, W.K. Reeves, and M.V. Orlova. 2016.
First record of *Chiroderma improvisum* Baker & Genoways, 1976
(Chiroptera: Phyllostomidae) from Saint Kitts, Lesser Antilles.
Check List 12:1854.

Bell, G.P. 1985. The sensory basis of prey location by the California
leaf-nosed bat *Macrotus californicus* (Chiroptera: Phyllostomidae).
Behavioral Ecology and Sociobiology 16:343–347.

Bell, G.P., G.A. Bartholomew, and K.A. Nagy. 1986. The role of energetics, water economy, foraging behavior and geothermal refugia
in the distribution of the bat, *Macrotus californicus*. Journal of
Comparative Physiology 156:441–450.

Bellard, C., C. Leclerc, and F. Courchamp. 2014. Impact of sea level
rise on the 10 insular biodiversity hotspots. Global Ecology and
Biogeography 23:203–212.

Benedict, K., and R. Mody. 2016. Epidemiology of histoplasmosis out-
breaks, United States, 1938–2013. Emerging Infectious Diseases
22:370–378.

Beolens, B., M. Watkins, and M. Grayson. 2009. The eponym dictionary of mammals. Baltimore: Johns Hopkins University Press.

Best, T.L., J.L. Hunt, L.A. McWilliams, and K.G. Smith. 2002. *Eumops
auripendulus*. Mammalian Species 708:1–5.

Blumenthal, L. 2007. Businesses for bats. Bats 25(3): 6–8.

Blumenthal, L. 2011. Bats and bat houses on the Cayman Islands:
Conservation ideas for the Caribbean and the tropics. https://
studylib.net/doc/9701664/.

Bohn, K., F. Montiel-Reyes, and I. Salazar. 2016. The complex songs of
two molossid species. *In* J. Ortega, ed., Sociality in bats, pp. 143–
160. New York: Springer.

Bond, R.M., and G.A. Seaman. 1958. Notes on a colony of *Brachyphylla cavernarum*. Journal of Mammalogy 39:150–151.

Boonman, A., Y. Bar-On, N. Cvikel, and Y. Yovel. 2013. It's not black
or white—On the range of vision and echolocation in echolocating
bats. Frontiers in Physiology 4(248): 1–12.

Borhidi, A. 1991. Phytogeography and vegetation ecology of Cuba. Budapest, Hungary: Akadémiai Kiadó.

Borroto-Páez, R., and C.A. Mancina, eds. 2011. Mamíferos en Cuba. Vasa, Finland: UPC Print.

Borroto-Páez, R., and C.A. Mancina. 2017. Biodiversity and conservation of Cuban mammals: Past, present, and invasive species. Journal of Mammalogy 98:964–985.

Boyles, J.G., P.M. Cryan, G.F. McCracken, and T.H. Kunz. 2011. Economic importance of bats in agriculture. Science 332:41–42.

Braun, J.K., and M.A. Mares. 1995. The mammals of Argentina: An etymology. Mastozoología Neotropical 2:173–206.

Briggs, J.C. 2014. Global biodiversity gain is concurrent with declining population sizes. Biodiversity Journal 5:447–452.

Brinkløv, S., and E. Warrant. 2017. Oilbirds. Current Biology 27: R1145–R1147.

Brown, E.E., D.D. Cashmore, N.B. Simmons, and R.J. Butler. 2019. Quantifying the completeness of the bat fossil record. Palaeontology 62:757–776.

Browne, P. 1789. The civil and natural history of Jamaica. London, United Kingdom: B. White and Son.

Buden, D.W. 1975. Taxonomic and zoogeographic appraisal of the big-eared bat (*Macrotus waterhousii* Gray) in the West Indies. Journal of Mammalogy 56:758–769.

Buden, D.W. 1976. A review of the bats of the endemic West Indian genus *Erophylla*. Proceedings of the Biological Society of Washington 89:1–30.

Buden, D.W. 1977. First records of bats of the genus *Brachyphylla* from the Caicos Islands, with notes on geographic variation. Journal of Mammalogy 58:221–225.

Calahorra-Oliart, A., S.M. Ospina-Garcés, and L. León-Paniagua. 2021. Cryptic species in *Glossophaga soricina* (Chiroptera: Phyllostomidae): Do morphological data support molecular evidence? Journal of Mammalogy 102:54–68.

Calderón-Acevedo, C.A., A. Rodríguez-Durán, and J.A. Soto-Centeno. 2021. Effect of land use, habitat suitability, and hurricanes on the population connectivity of an endemic insular bat. Science Reports 11(1): 9115. https://doi.org/10.1038/s41598-021-88616-7.

Campbell Wolfmeyer, M.T. 2010. Political implications of natural resources: A case study of guano on Monito and Mona islands. Ceiba 9:60–68.

Campos-Cerqueira, M., W.J. Arendt, and J.M. Wunderle, Jr. 2017. Have bird distributions shifted along an elevational gradient on a tropical mountain? Ecology and Evolution 7:9914–9924.

Cano-Torres, J., A. Olmedo-Reneaum, J. Esquivel-Sánchez, A. Camiro-Zúñiga, A. Pérez-Carrisoza, C. Madrigal-Iberri, R. Flores-Miranda, L. Ramírez-González, and P. Belaunzaran. 2019. Progressive disseminated histoplasmosis in Latin America and the Caribbean in people receiving highly active antiretroviral therapy for HIV infection: A systematic review. Medical Mycology 57:791–799.

Caraballo Ortiz, M. 2007. Mating system and fecundity of *Goetzea elegans*, an endangered tropical tree. MS thesis, University of Puerto Rico, Río Piedras.

Carew, J.L., and J.E. Mylroie. 1997. Geology of the Bahamas. *In* L. Vacher and T. Quinn, eds., Geology and hydrology of carbonate islands, pp. 91–139. Amsterdam: Elsevier.

Carlo, T.A, and J.M. Morales. 2016. Generalist birds promote tropical forest regeneration and increase plant diversity via rare-biased seed dispersal. Ecology 97:1819–1831.

Carstens, B.C., J. Sullivan, L.M. Dávalos, P.A. Larsen, and S.C. Pedersen. 2004. Exploring population genetic structure in three species of Lesser Antillean bats. Molecular Ecology 13:2557–2566.

Castillo-Figueroa, D. 2020. Why bats matter: A critical assessment of bat-mediated ecological processes in the Neotropics. European Journal of Ecology 6:77–101.

Castillo-Figueroa, D. 2022. Exploring a hidden structure in New World bats: The pollex. Tropical Natural History 22:1–11.

Catzeflis, F., G. Issartel, and J. Jemin. 2019. New data on the bats (Chiroptera) of Martinique Island (Lesser Antilles), with an emphasis on sexual dimorphism and sex ratios. Mammalia 83:501–514.

Cintrón, G., A.E. Lugo, D.J. Pool, and G. Morris. 1978. Mangroves of arid environments in Puerto Rico and adjacent islands. Biotropica 10:110–121.

Clairmont, L., E. Mora, and B. Fenton. 2014. Morphology, diet and flower-visiting by phyllostomid bats in Cuba. Biotropica 46:433–440.

Cláudio, V.C. 2019a. *Eptesicus fuscus. In* D.E. Wilson and R.A. Mittermeier, eds., Handbook of the mammals of the world. Volume 9, Bats, p. 843. Barcelona, Spain: Lynx Editions.

Cláudio, V.C. 2019b. *Nycticeius cubanus. In* D.E. Wilson and R.A. Mittermeier, eds., Handbook of the mammals of the world. Volume 9, Bats, p. 892. Barcelona, Spain: Lynx Editions.

Clement, M.J., and S.B. Castleberry. 2013. Tree structure and cavity microclimate: Implications for bats and birds. International Journal of Biometeorology 57:437–450.

Cooke, S.B., L.M. Dávalos, A.M. Mychajliw, S.T. Turvey, and N.S. Upham. 2017. Anthropogenic extinction dominates Holocene declines of West Indian mammals. Annual Review of Ecology and Systematics 48:301–327.

Corbett, R.J.M., C.L. Chambers, and M.J. Herder. 2008. Roosts and activity areas of *Nyctinomops macrotis* in northern Arizona. Acta Chiropterologica 10:323–329.

Corcoran, A.J., and T.J. Weller. 2018. Inconspicuous echolocation in hoary bats (*Lasiurus cinereus*). Proceedings of the Royal Society B 285:20180441.

Dávalos, L.M. 2004. Phylogeny and biogeography of Caribbean mammals. Biological Journal of the Linnaean Society 81:373–394.

Dávalos, L.M. 2009. Earth history and evolution of Caribbean bats. *In* T.H. Fleming and P.A. Racey, eds., Island bats: Evolution, ecology, and conservation, pp. 96–115. Chicago: University of Chicago Press.

Dávalos, L.M., and R. Eriksson. 2003. New and noteworthy records from ten Jamaican bat caves. Caribbean Journal of Science 39:140–144.

Dávalos, L.M., and A.L. Russell. 2012. Deglaciation explains bat extinction in the Caribbean. Ecology and Evolution 2:3045–3051.

Dávalos, L.M., and S.T. Turvey. 2012. West Indian mammals: The old, the new, and the recently extinct. *In* B.D. Patterson and L.P. Costa, eds., Bones, clones, and biomes: The history and geography of recent Neotropical mammals, pp. 157–202. Chicago: University of Chicago Press.

De la Cruz Mora, J.M. 2021. Caracterización del éxodo nocturno de *Eumops glaucinus* (Chiroptera; Molossidae) en zonas urbanas de Consolación del Sur, Cuba [Characterization of nightly exodus of *Eumops glaucinus* (Chiroptera: Molossidae) in urban areas of Consolación del Sur, Cuba]. Ecovida 11:157–164.

De la Cruz Mora, J.M., and L.Y. García Padrón. 2019. Aplicación de métodos no invasivos y convencionales en el estudio ecológico de *Natalus primus* (Chiroptera: Natalidae), uno de los murciélagos más amenazados de Cuba [Application of non-invasive and conventional methods in the study of the ecology of *Natalus primus* (Chiroptera: Natalidae), one of Cuba's most endangered species of bat]. Ecovida 19:24–38.

de la Torre, L. 1966. New bats of the genus *Sturnira* (Phyllostomidae) from the Amazonian lowlands of Peru and the Windward Islands, West Indies. Proceedings of the Biological Society of Washington 79:267–272.

de la Torre, L., and A. Schwartz. 1966. New species of *Sturnira* (Chiroptera: Phyllostomidae) from the islands of Guadeloupe and Saint Vincent, Lesser Antilles. Proceedings of the Biological Society of Washington 79:297–303.

Dinets, V. 2017. Roosting under tree leaves in Schwartz's myotis (Chiroptera). Neotropical Biology and Conservation 12:232–234.

Durocher, M., V. Nicolas, S. Perdikaris, D. Bonnissent, G. Robert, K. Debue, A. Evin, and S. Grouard. 2021. Archaeobiogeography of extinct rice rats (Oryzomyini) in the Lesser Antilles during the Ceramic Age (500 BCE–1500 CE). Holocene 31:433–445.

Egler, F.E. 1952. The natural vegetation of the Windward and Leeward islands: A review. Caribbean Forester 13:174–175.

Emrich, M.A., E.L. Clare, W.O.C. Symondson, S.E. Koenig, and M.B. Fenton. 2014. Resource partitioning by insectivorous bats in Jamaica. Molecular Ecology 23:3648–3656.

Fenton, M.B., P.A. Faure, and J.M. Ratcliffe. 2012. Evolution of high duty cycle echolocation in bats. Journal of Experimental Biology 215:2935–2944.

Fernández-Palacios, J.M., H. Kreft, S.D.H. Irl, S. Norder, C. Ah-Peng, P.A.V. Borges, K.C. Burns, L. de Náscimento, J.-Y. Meyer, E. Montes, and D.R. Drake. 2021. Scientists' warning—The outstanding biodiversity of islands is in peril. Global Ecology and Conservation 31:e01847. https://doi.org/10.1016/j.gecco.2021.e01847.

Fleming, T.H. 1982. Parallel trends in the species diversity of West Indian birds and bats. Oecologia 53:56–60.

Fleming, T.H., and K.L. Murray. 2009. Population and genetic conse-

quences of hurricanes for three species of West Indian phyllostomid bats. Biotropica 41:250–256.

Fleming, T.H., L.M. Dávalos, and M.A.R. Mello, eds. 2020. Phyllostomid bats: A unique mammalian radiation. Chicago: University of
Chicago Press.

Fleming, T.H., K.L. Murray, and B.C. Carstens. 2009. Phylogeogra-
phy and genetic structure of three evolutionary lineages of West
Indian phyllostomid bats. *In* T.H. Fleming and P.A. Racey, eds.,
Island bats: Evolution, ecology and conservation, pp. 116–150.
Chicago: University of Chicago Press.

Fooks, A.R., F. Cliquet, S. Finke, C. Freuling, T. Hemachudha, R.S.
Mani, T. Müller, S. Nadin-Davis, E. Picard-Meyer, H. Wilde, and
A.C. Banyard. 2017. Rabies. Nature Reviews Disease Primers
3(17091): 1–19. https://doi.org/10.1038/nrdp.2017.91.

Frank, E.F. 1998. History of the guano mining industry, Isla de
Mona, Puerto Rico. Journal of Cave and Karst Studies 60:
121–125.

Fraser, E., A. Silvis, R.M. Brigham, and Z.J. Czenze, eds. 2020. Bat
echolocation research: A handbook for planning and conduct-
ing acoustic studies. 2nd ed. Austin, Texas: Bat Conservation
International.

Frick, W.F., T. Kingston, and J. Flanders. 2020. A review of the major
threats and challenges to global bat conservation. Annals of the
New York Academy of Science 1469:5–25.

Friedenberg, N.A., and W.F. Frick. 2021. Assessing fatality minimi-
zation for hoary bats amid continued wind energy development.
Biological Conservation 262(109309). https://doi.org/10.1016/j
.biocon.2021.109309.

Fundora Caballero, D., A. Espinosa Lima, and A. Espinosa Romo.
2021. Quiropterofauna del área protegida Loma de Santa María,
Ciego de Ávila, Cuba [The bat fauna of the Loma de Santa María
protected area, Ciego de Ávila, Cuba]. Revista Cubana de Ciencias
Biológicas 9:1–4.

Furey, N.M., and P.A. Racey. 2016. Conservation ecology of cave bats.
In C.C. Voigt and T. Kingston, eds., Bats in the Anthropocene:
Conservation of bats in a changing world, pp. 463–500. New York:
SpringerOpen.

Gallant, L.R., M.B. Fenton, C. Grooms, W. Bogdanowicz, R.S. Stewart, E.L. Clare, J.P. Smol, and J.M. Blais. 2021. A 4,300-year history of dietary changes in a bat roost determined from a tropical

guano deposit. Journal of Geophysical Research: Biosciences 126:e2020JG006026. https://doi.org/10.1029/2020JG006026.

Gallant, L.R, C. Grooms, L.E. Kimpe, J.P. Smol, W. Bogdanowicz, R.S. Stewart, E.L. Clare, M.B. Fenton, and J.M. Blais. 2019. A bat guano deposit in Jamaica recorded agricultural changes and metal exposure over the last >4300 years. Palaeogeography, Palaeoclimatology, Palaeoecology 538:e109470. https://doi.org/10.1016/j.palaeo.2019.109470.

Gannon, M.R. 1991. Foraging ecology, reproductive biology, and systematics of the red fig-eating bat (*Stenoderma rufum*) in the tabonuco rain forest of Puerto Rico. PhD dissertation, Texas Tech University, Lubbock.

Gannon, M.R., and M.R. Willig. 1994. The effects of Hurricane Hugo on bats of the Luquillo Experimental Forest of Puerto Rico. Biotropica 26:320–331.

Gannon, M.R., and M.R. Willig. 2009. Island in the storm: Disturbance ecology of plant-visiting bats on the hurricane-prone island of Puerto Rico. *In* T.H. Fleming and P.A. Racey, eds., Island bats: Evolution, ecology and conservation, pp. 281–301. Chicago: University of Chicago Press.

Gannon, M.R., A. Kurta, A. Rodríguez-Durán, and M.R. Willig. 2005. Bats of Puerto Rico: An island focus and a Caribbean perspective. Lubbock: Texas Tech University Press.

García-Arévalo, M.A. 2019. Tainos, arte y sociedad [Tainos, art and society]. Santo Domingo, Dominican Republic: Banco Popular Dominicano.

García Martinó, A.R., G.S. Warner, F.N. Scatena, and D.L. Civico. 1996. Rainfall and elevation relationships in the Luquillo Mountains of Puerto Rico. Caribbean Journal of Science 32:413–424.

García-Rivera, L., and C.A. Mancina. 2011. Murciélagos insectívoros [Insectivorous bats]. *In* R. Borroto-Páez and C.A. Mancina, eds., Mamíferos en Cuba. [Mammals in Cuba], pp. 149–165. Vasa, Finland: UPC Print.

Geiling, N. 2019. The best places around the world to see bats (by the millions). Smithsonian Magazine. https://www.smithsonianmag.com/travel/best-places-around-world-see-bats-180953185.

Geiser, F., and C. Stawski. 2011. Hibernation and torpor in tropical and subtropical bats in relation to energetics, extinction, and the evolution of endothermy. Integrative and Comparative Biology 51:337–348.

Genoways, H.H. 2001. Review of Antillean bats of the genus *Ariteus*. Occasional Papers of the Museum, Texas Tech University 206:1–11.

Genoways, H.H., and R.J. Baker. 1975. A new species of *Eptesicus* from Guadeloupe, Lesser Antilles (Chiroptera: Vespertilionidae). Occasional Papers of the Museum, Texas Tech University 34:1–7.

Genoways, H.H., and J.K. Jones, Jr. 1975. Additional records of the stenodermine bat, *Sturnira thomasi*, from the Lesser Antillean island of Guadeloupe. Journal of Mammalogy 56:924–925.

Genoways, H.H., C.J. Phillips, and R.J. Baker. 1998. Bats of the Antillean island of Grenada: A new zoogeographic perspective. Occasional Papers of the Museum, Texas Tech University 177:1–28.

Genoways, H.H., R.J. Baker, J.W. Bickham, and C.J. Phillips. 2005. Bats of Jamaica. Special Publications of the Museum, Texas Tech University 48:1–154.

Genoways, H.H., P.A. Larsen, S.C. Pedersen, and J.J. Huebschman. 2009. Bats of Saba, Netherlands Antilles: A zoogeographic perspective. Acta Chiropterologica 9:97–114.

Genoways, H.H., S.C. Pedersen, C.J. Phillips, and L.K. Gordon. 2007b. Bats of Anguilla, Northern Lesser Antilles. Occasional Papers of the Museum, Texas Tech University 270:1–12.

Genoways, H.H., R.J. Larsen, S.C. Pedersen, G.G. Kwiecinski, and P.A. Larsen. 2011. Bats of Barbados. Chiroptera Neotropical 17:1029–1054.

Genoways, H.H., S.C. Pedersen, P.A. Larsen, G.G. Kwiecinski, and J. Huebschman. 2007a. Bats of Saint Martin, French West Indies/ Sint Maarten, Netherlands Antilles. Mastozoología Neotropical 14:169–188.

Genoways, H.H., R.M. Timm, R.J. Baker, C.J. Phillips, and D.A. Schlitter. 2001. Bats of the West Indian island of Dominica: Natural history, areography, and trophic structure. Special Publications of the Museum, Texas Tech University 43:1–42.

Genoways, H.H., G.G. Kwiecinski, P.A. Larsen, S.C. Pedersen, R.J. Larsen, J.D. Hoffman, M. de Silva, C.J. Phillips, and R.J. Baker. 2010. Bats of the Grenadine Islands, West Indies, and placement of Koopman's Line. Chiroptera Neotropical 16:501–521.

Gessinger, G., R. Page, L. Wilfert, A. Surlykke, S. Brinkløv, and M. Tschapka. 2021. Phylogenetic patterns in mouth posture and echolocation emission behavior of phyllostomid bats. Frontiers in Ecology and Evolution. https://doi.org/10.3389/fevo.2021.630481.

Ghanem, S.H., and C.C. Voigt. 2012. Increasing awareness of ecosystem services provided by bats. Advances in the Study of Behavior 44:279–302.

Gile, P.L., and J.O. Carrero. 1918. The bat guanos of Porto Rico and their fertilizing value. Porto Rico Agricultural Experiment Station Bulletin 25:1–66.

Gleason, H.A., and M.T. Cook. 1926. Plant ecology of Porto Rico. Scientific Survey of Porto Rico and the Virgin Islands. New York: New York Academy of Science.

Gomes, G.A., and F.A. Reid. 2015. Bats of Trinidad and Tobago. A field guide and natural history. Port of Spain, Trinidad and Tobago: Trinibat.

Gonçalves, F., L.P. Sales, M. Galetti, and M.M. Pires. 2021. Combined impacts of climate and land use change and the future restructuring of Neotropical bat diversity. Perspectives in Ecology and Conservation 19:454–463.

González-Gutiérrcz, K., J.H. Castaño, J. Pérez-Torres, and H.R. Mosquera-Mosquera. 2022. Structure and roles in pollination networks between phyllostomid bats and flowers: A systematic review for the Americas. Mammalian Biology. https://doi.org/10.1007/s42991-021-00202-6.

Good, R. 1953. The geography of flowering plants. London, United Kingdom: Longmans, Green and Co.

Goodwin, G.G., and A.M. Greenhall. 1961. A review of the bats of Trinidad and Tobago: Descriptions, rabies infection, and ecology. Bulletin of the American Museum of Natural History 122:187–302.

Goodwin, R.E. 1970. The ecology of Jamaican bats. Journal of Mammalogy 51:571–579.

Gorresen, P.M., P.M. Cryan, D.C. Dalton, S. Wolf, and F. Bonaccorso. 2015. Ultraviolet vision may be widespread in bats. Acta Chiropterologica 17:193–198.

Gosse, P.H. 1851. A naturalist's sojourn in Jamaica. London, United Kingdom: Longman, Brown, Green, and Longman.

Gotch, A.F. 1996. Latin names explained: A guide to the scientific classification of reptiles, birds and mammals. New York: Facts on File.

Griffiths, T.A., and Klingener, D. 1998. On the distribution of Greater Antillean bats. Biotropica 20:240–251.

Grindal, S.D. 2004. Notes on the natural history of bats on Barbados. Journal of the Barbados Museum and Historical Society 50:9–27.

Gunnell, G.F., and N.B. Simmons. 2005. Fossil evidence and the origin of bats. Journal of Mammalian Evolution 12:209–246.

Gustin, M.K., and G.F. McCracken. 1987. Scent recognition between females and pups in the bat *Tadarida brasiliensis mexicana*. Animal Behaviour 35:13–19.

Handley, C.O., Jr. 1976. Mammals of the Smithsonian Venezuelan Project. Brigham Young University Science Bulletin 20:1–89.

Harper, C.J., S.M. Swartz, and E.L. Brainerd. 2013. Specialized bat tongue is a hemodynamic nectar mop. Proceedings of the National Academy of Sciences of the United States of America 110:8852–8857.

Hayward, C.E. 2013. DNA barcoding expands dietary identification and reveals dietary similarity in Jamaican frugivorous bats. MS thesis, University of Western Ontario, London, Ontario, Canada.

Hedges, S.B., R. Powell, R.W. Henderson, S. Hanson, and J.C. Murphy. 2019. Definition of the Caribbean Islands biogeographic region, with checklist and recommendations for standardized common names of amphibians and reptiles. Caribbean Herpetology 67:1–53.

Hein, C.D., and M.R. Schirmacher. 2016. Impact of wind energy on bats: A summary of our current knowledge. Human-Wildlife Interactions 10:19–27.

Heinicke, M.P., W.E. Duellman, and S.B. Hedges. 2007. Major Caribbean and Central American frog faunas originated by ancient oceanic dispersal. Proceedings of the National Academy of Sciences of the United States of America 104:10092–10097.

Henareh Khalyani, A., W.A. Gould, E. Harmsen, A. Terando, M. Quiñones, and J.A. Collazo. 2016. Climate change implications for tropical islands: Interpolating and interpreting statistically downscaled GCM projections for management and planning. Journal of Applied Meteorology and Climatology 55:265–282.

Henson, O.W., Jr., and A. Novick. 1966. An additional record of the bat, *Phyllonycteris aphylla*. Journal of Mammalogy 47:351–352.

Hernández Ayala, J.J., and M. Heslar. 2019. Examining the spatio-temporal characteristics of droughts in the Caribbean using the standardized precipitation index (SPI). Climate Research 78:102–116.

Higman, B.W. 2011. A concise history of the Caribbean. New York: Cambridge University Press.

Hill, J.E., and P.G.H. Evans. 1985. A record of *Eptesicus fuscus* (Chiroptera: Vespertilionidae) from Dominica, West Indies. Mammalia 49:133–136.

Hirsbrunner, A., A. Rodríguez-Durán, J. Jarvis, R. Rudd, and A. Davis. 2020. Detection of rabies viral neutralizing antibodies in the Puerto Rican *Brachyphylla cavernarum*. Infection Ecology and Epidemiology 10:1840773. https://doi.org/10.1080/20008686.2020.1840773.

Hoffman, J.D., G.D. Kadlubar, S.C. Pedersen, R.J. Larsen, P.A. Larsen, C.J. Phillips, G.G. Kwiecinski, and H.H. Genoways. 2019. Predictors of bat species richness within the islands of the Caribbean basin. Special Publications of the Museum, Texas Tech University 71:337–350.

Hoffmann, S., D. Genzel, S. Prosch, L. Baier, S. Weser, L. Wiegrebe, and U. Firzlaff. 2015. Biosonar navigation above water I: Estimating flight height. Journal of Neurophysiology 113:1135–1145.

Holdridge, L.R. 1967. Life zone ecology. San José, Costa Rica: Tropical Science Center.

Horowitz, M.M., ed. 1971. Peoples and cultures of the Caribbean. An anthropological reader. Garden City, New York: The Natural History Press.

Ibéné, B., F. Leblanc, and C. Pentier. 2007. Contribution à l'étude des chiroptères de la Guadeloupe [Contribution to the study of bats from Guadeloupe]. Rapport final 2006. Diren, Guadeloupe: L'ASFA et Groupe Chiroptères Guadeloupe. https://docplayer.fr/64236417-Contribution-a-l-etude-des-chiropteres-de-la-guadeloupe.html.

Instituto de Investigación de Recursos Biológicos Alexander von Humboldt. 2021. Expedición Cangrejo Negro, Isla de Providencia, 17 al 24 Enero 2021 [Expedition Black Crab, Providence Island, 17 to 24 January 2021]. http://repository.humboldt.org.co/handle/20.500.11761/35734?show=full.

International Union for the Conservation of Nature. 2012. Threats classification scheme. Version 3.2. https://www.iucnredlist.org/resources/threat-classification-scheme. Accessed 27 November 2021.

International Union for the Conservation of Nature. 2016. Global standard for the identification of Key Biodiversity Areas. https://portals.iucn.org/library/node/46259. Accessed 27 November 2021.

International Union for the Conservation of Nature. 2019. Guidelines for using the IUCN Red List categories and criteria. Version 14. http://www.iucnredlist.org/documents/RedListGuidelines.pdf. Accessed 27 November 2021.

Iturralde-Vinent, M.A., and R.D.E. MacPhee. 1999. Paleogeography of the Caribbean region: Implications for Cenozoic biogeography. Bulletin of the American Museum of Natural History 95:1–95.

Jakobsen, L., S. Brinkløv, and A. Surlykke. 2013. Intensity and directionality of bat echolocation signals. Frontiers in Physiology 4(89): 1–9. https://doi.org/10.3389/fphys.2013.00089.

Jennings, L.N., J. Douglas, E. Treasure, and G. González. 2014. Climate change effects in El Yunque National Forest, Puerto Rico, and the Caribbean region. General Technical Report SRS-193. Asheville, North Carolina: Southern Research Station.

Jones, G., and M. Holderied. 2007. Bat echolocation calls: Adaptation and convergent evolution. Proceedings of the Royal Society B 274:905–912.

Jones, J.K., Jr. 1978. A new bat of the genus *Artibeus* from the Lesser Antillean island of Saint Vincent. Occasional Papers of the Museum, Texas Tech University 5:1–6.

Jones, J.K., Jr., and R.J. Baker. 1978. *Chiroderma improvisum*. Mammalian Species 134:1–2.

Jones, J.K., Jr., and R.J. Baker. 1979. Notes on a collection of bats from Montserrat, Lesser Antilles. Occasional Papers of the Museum, Texas Tech University 60:1–6.

Jones, J.K., Jr., and H.H. Genoways. 1973. *Ardops nichollsi*. Mammalian Species 24:1–2.

Jones, K.E., and A. Purvis. 1997. An optimum body size for mammals? Comparative evidence from bats. Functional Ecology 11:751–756.

Jones, K.E., K.E. Barlow, N. Vaughan, A. Rodríguez-Durán, and M.R. Gannon. 2001. Short-term impacts of extreme environmental disturbance on the bats of Puerto Rico. Animal Conservation 4:59–66.

Jones, T.S. 1951. Bat records from the islands of Grenada and Tobago, British West Indies. Journal of Mammalogy 32:223–224.

Jung, K., J. Molinari, and E.K.V. Kalko. 2014. Driving factors for the evolution of species-specific echolocation call design in New World free-tailed bats (Molossidae). PLoS One 9(1): e85279. https://doi.org/10.1371/journal.pone.0085279.

Kairo, M., B. Ali, O. Cheesman, K. Haysom, and S. Murphy. 2003. Invasive species threats in the Caribbean region. Report to the Nature Conservancy. Curepe, Trinidad and Tobago: CAB International.

Kauffman, C. 2007. Histoplasmosis: A clinical and laboratory update. Clinical Microbiology Reviews 20:115–32.

Keeley, A.T.H., and B.W. Keeley. 2004. The mating system of *Tadarida brasiliensis* (Chiroptera: Molossidae) in a large highway bridge colony. Journal of Mammalogy 85:113–119.

Kerridge, D.C., and R.J. Baker. 1978. *Natalus micropus*. Mammalian Species 114:1–3.

Klingener, D., H.H. Genoways, and R.J. Baker. 1978. Bats from southern Haiti. Annals of the Carnegie Museum 47:81–99.

Koenig, S., and L. Dávalos. 2015. *Phyllonycteris aphylla*. The IUCN Red List of Threatened Species 2015:e.T17173A22133396. https://dx.doi.org/10.2305/IUCN.UK.2015-4.RLTS.T17173A22133396.en. Accessed 15 December 2021.

Koopman, K.F. 1959. The zoogeographical limits of the West Indies. Journal of Mammalogy 40:236–240.

Koopman, K.F. 1968. Taxonomic and distributional notes on Lesser Antillean bats. American Museum Novitates 2333:1–13.

Koopman, K.F. 1975. Bats of the Virgin Islands in relation to those of the Greater and Lesser Antilles. American Museum Novitates 2581:1–7.

Koopman, K.F. 1977. Zoogeography. *In* R.J. Baker, J.K. Jones, Jr., and D.C. Carter, eds., Biology of bats of the New World Family Phyllostomidae, Part 1, pp. 39–47. Lubbock, Texas: The Museum, Texas Tech University.

Koopman, K.F. 1989. A review and analysis of the bats of the West Indies. *In* C.A. Woods, ed., Biogeography of the West Indies: Past, present, and future, pp. 635–644. Gainesville, Florida: Sandhill Crane Press.

Koopman, K.F., and E.E. Williams. 1951. Fossil Chiroptera collected by H.E. Anthony in Jamaica, 1919–1920. American Museum Novitates 1519:1–28.

Koopman, K.F., M.K. Hecht, and E. Ledecky-Janecek. 1957. Notes on the mammals of the Bahamas with special reference to the bats. Journal of Mammalogy 38:164–174.

Kunz, T.H. 1982. Roosting ecology of bats. *In* T.H. Kunz, ed., Ecology of bats, pp. 1–55. New York: Plenum Press.

Kunz, T.H., E. Braun de Torrez, D.M. Bauer, T.A. Lobova, and T.H. Fleming. 2011. Ecosystem services provided by bats. Annals of the New York Academy of Science 1223:1–38.

Kunz, T.H., S.A. Gauthreaux, Jr., N.I. Hristov, J.W. Horn, G. Jones, E.K.V. Kalko, R.P. Larkin, G.F. McCracken, S.M. Swartz, R.B. Srygley, R. Dudley, J.K. Westbrook, and M. Wikelski. 2008. Aero-ecology: Probing and modeling the aerosphere. Integrative and Comparative Biology 48:1–11.

Kurta, A. 1985. External insulation available to a non-nesting mammal, the little brown bat (*Myotis lucifugus*). Comparative Biochemistry and Physiology 82A:413–420.

Kurta, A., and S.M. Smith. 2014. Hibernating bats and abandoned mines in the Upper Peninsula of Michigan. Northeastern Naturalist 21:587–605.

Kurta, A., G.P. Bell, K.A. Nagy, and T.H. Kunz. 1989. Energetics of pregnancy and lactation in free-ranging little brown bats (*Myotis lucifugus*). Physiological Zoology 62:804–818.

Kurta, A., J.O. Whitaker, Jr., W. Wrenn, and J.A. Soto-Centeno. 2007. Ectoparasitic assemblages on mormoopid bats from Puerto Rico. Journal of Medical Entomology 44:953–958.

Kwiecinski, G.G., and W.C. Coles. 2007. Presence of *Stenoderma rufum* beyond the Puerto Rican Bank. Occasional Papers of the Museum, Texas Tech University 266:1–9.

Kwiecinski, G.G., J.-P. Bacle, K.C. Lindsay, and H.H. Genoways. 2010. New records of bats from the British Virgin Islands. Caribbean Journal of Science 46:64–70.

Kwiecinski, G.G., S.C. Pedersen, H.H. Genoways, P.A. Larsen, R.J. Larsen, J.D. Hoffman, F. Springer, C.J. Phillips, and R.J. Baker. 2018. Bats of St. Vincent, Lesser Antilles. Special Publications of the Museum, Texas Tech University 68:1–68.

Ladle, R.J., J.V.L. Firmino, A.C.M. Malhado, and A. Rodríguez-Durán. 2012. Unexplored diversity and conservation potential of Neotropical hot caves. Conservation Biology 26:978–982.

Laird, T. 2018. Bats. London, United Kingdom: Reaction Books.

Lambeck, K., and J. Chappell. 2001. Sea level change through the last glacial cycle. Science 292:679–686.

Larsen, P.A., H.H. Genoways, and S.C. Pedersen. 2006. New records of bats from Saint Barthélemy, French West Indies. Mammalia 70:321–325.

Larsen, P.A., S.R. Hoofer, M.C. Bozeman, S.C. Pedersen, H.H. Genoways, C.J. Phillips, D.E. Pumo, and R.J. Baker. 2007. Phylogenetics and phylogeography of the *Artibeus jamaicensis* complex based on cytochrome-b DNA sequences. Journal of Mammalogy 88:712–727.

Larsen, P.A., M.R. Marchán-Rivadeneira, and R.J. Baker. 2010. Natural hybridization generates mammalian lineage with species characteristics. Proceedings of the National Academy of Sciences of the United States of America 107:11447–11452.

Larsen, P.A., L. Siles, S.C. Pedersen, and G.G. Kwiecinski. 2011. A new species of *Micronycteris* (Chiroptera: Phyllostomidae) from Saint Vincent, Lesser Antilles. Mammalian Biology 76:687–700.

Larsen, R.J., K.A. Boegler, H.H. Genoways, W.P. Masefield, R.A. Kirsch, and S.C. Pedersen. 2007. Mist netting bias, species accumulation curves, and the rediscovery of two bats on Montserrat (Lesser Antilles). Acta Chiropterologica 9:423–435.

Larsen, R.J., P.A. Larsen, H.H. Genoways, F.M. Catzeflis, K. Geluso, G.G. Kwiecinski, S.C. Pedersen, F. Simal, and R.J. Baker. 2012. Evolutionary history of Caribbean species of *Myotis*, with evidence of a third Lesser Antillean endemic. Mammalian Biology 77:124–134.

Larsen, R.J., P.A. Larsen, C.D. Phillips, H.H. Genoways, G.G. Kwiecinski, S.C. Pedersen, C.J. Phillips, and R.J. Baker. 2017. Patterns of morphological and molecular evolution in the Antillean tree bat, *Ardops nichollsi* (Chiroptera: Phyllostomidae). Occasional Papers of the Museum, Texas Tech University 345:1–28.

Lavery, T., and J. Fasi. 2017. Buying through your teeth: Traditional currency and conservation of flying foxes *Pteropus* spp. in Solomon Islands. Oryx 53:1–8.

Ledrú, A.P. 1810. Voyage aux îles de Ténériffe, la Trinité, Saint-Thomas, Sainte-Croix et Porto Rico, exécuté par ordre du gouvernement Français, depuis le 30 Septembre 1796 jusqu'au 7 Juin 1798, sous la direction du Capitaine Baudin, pour faire des recherches et des collections relatives à l'histoire naturelle [Voyage to the islands of Tenerife, Trinidad, Saint Thomas, Saint Croix and Puerto Rico, executed under orders of the French government, from 30 September 1796 to 7 June 1798, under the direction of Captain Baudin, to carry out research and collections relating to natural history]. Paris, France: Arthus Bertrand.

Lee, H. 1960. To kill a mockingbird. Philadelphia: J.B. Lippincott.

Leiser-Miller, L.B, and S.E. Santana. 2020. Morphological diversity in the sensory system of phyllostomid bats: Implications for acoustic and dietary ecology. Functional Ecology 34:1416–1427.

Lenoble, A. 2019. The past occurrence of the Guadeloupe big-eyed bat *Chiroderma improvisum* Baker and Genoways, 1976 on Marie-Galante (French West Indies) with comments on bat remains from pre-Columbian sites in the eastern Caribbean. Acta Chiropterologica 21:299–308.

Lenoble, A., B. Angin, J.P. Huchet, and A. Royer. 2014a. Seasonal insectivory of the Antillean fruit-eating bat (*Brachyphylla cavernarum*). Caribbean Journal of Science 48:127–131.

Lenoble, A., B. Corentin, T. Bos, E. Discamps, and A. Queffelec. 2014b. Predation of lesser naked-backed bats (*Pteronotus davyi*) by a pair of American kestrels (*Falco sparverius*) on the island of Marie-Galante, French West Indies. Journal of Raptor Research 48:78–81.

Lim, B.K., L.O. Loureiro, and G.S.T. Garbino. 2020. Cryptic diversity and range extension in the big-eyed bat genus *Chiroderma* (Chiroptera, Phyllostomidae). ZooKeys 918:41–63.

Lindsay, K.C., J.-P. Bacle, and G.G. Kwiecinski. 2009. A bat conservation and management plan for St. Thomas and St. John, U.S. Virgin Islands. Saint Thomas, U.S. Virgin Islands: Island Resources Foundation.

Lindsay, K.C., G.G. Kwiecinski, S.C. Pedersen, J.-P. Bacle, and H.H. Genoways. 2010. First record of *Ardops nichollsi* from Antigua, Lesser Antilles. Mammalia 74:93–95.

Lobova, T., C.K. Geiselman, and S.A. Mori. 2009. Seed dispersal by bats in the Neotropics. Memoirs of the New York Botanical Garden 101:1–471.

López-González, C., and C. Ocampo-Ramírez. 2021. External ears in Chiroptera: Form-function relationships in an ecological context. Acta Chiropterologica 23:525–545.

López Marrero, T., K. Yamane, T. Heartsill Scalley, and N. Villanueva-Colón. 2012. The various shapes of the insular Caribbean: Population and environment. Caribbean Studies 40:17–37.

Loureiro, L., B.K. Lim, and M.D. Engstrom. 2018. A new species of mastiff bat (Chiroptera, Molossidae, *Molossus*) from Guyana and Ecuador. Mammalian Biology 90:10–21.

Loureiro, L.O., M.D. Engstrom, and B.K. Lim. 2019a. Comparative phylogeography of mainland and insular species of Neotropical molossid bats (*Molossus*). Ecology and Evolution 10:389–409.

Loureiro, L.O., M.D. Engstrom, and B.K. Lim. 2019b. Not all *Molossus* are created equal: Genetic variation in the mastiff bat reveals diversity masked by conservative morphology. Acta Chiropterologica 21:51–64.

Loureiro, L.O., M.D. Engstrom, and B.K. Lim. 2020. Does evolution of echolocation calls and morphology in *Molossus* result from convergence or stasis? PLoS One 15(9): e0238261. https://doi.org/10.1371/journal.pone.0238261.

Low, M.-R., W.Z. Hoong, Z. Shen, B. Murugavel, N. Mariner, I. M. Paguntalan, K. Tanalgo, M.M. Aung, Sheherazade, L.A. Bansa, T. Sritongchuay, J.H. Preble, and S.A. Aziz. 2021. Bane or blessing? Reviewing cultural values of bats across the Asia-Pacific region. Journal of Ethnobiology 41:18–34.

Lugo, A.E. 1991. Dominancia y diversidad de plantas en Isla de Mona [Dominance and diversity of plants on Mona Island]. Acta Científica 5:65–71.

Lugo, A.E. 1996. Caribbean island landscapes: Indicators of the effects of economic growth on the region. Environment and Development Economics 1:128–136.

Lugo, A.E. 2019. Social-ecological-technological effects of Hurricane María on Puerto Rico. Planning for resilience under extreme events. Cham, Switzerland: Springer.

Lugo, A.E. 2020. Effects of extreme disturbance events: From ecesis to social-ecological-technological systems. Ecosystems 23:1726–1747.

Lugo, A.E., T.A. Carlo, and J.M. Wunderle, Jr. 2012a. Natural mixing of species: Novel plant-animal communities on Caribbean islands. Animal Conservation 15:233–241.

Lugo, A.E., E.H. Helmer, and E. Santiago-Valentín. 2012b. Caribbean landscapes and their biodiversity. Interciencia 37:705–710.

Lugo, A.E., and E. Medina. 2014. Mangrove forests. *In* Y. Wang, ed., Encyclopedia of natural resources. Volume 1, Land, pp. 343–352. New York: Taylor and Francis.

Lugo, A.E., E. Medina, E. Cuevas, and O. Ramos-González. 2019. Ecological and physiological aspects of Caribbean shrublands. Caribbean Naturalist 58:1–35.

Lugo, A.E., M. Quiñones, and O.M.R. González. 2016. Islas, islotes y cayos del archipiélago puertorriqueño [Islands, islets and keys of the Puerto Rican archipelago]. Corriente Verde 7:24–28.

Lugo, A.E., K.M. Winchell, and T.A. Carlo. 2018. Novelty in ecosystems. *In* D.A. DellaSala and M.I. Goldstein, eds., The Encyclopedia of the Anthropocene, pp. 259–271. Oxford, United Kingdom: Elsevier.

Lundberg, J., and D.A. McFarlane. 2009. Bats and bell holes: The microclimatic impact of bat roosting, using a case study from Runaway Bay Caves, Jamaica. Geomorphology 106:78–85.

Luo, B., L. Leiser-Miller, S.E. Santana, L. Zhang, L. Tong, Y. Xiao, Y. Liu, and J. Feng. 2019. Echolocation call divergence in bats: A comparative analysis. Behavioral Ecology and Sociobiology. 73(154). https://doi.org/10.1007/s00265-019-2766-9.

MacArthur, R.H., and E.O. Wilson. 1963. An equilibrium theory of insular zoogeography. Evolution 17:373–387.

MacArthur, R.H., and E.O. Wilson. 1967. The theory of island biogeography. Princeton, New Jersey: Princeton University Press.

Macías, S., and E.C. Mora. 2003. Variation of echolocation calls of *Pteronotus quadridens* (Chiroptera: Mormoopidae) in Cuba. Journal of Mammalogy 84:1428–1436.

Macías, S., E.C. Mora, and A. García. 2006a. Acoustic identification of mormoopid bats: A survey during the evening exodus. Journal of Mammalogy 87:324–330.

Macías, S., E.C. Mora, A. García, and Y. Macías. 2006b. Echolocation behavior of *Brachyphylla nana* (Chiroptera: Phyllostomidae) under laboratory conditions. Caribbean Journal of Science 42:114–120.

Macías, S., E.C. Mora, C. Koch, and O. von Helversen. 2009. Echolocation behaviour of *Phyllops falcatus* (Chiroptera: Phyllostomidae): Unusual frequency range of the first harmonic. Acta Chiropterologica 7:275–283.

MacSwiney González, M.C., F.M. Clarke, and P.A. Racey. 2008. What you see is not what you get: The role of ultrasonic detectors in increasing inventory completeness in Neotropical bat assemblages. Journal of Applied Ecology 45:1364–1371.

Maina, J.N. 2000. What it takes to fly: The structural and functional respiratory refinements in birds and bats. Journal of Experimental Biology 203:3045–3064.

Maine, J.J., and J.G. Boyles. 2015. Bats initiate vital agroecological interactions in corn. Proceedings of the National Academy of Sciences of the United States of America 112:12438–12443.

Mancina, C.A. 2008. Effect of moonlight on nocturnal activity of two Cuban nectarivores: The Greater Antillean long-tongued bat (*Monophyllus redmani*) and Poey's flower bat (*Phyllonycteris poeyi*). Bat Research News 49:71–74.

Mancina, C.A. 2010. *Phyllonycteris poeyi* (Chiroptera: Phyllostomidae). Mammalian Species 852:41–48.

Mancina, C.A. 2012. Mamíferos. *In* González Alonso, H., L. Rodríguez Schettino, A. Rodríguez, C.A. Mancina, and I. Ramos García, eds. Libro rojo de los vertebrados de Cuba [Red book of Cuban vertebrates], pp. 268–291. Havana, Cuba: Editorial Academia.

Mancina, C.A., and I. Castro-Arellano. 2013. Unusual temporal niche overlap in a phytophagous bat ensemble of western Cuba. Journal of Tropical Ecology 29:511–521.

Mancina, C.A, and L. García-Rivera. 2000. Notes on the natural history of *Phyllops falcatus* (Gray, 1893) (Phyllostomidae: Stenodermatinae) in Cuba. Chiroptera Neotropical 6:123–125.

Mancina, C.A., L. García-Rivera, and R.T. Capote. 2007a. Habitat use by phyllostomid bat assemblages in secondary forests of the "Sierra del Rosario" Biosphere Reserve, Cuba. Acta Chiropterologica 9:203–218.

Mancina, C.A., L. García-Rivera, and B.W. Miller. 2012. Wing morphology, echolocation, and resource partitioning in syntopic Cuban mormoopid bats. Journal of Mammalogy 93:1308–1317.

Mancina, C.A., L.M. Echenique-Díaz, A. Tejedor, L. García, Á. Daniel-Alvarez, and M.A. Ortega-Huerta. 2007b. Endemics under threat: An assessment of the conservation status of Cuban bats. Hystrix, 18:3–15.

Martin, P.H., T.J. Fahey, and R.E. Sherman. 2011. Vegetation zonation in a Neotropical montane forest: Environment, disturbance and ecotones. Biotropica 43:533–543.

Martin, P.H., R.E. Sherman, and T.J. Fahey. 2004. Forty years of tropical forest recovery from agriculture: Structure and floristics of secondary and old growth riparian forests in the Dominican Republic. Biotropica 36:297–317.

Martínez-Ferreira, S.R., M.Y. Alvarez-Añorve, A.E. Bravo-Monzón, C. Montiel-González, J.I. Flores-Puerto, S.P. Morales-Díaz, X. Chiappa-Carrara, K. Oyama, and L.D. Avila-Cabadilla. 2020. Taxonomic and functional diversity and composition of bats in a regenerating Neotropical dry forest. Diversity 12:332. https://doi.org/10.3390/d12090332.

Martino, A.M.G., D. Borges, and J.M. Nassar. 2019. Activity records of the endangered Paraguaná moustached bat, *Pteronotus paraguanensis*, in the main vegetation types of the Paraguaná Peninsula, Venezuela. Acta Chiropterologica 21:165–174.

Masson, D., and M. Breuil. 1992. Un *Myotis* (Chiroptera, Vespertilionidae) en Guadeloupe [A *Myotis* (Chiroptera, Vespertilionidae) in Guadeloupe]. Mammalia 56:473–475.

Masson, D., M. Breuil, and A. Breuil. 1990. Premier inventaire des chauves-souris de l'île de Marie-Galante (Antilles Françaises) [First survey of bats from the island of Marie-Galante (French Antilles)]. Mammalia 54:656–658.

Masson, D., A. Breuil, M. Breuil, F. Leboulenger, F. Leugé, and C. Masson. 1994. La place des chiroptères dans la dissémination, par endophytosporie, des plantes forestières de la Guadeloupe. Rapport final [The role of bats in the dispersion of the forest plants of Guadeloupe by means of seed ingestion. Final report]. Paris, France: Ministere de L'Environnement.

Mathis, V.L., and D.L. Reed. 2021. Two new recent country records of mormoopid bats (Chiroptera: Mormoopidae) from Long Island, The Bahamas. Florida Field Naturalist 49:13–17.

McCarthy, T.J., and R.W. Henderson. 1992. Confirmation of *Ardops nichollsi* on Marie Galante, Lesser Antilles, and comments on other bats. Caribbean Journal of Science 28:106–107.

McCracken, G.F., J.P. Hayes, J. Cevallos, S.Z. Guffey, and F.C. Romero. 1997. Observations on the distribution, ecology, and behaviour of bats on the Galapagos Islands. Journal of Zoology 243:757–770.

McCracken, G.F., K. Safi, T.H. Kunz, D.K. Dechmann, S.M. Swartz, and M. Wikelski. 2016. Airplane tracking documents the fastest flight speeds recorded for bats. Royal Society Open Science 3:160398. https://doi.org/10.1098/rsos.160398.

McDonough, M.M., L.K. Ammerman, R.M. Timm, and H.H. Genoways. 2008. Speciation within bonneted bats (genus *Eumops*): The complexity of morphological, mitochondrial, and nuclear data sets in systematics. Journal of Mammalogy 89:1306–1315.

McFarlane, D.A. 1985. The rat-bat caves of Jamaica. Terra 23:14–17.

McFarlane, D.A. 1986. Cave bats in Jamaica. Oryx 20:27–30.

McFarlane, D.A., and K.L. Garrett. 1989. The prey of common barn-owls (*Tyto alba*) in dry limestone scrub forest of southern Jamaica. Caribbean Journal of Science 25:1–23.

McNab, B.K. 1976. Seasonal fat reserves of bats in two tropical environments. Ecology 57:332–338.

McNab, B.K. 2009. Physiological adaptations of bats and birds to island life. *In* T.H. Fleming and P.A. Racey, eds., Island bats: Evolution, ecology, and conservation, pp. 153–175. Chicago: University of Chicago Press.

Medellín, R.A. 2019. Family Noctilionidae (bulldog bats). *In* D.E. Wilson, and R.A. Mittermeier, eds., Handbook of the mammals of the world. Volume 9, Bats, pp. 404–411. Barcelona, Spain: Lynx Edicions.

Mendes, P., and A.C. Srbek Araujo. 2020. Effects of land-use changes on Brazilian bats: A review of current knowledge. Mammal Review 51:127–142.

Miller, G.S. 1931. The red bats of the Greater Antilles. Journal of Mammalogy 12:409–410.

Mollentze, N., and D.G. Streicker. 2020. Viral zoonotic risk is homogenous among taxonomic orders of mammalian and avian reservoir hosts. Proceedings of the National Academy of Sciences of the United States of America 117:9423–9430.

Mora, E.C., and S. Macías. 2007. Echolocation calls of Poey's flower bat (*Phyllonycteris poeyi*) unlike those of other phyllostomids. Naturwissenschaften 94:380–383.

Mora, E.C., and L. Torres. 2008. Echolocation in the large molossid bats *Eumops glaucinus* and *Nyctinomops macrotis*. Zoological Science 25:1–6.

Mora, E.C., A. Rodríguez, S. Macías, I. Quiñonez, and M.M. Mellado. 2005. The echolocation behaviour of *Nycticeius cubanus* (Chiroptera: Vespertilionidae): Inter- and intra-individual plasticity in vocal signatures. Bioacoustics 15:175–193.

Mora, E.C., C. Ibáñez, S. Macías, J. Juste, I. López, and L. Torres. 2011. Plasticity in the echolocation inventory of *Mormopterus minutus* (Chiroptera, Molossidae). Acta Chiropterologica 13:179–187.

Morales, A.E., M. De La Mora, and D. Piñeiro. 2018. Spatial and environmental factors predict skull variation and genetic structure in the cosmopolitan bat *Tadarida brasiliensis*. Journal of Biogeography 45:1529–1540.

Morales-Martínez, D.M., H.F. López-Arévalo, M. Vargas-Ramírez. 2021. Beginning the quest: Phylogenetic hypothesis and identification of evolutionary lineages in bats of the genus *Micronycteris* (Chiroptera, Phyllostomidae). ZooKeys 1028:135–159.

Moratelli, R., D.E. Wilson, R.L.M. Novaes, K.F. Helgen, and E.E. Gutiérrez. 2017. Caribbean *Myotis* (Chiroptera, Vespertilionidae), with description of a new species from Trinidad and Tobago. Journal of Mammalogy 98:1–15.

Moratelli, R., C. Burgin, V.C. Cláudio, R.L.M. Novaes, A. López-Baucells, and R. Haslauer. 2019. Family Vespertilionidae (vesper bats). *In* D.E. Wilson and R.A. Mittermeier, eds., Handbook of the mammals of the world. Volume 9, Bats, pp. 855–996. Barcelona, Spain: Lynx Edicions.

Morgan, C.N., R.M. Wallace, A. Vokaty, J.F.R Seetahal, and Y.J. Nakazawa. 2020. Risk modeling of bat rabies in the Caribbean islands. Tropical Medicine and Infectious Disease 5(35): 1–17. https://doi.org/10.3390/tropicalmed5010035.

Morgan, G.S. 1994. Mammals of the Cayman Isands. *In* M.A. Brunt and J.E. Davies, eds., The Cayman Islands: Natural history and biogeography, pp. 435–463. Boston: Kluwer Academic.

Morgan, G.S., and N.J. Czaplewski. 2003. A new bat (Chiroptera: Natalidae) from the early Miocene of Florida, with comments on natalid phylogeny. Journal of Mammalogy 84:729–752.

Muñoz, E., A.J. Busalacchi, S. Nigam, and A. Ruiz-Barradas. 2008. Winter and summer structure of the Caribbean low-level jet. Journal of Climate 21:1260–1276.

Muñoz-Romo, M., E. Herrera, and T.H. Kunz. 2008. Roosting behavior and group stability of the big fruit-eating bat *Artibeus lituratus* (Chiroptera: Phyllostomidae). Mammalian Biology 73:214–221.

Murray, K.L., and T.H. Fleming. 2008. Social structure and mating system of the buffy flower bat, *Erophylla sezekorni* (Chiroptera, Phyllostomidae). Journal of Mammalogy 89:1391–1400.

Murray, K.L., E. Fraser, C. Davy, T.H. Fleming, and M.B. Fenton. 2009. Characterization of the echolocation calls of bats from Exuma, Bahamas. Acta Chiropterologica 11:415–424.

Muscarella, R.A., K.L. Murray, D. Ortt, A.L. Russell, and T.H. Fleming. 2011. Exploring demographic, physical, and historical explanations for the genetic structure of two lineages of Greater Antillean bats. PLoS One 6(3): e17704. https://doi.org/10.1371/journal.pone.0017704.

Myers, N., R.A. Mittermeir, C.G. Mittermeir, G.A.B. da-Fonseca, and J. Kent. 2000. Biodiversity hotspots for conservation priorities. Nature 403:853–858.

Nassar, J.M., H. Beck, L. Sternberg, and T.H. Fleming. 2003. Dependence on cacti and agaves in nectar-feeding bats from Venezuelan arid zones. Journal of Mammalogy 84:106–116.

National Environment and Planning Agency. 2011. Bat management plan for Jamaica. Kingston, Jamaica: Ecosystems Management Branch, National Environment and Planning Agency.

Nellis, D.W., and C.P. Ehle. 1977. Observations on the behavior of *Brachyphylla cavernarum* in Virgin Islands. Mammalia 41:403–409.

Neumann, C.J., G.W. Cry, E.L. Caso, and B.R. Jarvinen. 1978. Tropical cyclones of the North Atlantic Ocean, 1871–1977. Asheville, North Carolina: National Climatic Center, U.S. Department of Commerce, National Oceanic and Atmospheric Administration.

Newman, B.C., S.E. Henke, S.E. Koenig, and R.L. Powell. 2016. Distribution and general habitat use analysis of the Jamaican boa (*Chilabothrus subflavus*). South American Journal of Herpetology 11:228–234.

Nicholson, K.E., R.E. Glor, J.J. Kolbe, A. Larson, S.B. Hedges, and J.B. Losos. 2005. Mainland colonization by island lizards. Journal of Biogeography 32:929–938.

Nieves-Rivera, A.M., C. Santos, F. Dugan, and E. Miller. 2009. Guanophilic fungi in three caves of southwestern Puerto Rico. International Journal of Speleology 38:61–70.

Norberg, U., and J. Rayner. 1987. Ecological morphology and flight in bats (Mammalia; Chiroptera): Wing adaptations, flight performance, foraging strategy and echolocation. Philosophical Transactions of the Royal Society B 316:335–427.

Norman, L.J., C. Dodsworth, D. Foresteire, and L. Thaler. 2021. Human click-based echolocation: Effects of blindness and age, and real-life implications in a 10-week training program. PLoS One 16(6): e0252330. https://doi.org/10.1371/journal.pone.0252330.

Núñez-Novas, M.S., and Y.M. León. 2011. Análisis de la colección de murciélagos (Mammalia: Chiroptera) del Museo Nacional de Historia Natural de Santo Domingo [Analysis of the collection of bats (Mammalia: Chiroptera) at the National Museum of Natural History of Santo Domingo]. Novitates Caribaea 4:109–119.

Núñez-Novas, M.S., M.S.R. Guzmán-Pérez, and A. Contreras-Puigbó. 2019. Guía de identificación de los murciélagos de La Española y anotaciones sobre las familias y especies [Guide to the identification of the bats from Hispaniola with notes on the families and species]. Novitates Caribaea 13:39–63.

Núñez-Novas, M.S., Y.M. León, J. Mateo, and L.M. Dávalos. 2014. Horas de éxodo y estacionalidad de los murciélagos en cuatro cuevas de República Domincana [Exit times and seasonality in bats at four caves in the Dominican Republic]. Novitates Caribaea 7:83–94.

Núñez-Novas, M.S., Y.M. León, J. Mateo, and L.M. Dávalos. 2016. Records of the cave-dwelling bats (Mammalia: Chiroptera) of Hispaniola with an examination of seasonal variation in diversity. Acta Chiropterologica 18:269–278.

Núñez-Novas, M.S., R. Torres, A. Rodríguez-Durán, and J. Zorilla. 2021. Spatial distribution of bat species on Hispaniola Island, the Greater Antilles. Acta Chiropterologica 23:443–454.

O'Brien, J. 2011. Bats of the western Indian Ocean islands. Animals 1:259–290.

Odum, H.T. 1970. Summary: An emerging view of the ecological systems at El Verde. In H.T. Odum and R.F. Pigeon, eds., A tropical rain forest, pp. 191–289. Washington, D.C.: Division of Technical Information, U.S. Atomic Energy Commission.

O'Farrell, M.J., C. Corben, and W.L. Gannon. 2000. Geographic variation in the echolocation calls of the hoary bat (Lasiurus cinereus). Acta Chiropterologica 2:185–196.

Oliver, J., and Y. Narganes. 2003. The zooarcheological remains from Juan Miguel Cave and Finca de Doña Rosa, Barrio Caguana,

Puerto Rico. Ritual edibles or quotidian meals? Proceedings of the International Congress on Caribbean Archaeology 20:227–242.

Orihuela, J., and A. Tejedor. 2012. Peter's ghost-faced bat *Mormoops megalophylla* (Chiroptera: Mormoopidae) from a pre-Columbian archeological deposit in Cuba. Acta Chiropterologica 14:63–72.

Orihuela, J., L.W. Viñola, and R.A. Viera. 2020a. New bat locality records from Cuba with emphasis on the province of Matanzas. Novitates Caribaea 15:96–116.

Orihuela J., L.W. Viñola, O. Jiménez Vázquez, A.M. Mychajliw, O. Hernández de Lara, L. Lorenzo, and J.A. Soto-Centeno. 2020b. Assessing the role of humans in Greater Antillean land vertebrate extinctions: New insights from Cuba. Quaternary Science Reviews. 249:106597. https://doi.org/10.1016/j.quascirev.2020.106597.

Orr, R.T., and G. Silva-Taboada. 1960. A new species of bat of the genus *Antrozous* from Cuba. Proceedings of the Biological Society of Washington 73:83–86.

Ortega, J., and I. Castro-Arellano. 2001. *Artibeus jamaicensis.* Mammalian Species 662:1–9.

Ortega J., B. Hernández-Chávez, A. Rizo-Aguilar, and J.A. Guerrero. 2010. Estructura social y composición temporal en una colonia de *Nyctinomops laticaudatus* (Chiroptera: Molossidae) [Social structure and temporal composition of a colony of *Nyctinomops laticaudatus* (Chiroptera: Molossidae)]. Revista Mexicana de Biodiversidad 81:853–862.

Ortega J., J.E. Maldonado, G.S. Wilkinson, H.T. Arita, and R.C. Fleischer. 2003. Male dominance, paternity, and relatedness in the Jamaican fruit-eating bat (*Artibeus jamaicensis*). Molecular Ecolgy 12:2409–2415.

Osburn, W. 1865. Notes on the Chiroptera of Jamaica. Proceedings of the Zoological Society of London 1865:61–85.

Panyutina, A.A., A.N. Kuznetsov, I.A. Volodin, A.V. Abramov, and I.B. Soldatova. 2017. A blind climber: The first evidence of ultrasonic echolocation in arboreal mammals. Integrative Zoology 12:172–184.

Pavan, A.C. 2019. Family Mormoopidae (ghost-faced bats, naked-backed bats and mustached bats). *In* D.E. Wilson and R.A. Mittermeier, eds., Handbook of the mammals of the world. Volume 9, Bats, pp. 424–435. Barcelona, Spain: Lynx Edicions.

Pavan A.C., and G. Marroig. 2016. Integrating multiple evidences in taxonomy: Species diversity and phylogeny of mustached bats (Mormoopidae: *Pteronotus*). Molecular Phylogenetics and Evolution 103:184–198.

Pedersen, S.C., H.H. Genoways, and P.W. Freeman. 1996. Notes on bats from Montserrat (Lesser Antilles) with comments concerning the effects of Hurricane Hugo. Caribbean Journal of Science 32:206–213.

Pedersen, S.C., T.E. Popowics, G.G. Kwiecinski, and D.E.B. Knudsen. 2012. Sublethal pathology in bats associated with stress and volcanic activity on Montserrat, West Indies. Journal of Mammalogy 93:1380–1392.

Pedersen, S.C., H.H. Genoways, M.N. Morton, J.W. Johnson, and S. Courts. 2003. Bats of Nevis, Northern Lesser Antilles. Acta Chiropterologica 5:251–267.

Pedersen, S.C., H.H. Genoways, M.N. Morton, G.G. Kwiecinski, and S.E. Courts. 2005. Bats of St. Kitts (St. Christopher), northern Lesser Antilles, with comments regarding capture rates of Neotropical bats. Caribbean Journal of Science 41:744–760.

Pedersen, S.C., H.H. Genoways, M.N. Morton, K.C. Lindsay, and J. Cindric. 2007. Bats of Barbuda, northern Lesser Antilles. Occasional Papers of the Museum, Texas Tech University 271:1–19.

Pedersen, S.C., G.G. Kwiecinski, H.H. Genoways, R.J. Larsen, P.A. Larsen, C.J. Phillips, and R.J. Baker. 2018a. Bats of Saint Lucia, Lesser Antilles. Special Publications of the Museum, Texas Tech University 69:1–61.

Pedersen, S.C., G.G. Kwiecinski, P.A. Larsen, M.N. Morton, R.A. Adams, H.H. Genoways, and V.J. Swier. 2009. Bats of Montserrat: Population fluctuation and response to hurricanes and volcanoes, 1978–2005. *In* T.H. Fleming and P.A. Racey, eds., Island bats: Evolution, ecology, and conservation, pp. 302–340. Chicago: University of Chicago Press.

Pedersen, S.C., P.A. Larsen, S.A. Westra, E. van Norren, W. Overman, G.G. Kwiecinski, and H.H. Genoways. 2018b. Bats of Sint Eustatius, Caribbean Netherlands. Occasional Papers of the Museum, Texas Tech University 353:1–24.

Pedersen, S.C., H.H. Genoways, M.N. Morton, V.J. Swier, P.A. Larsen, K.C. Lindsay, R.A. Adams, and J.D. Appino. 2006. Bats of Antigua, northern Lesser Antilles. Occasional Papers of the Museum, Texas Tech University 249:1–18.

Pelletier, M., E. Stoetzel, D. Cochard, and A. Lenoble. 2017. Sexual dimorphism in the pelvis of Antillean fruit-eating bat (*Brachyphylla cavernarum*) and its application to a fossil accumulation from the Lesser Antilles. Geobios 50:311–318.

Petit, S. 1997. The diet and reproductive schedules of *Leptonycteris curasoe curasoe* and *Glossophaga longirostris elongata* (Chiroptera: Glossophaginae) on Curaçao. Biotropica 29:214–233.

Picard, R., and F. Catzeflis. 2013. Première étude des chauves-souris dans les goyaveraies de Martinique [First study of bats in the guava orchards of Martinique]. *In* L. Vernier and M. Burac, eds., Biodiversité insulaire: La flore, la faune et l'homme dans les Petites Antilles [Island biodiversity: Flora, fauna, and man in the Lesser Antilles]. France: Direction de l'Environnement, de l'Aménagement et du Logement de Martinique et Université des Antilles et de la Guyane. https://www.biodiversite-martinique.fr/document /premiere-etude-des-chauves-souris dans-les-goyaveraies-de -martinique.

Pierson, E.D., W.E. Rainey, R.M. Warner, and C.C. White-Warner. 1986. First record of *Monophyllus* from Montserrat, West Indies. Mammalia 50:269–271.

Pietsch, T.W., and B. Marx. 2021. Charles Plumier's (1646–1704) *Vespertilio maximus ex insula Sancti Vincentii*: A previously unpublished description and drawings of the greater bulldog bat, *Noctilio leporinus* (Linnaeus, 1758). Proceedings of the Biological Society of Washington 134:29–41.

Pindell, J., and L. Kennan. 2009. Tectonic evolution of the Gulf of Mexico, Caribbean and northern South America in the mantle reference frame: An update. Special Publication of the Geological Society of London 328:1–55.

Pio, D.V., F.M. Clarke, I. MacKie, and P.A. Racey. 2010. Echolocation calls of the bats of Trinidad, West Indies: Is guild membership reflected in echolocation signal design? Acta Chiropterologica 12:217–229.

Presley, S.J., J. Graf, A.F. Hassan, A.R. Sjodin, and M.R. Willig. 2021. Effects of host species identity and diet on the biodiversity of the microbiota in Puerto Rican Bats. Current Microbiology 78:3526–3540.

Radeloff, V.C., J.W. Williams, B.L. Bateman, K.D. Burke, S.K. Carter, E.S. Childress, K.J. Cromwell, C. Gratton, A.O. Hasley, B.M. Kraemer, et al. 2015. The rise of novelty in ecosystems. Ecological Applications 25:2051–2068.

Ramírez-Fráncel, L.A., L.V. García-Herrera, S. Losada-Prado, G. Reinoso-Flórez, A. Sánchez-Hernández, S. Estrada-Villegas, B. Lim, and G. Guevara. 2022. Bats and their vital ecosystem services: A global review. Integrative Zoology 17:2–23.

Recht, J., V.J. Schuenemann, and M.R. Sánchez-Villagra. 2020. Host diversity and origin of zoonoses: The ancient and the new. Animals 10(9): 1672. https://doi.org/10.3390/ani10091672.

Red Latinoamericana y del Caribe para la Conservación de los Murciélagos [Latin American and Caribbean Network for the Conservation of Bats]. 2021. Conservación: AICOMs & SICOMs [Conservation: AICOMs & SICOMs]. https://relcomlatinoamerica .net. Accessed 20 June 2022.

Reeves, W.K., J. Beck, M.V. Orlova, J.L. Daly, K. Pippin, F. Revan, and A.D. Loftis. 2016. Ecology of bats, their ectoparasites, and associated pathogens on Saint Kitts Island. Journal of Medical Entomology 53:1218–1225.

Riccucci, M. 2013. Bats as materia medica: An ethnomedical review and implications for conservation. Vespertilio 16:249–270.

Ricklefs, R., and E. Bermingham. 2008. The West Indies as a laboratory of biogeography and evolution. Philosophical Transactions of the Royal Society B 363:2393–2413.

Ricklefs, R.E., and I.J. Lovette. 1999. The roles of island area per se and habitat diversity in the species-area relationships of four Lesser Antillean faunal groups. Journal of Animal Ecology 68:1142–1160.

Riskin, D.K., and M.B. Fenton. 2001. Sticking ability in Spix's diskwinged bat, *Thyroptera tricolor* (Microchiroptera: Thyropteridae). Canadian Journal of Zoology 79:2261–2267.

Riskin, D.K., J.E.A. Bertram, and J.W. Hermanson. 2016. The evolution of terrestrial locomotion in bats: The bad, the ugly, and the good. *In* J.E.A. Bertram, ed. Understanding mammalian locomotion: Concepts and applications, pp. 307–323. Hoboken, New Jersey: Wiley Blackwell.

Rivas-Camo, N.A., P.A. Sabido-Villanueva, C.R. Peralta-Muñoz, and R.A. Medellín. 2020. Cuba in Mexico: First record of *Phyllops falcatus* (Gray, 1839) (Chiroptera, Phyllostomidae) for Mexico and other new records of bats from Cozumel, Quintana Roo. ZooKeys 973:153–162.

Robertson, R.E.A. 2009. Antilles geology. *In* R.G. Gillespie and D.A. Clague, eds., Encyclopedia of islands, pp. 29–35. Berkeley: University of California Press.

Rocha, R., A. López-Baucells, and A. Fernández-Llamazares. 2021. Ethnobiology of bats: Exploring human-bat inter-relationships in a rapidly changing world. Journal of Ethnobiology 41:3–17.

Rodríguez, A., and E.C. Mora. 2006. The echolocation repertoire of *Eptesicus fuscus* (Chiroptera: Vespertilionidae) in Cuba. Caribbean Journal of Science 42:121–128.

Rodríguez-Durán, A. 1995. Metabolic rates and thermal conductance in four species of Antillean bats roosting in hot caves. Comparative Biochemistry and Physiology 110A:347–355.

Rodríguez-Durán, A. 1996. Foraging ecology of the Puerto Rican boa (*Epicrates inornatus*): Bat predation, carrion feeding, and piracy. Journal of Herpetology 30:533–536.

Rodríguez-Durán, A. 1998. Distribution and nonrandom aggregations of the cave bats of Puerto Rico. Journal of Mammalogy 79:141–146.

Rodríguez-Durán, A. 1999. First record of reproductive *Lasiurus borealis minor* (Miller) from Puerto Rico (Chiroptera). Caribbean Journal of Science 35:143–144.

Rodríguez-Durán, A. 2002. Los murciélagos en las culturas pre-colombinas de Puerto Rico [Bats in pre-Columbian societies of Puerto Rico]. Focus 1:15–18.

Rodríguez-Durán, A. 2009. Bat assemblages in the West Indies: The role of caves. *In* T.H. Fleming and P.A. Racey, eds., Island bats: Evolution, ecology, and conservation, pp. 265–280. Chicago: University of Chicago Press.

Rodríguez-Durán, A. 2020. Roosting ecology: The importance of detailed description. *In* T.H. Fleming, L.M. Dávalos, and M. Mello, eds., Phyllostomid bats: A unique mammalian radiation, pp. 311–324. Chicago: University of Chicago Press.

Rodríguez-Durán, A. 2022. First census and monitoring of bat colonies in hot caves in Puerto Rico. Final report. State Wildlife Action Plan Contract 2020-000089. San Juan, Puerto Rico: Department of Natural and Environmental Resources.

Rodríguez-Durán, A., and W. Feliciano-Robles. 2015. Impact of wind facilities on bats in the Neotropics. Acta Chiropterologica 17:365–370.

Rodríguez-Durán, A., and W. Feliciano-Robles. 2016. Conservation value of remnant habitat for Neotropical bats on islands. Caribbean Naturalist 35:1–10.

Rodríguez-Durán, A., and T.H. Kunz. 2001. Biogeography of West Indian bats: An ecological perspective. *In* C.A. Woods and F. Sergile, eds., Biogeography of the West Indies: Patterns and perspectives, pp. 355–368. Boca Raton, Florida: CRC Press.

Rodríguez-Durán, A., and A.R. Lewis. 1985. Seasonal predation by merlins on sooty mustached bats in western Puerto Rico. Biotropica 17:71–74.

Rodríguez-Durán, A., and A.R. Lewis. 1987. Patterns of population size, diet, and activity for a multispecies assemblage of bats. Caribbean Journal of Science 23:352–360.

Rodríguez-Durán, A., and W. Otero. 2011. Species richness and diversity of a West Indian bat assemblage in a fragmented ecosystem. Acta Chiropterologica 13:439–445.

Rodríguez-Durán, A., and E. Padilla-Rodríguez. 2008. Oxygen transport and wing morphology of Antillean bats. Caribbean Journal of Science 44:375–379.

Rodríguez-Durán, A., and E. Padilla-Rodríguez. 2010. New Records for the bat fauna of Mona Island, Puerto Rico, with notes on their natural history. Caribbean Journal of Science 46:102–105.

Rodríguez-Durán, A., and J. Rosa. 2020. Remarkable variation in the diet of *Noctilio leporinus* in Puerto Rico: The fishing bat turns carnivorous. Acta Chiropterologica 22:175–178.

Rodríguez-Durán, A., and E. Santiago-Valentín. 2014. Una historia de la mastozoología en el Caribe insular [A history of mammalogy in the insular Caribbean]. *In* J. Ortega, J.L. Martínez, and D.G. Tirira, eds., Historia de la mastozoología en Latinoamérica, las Guayanas y el Caribe [History of mammalogy in Latin America, the Guianas, and the Caribbean], pp. 129–142. Quito, Ecuador: Editorial Murciélago Blanco and Asociación Ecuatoriana de Mastozoología.

Rodríguez-Durán, A., and J.A. Soto-Centeno. 2003. Temperature selection by tropical bats roosting in caves. Journal of Thermal Biology 28:465–468.

Rodríguez-Durán, A., A.R. Lewis, and Y. Montes. 1993. Skull morphology and diet of Antillean bat species. Caribbean Journal of Science 29:258–261.

Rodríguez-Durán, A., J. Pérez, M.A. Montalbán, and J.M. Sandoval. 2010. Predation by free-roaming cats on an insular population of bats. Acta Chiropterologica 12:359–362.

Rodríguez-Herrera, B., R.A. Medellín, and R.M. Timm. 2007. Neotropical tent-roosting bats—field guide / Murciélagos Neotropicales que acampan en hojas—Guía de campo. Santo Domingo de Heredia, Costa Rica: Editorial INBIO.

Rodríguez-Herrera, B., L. Viquez-R, E. Cordero-Schmidt, J.M. Sandoval, and A. Rodríguez-Durán. 2016. Energetics of tent roosting bats: The case of *Ectophylla alba* and *Uroderma bilobatum* (Chiroptera: Phyllostomidae). Journal of Mammalogy 97:246–252.

Rodríguez-Ramos, R., J.R. Pagán-Jimenez, Y. Narganes Storde, and M.J. Lace. 2019. Guácaras in early precolonial Puerto Rico: The case of Cueva Ventana. *In* C.L. Hofman and A.T. Antczak, eds., Early settlers of the insular Caribbean: Dearchaizing the archaic, pp. 201–214. Leiden, The Netherlands: Sidestone Press.

Rolfe, A.K., and A. Kurta. 2012. Diet of mormoopid bats on the Caribbean island of Puerto Rico. Acta Chiropterologica 14:369–377.

Rolfe, A.K., A. Kurta, and D.L. Clemans. 2014. Species-level analysis of diets of two mormoopid bats from Puerto Rico. Journal of Mammalogy 95:587–596.

Rose, A., S. Wöhl, J. Bechler, M. Tschapka, and M. Knörnschild. 2019. Maternal mouth-to-mouth feeding behaviour in flower-visiting bats, but no experimental evidence for transmitted dietary preferences. Behavioural Processes 165:29–35.

Rowse, E.G., D. Lewanzik, E.L. Stone, S. Harris, G. Jones. 2016. Dark matters: The effects of artificial lighting on bats. *In* C.C. Voigt and T. Kingston, eds., Bats in the Anthropocene: Conservation of bats in a changing world, pp. 187–213. New York: SpringerOpen.

Runkel, V., G. Gerding, and U. Marckmann. 2021. The handbook of acoustic bat detection. Exeter, United Kingdom: Pelagic Publishing.

Russell, A.L., C.A. Pinzari, M.J. Vonhof, K.J. Olival, and F.J. Bonaccorso. 2015. Two tickets to paradise: Multiple dispersal events in the founding of hoary bat populations in Hawai`i. PloS One 10(6): e0127912. https://doi.org/10.1371/journal.pone.0127912.

Russello, M.A., and G. Amato. 2004. A molecular phylogeny of

Amazona: Implications for Neotropical parrot biogeography, taxonomy, and conservation. Molecular Phylogenetics and Evolution 30:421–437.

Russo, D., L. Ancillotto, and G. Jones. 2017. Bats are still not birds in the digital era: Echolocation call variation and why it matters for bat species identification. Canadian Journal of Zoology 96:63–78.

Sanchez, L., C.R. Moreno, and E.C. Mora. 2017. Echolocation calls of *Natalus primus* (Chiroptera: Natalidae): Implications for conservation monitoring of this species. Cogent Biology 3: 11355027. https://doi.org/10.1080/23312025.2017.1355027.

Sánchez, O., and D.E. Wilson. 2016. Food items of *Macrotus waterhousii* (Chiroptera: Phyllostomidae) in central Mexico. Therya 7:161–177.

Sánchez-Losada, M., and C.A. Mancina. 2020. Diet segregation between sexes by a gregarious Greater Antillean bat, *Phyllonycteris poeyi* (Chiroptera: Phyllostomidae). Acta Chiropterologica 21:385–393.

Sánchez-Lozada, M., H. Vela, H.M. Perdomo, J. Mozón, A.H. de la Cruz, A. Hernández, A. Longueira, A. Espinosa, T.M. Rodríguez-Cabrera, A. Vidal, and C.A. Mancina. 2018. Datos de distribución de murciélagos en Cuba: Un acercamiento a través de inventarios biológicos rápidos [Data on the distribution of bats in Cuba based on biological quick-surveys]. Poeyana 507:76–81.

Santana, S.E., T.O. Dial, T.P. Eiting, and M.E. Alfaro. 2011. Roosting ecology and the evolution of pelage markings in bats. PloS One 6(10): e25845. https://doi.org/10.1371/journal.pone.0025845.

Scatena, F.N., and A.E. Lugo. 1995. Geomorphology, disturbance, and the soil and vegetation of two subtropical wet steepland watersheds of Puerto Rico. Geomorphology 13:199–213.

Schaetz, B.A., A. Kurta, A. Rodríguez-Durán, O.M. Münzer, and R. Foster. 2009. Identification of bats on Puerto Rico using the scanning electron microscope to examine external hairs. Caribbean Journal of Science 45:125–130.

Schnitzler, H.-U. 2009. Echolocation by insect-eating bats. BioScience 51:557–569.

Schnitzler, H.-U., E.K.V. Kalko, I. Kaipf, and A.D. Grinnell. 1994. Fishing and echolocation behavior of the greater bulldog bat, *Noctilio leporinus*, in the field. Behavioral Ecology and Sociobiology 35:327–345.

Seetahal, J.F.R., A. Vokaty, M.A.N. Vigilato, C.V.E. Carrington, J. Pradel, B. Louison, A. Van Sauers, R. Roopnarine, J.C. González Arrebato, M.E. Millien, C. James, and C.E. Rupprecht. 2018. Rabies in the Caribbean: A situational analysis and historic review. Tropical Medicine and Infectious Disease 3(89): 1–21. https://doi.org/10.3390/tropicalmed3030089.

Shamel, H.H. 1945. A new *Eptesicus* from Jamaica. Proceedings of the Biological Society of Washington 58:107–110.

Sherwin, R.E., and W.L. Gannon. 2005. *Ariteus flavescens*. Mammalian Species 787:1–3.

Siemers, B.M., G. Schauermann, H. Turni, and S. von Merten. 2009. Why do shrews twitter? Communication or simple echo-based orientation. Biology Letters 5:593–596.

Silander, S.R. 1979. A study of the ecological life history of *Cecropia peltata* L., an early secondary successional species in the rain forest of Puerto Rico. MS thesis, University of Tennessee, Knoxville.

Silva-Taboada, G. 1976. Historia y actualización taxonómica de algunas especies antillanas de murciélagos de los géneros *Pteronotus*, *Brachyphylla*, *Laisurus*, y *Antrozous* (Mammalia:Chiroptera) [History and taxonomic update of some Antillean species of bats in the genera *Pteronotus*, *Brachyphylla*, *Laisurus*, and *Antrozous* (Mammalia: Chiroptera)]. Poeyana 153:1–24.

Silva-Taboada, G. 1979. Los murciélagos de Cuba [The bats of Cuba]. La Habana, Cuba: Editorial Academia.

Silva-Taboada, G., W. Suarez Duque, and S. Díaz Franco. 2007. Compendio de los mamíferos terrestres autóctonos de Cuba vivientes y extinguidos [Compendium of living and extinct autochthonous terrestrial mammals of Cuba]. Havana, Cuba: Museo Nacional de Historia Natural.

Simal, F., C. de Lannoy, L. García-Smith, O. Doest, J.A. Freitas, F. Franken, I. Zaandam, A. Martino, J.A. González-Carcacía, C.L. Peñaloza, P. Bertuol, D. Simal, and J.F. Nassar. 2015. Island-island and island-mainland movements of the Curaçaoan long-nosed bat, *Leptonycteris curasoae*. Journal of Mammalogy 96:579–590.

Simmons, N.B. 2005. Order Chiroptera. *In* D.E. Wilson and D.M. Reeder, eds., Mammal species of the world: A taxonomic and geographic reference, 3rd ed., pp. 312–529. Washington, D.C.: Smithsonian Institution Press.

Simmons, N.B., and T.H. Quinn. 1994. Evolution of the digital tendon locking mechanism in bats and dermopterans: A phylogenetic perspective. Journal of Mammalian Evolution 2:231–254.

Simmons, N.B., K. Seymour, J. Habersetzer, and G. Gunnell. 2008. Primitive early Eocene bat from Wyoming and the evolution of flight and echolocation. Nature 451:818–821.

Snyder, N.F.R., J.W. Wiley, and C.B. Kepler. 1987. The parrots of Luquillo: Natural history and conservation of the Puerto Rican parrot. Los Angeles: Western Foundation of Vertebrate Zoology.

Solari, S., R.L. Medellín, B. Rodríguez-Herrera, E.R. Dumont, and S.F. Burneo. 2019. Family Phyllostomidae (New World Leaf-Nosed bats). *In* D.E. Wilson and R.A. Mittermeier, eds., Handbook of the mammals of the world. Volume 9, Bats, pp. 444–487. Barcelona, Spain: Lynx Edicions.

Soto-Centeno, J.A. 2013. Extinction and phylogeography of Caribbean bats during the late Quaternary. PhD dissertation, University of Florida, Gainesville.

Soto-Centeno, J.A., and C.A. Calderón-Acevedo. 2022. Biogeography and validation of species limits in Caribbean red bats (Vespertilionidae: *Lasiurus*). BioRxiv doi: https://doi.org/10.1101/2022.02.11.479705.

Soto-Centeno, J.A., and A. Kurta. 2003. Description of fetal and newborn brown flower bats, *Erophylla sezekorni* (Chiroptera: Phyllostomidae). Caribbean Journal of Science 39:233–234.

Soto-Centeno, J.A., and A. Kurta. 2006. Diet of two nectarivorous bats, the brown flower bat (*Erophylla sezekorni*) and the Greater Antillean long-tongued bat (*Monophyllus redmani*), on Puerto Rico. Journal of Mammalogy 87:19–26.

Soto-Centeno, J.A., and N.B. Simmons. Environment drives phenotypic convergence in the most widely distributed New World bat. Journal of Biogeography (in review).

Soto-Centeno, J.A., and D.W. Steadman. 2015. Fossils reject climate change as the cause of extinction of Caribbean bats. Science Reports 5(7971): 1–7. https://doi.org/10.1038/srep07971.

Soto-Centeno, J.A., M. O'Brien, and N.B. Simmons. 2015. The importance of late Quaternary climate change and karst on distributions of Caribbean mormoopid bats. American Museum Novitates 3847:1–32.

Soto-Centeno, J.A., N.B. Simmons, and D.W. Steadman. 2017. The bat community of Haiti and evidence for its long-term persistence at high elevations. PLoS One 12(6): e0178066. https://doi.org/10.1371/journal.pone.0178066.

Soto-Centeno, J.A., D. Phillips, A. Kurta, and K. Hobson. 2014. Food resource partitioning in syntopic nectarivorous bats on Puerto Rico. Journal of Tropical Ecology 30:359–369.

Speer, K.A., J.A. Soto-Centeno, N.A. Albury, Z. Quicksall, M.G. Marte, and D.L. Reed. 2015. Bats of the Bahamas: Natural history and conservation. Bulletin of the Florida Museum of Natural History 53:45–95.

Speer, K.A., B.J. Petronio, N.B. Simmons, R. Richey, K. Magrini, J.A. Soto-Centeno, and D.L. Reed. 2017. Population structure of a widespread bat (*Tadarida brasiliensis*) in an island system. Ecology and Evolution 7:7585–7598.

Speer, K.A., E. Luetke, E. Bush, B. Sheth, A. Gerace, Z. Quicksall, M. Miyamoto, C.W. Dick, K. Dittmar, N. Albury, and D.L. Reed. 2019. A fly on the cave wall: Parasite genetics reveal fine-scale dispersal patterns of bats. Journal of Parasitology 105:555–566.

Stanchak, K.E., J.H. Arbour, and S.E. Santana. Anatomical diversification of a skeletal novelty in bat feet. Evolution 73:1591–1603.

Steadman, D.W., N.A. Albury, B. Kakuk, J.I. Mead, J.A. Soto-Centeno, H.M. Singleton, and J. Franklin. 2015. Vertebrate community on an ice-age Caribbean island. Proceedings of the National Academy of Sciences of the United States of America 112(44):E5963-E5071.

Stehlé, H. 1945. Forest types of the Caribbean islands. Caribbean Forester 6 (supplement): 273–408.

Stevens-Arroyo, A.M. 2006. Cave of the Jagua: The mythological world of the Taínos. Scranton, Pennsylvania: University of Scranton Press.

Stoetzel, E., A. Royer, D. Cochard, and A. Lenoble. 2016. Late Quaternary changes in bat palaeobiodiversity and palaeobiogeography under climatic and anthropogenic pressure: New insights from Marie-Galante, Lesser Antilles. Quaternary Science Review 143:150–174.

Stoetzel, E., C. Bochaton, S. Bailon, D. Cochard, M. Gala, and V. Laroulandie. 2021. Multi-taxa neo-taphonomic analysis of bone remains from barn owl pellets and cross-validation of observations: A case study from Dominica (Lesser Antilles). Quaternary 4(4): 38. https://doi.org/10.3390/quat4040038.

Surlykke, A., S.B. Pedersen, and L. Jakobsen. 2009. Echolocating bats emit a highly directional sonar sound beam in the field. Proceedings of the Royal Society B 276:853–860.

Surlykke, A., L. Jakobsen, E.K.V. Kalko, and R. Page. 2013. Echolocation intensity and directionality of perching and flying fringe-lipped bats, *Trachops cirrhosus* (Phyllostomidae). Frontiers in Physiology 4(143): 1–9. https://doi.org/10.3389/fphys.2013.00143.

Swartz, S.M., and J.J. Allen. 2020. Structure and function of bat wings: A view from the Phyllostomidae. *In* T.H. Fleming, L.M. Dávalos, and M.A.R. Mello, eds., Phyllostomid bats: A unique mammalian radiation, pp. 151–168. Chicago: University of Chicago Press.

Swartz, S.M., J. Iriarte-Díaz, D.K. Riskin, and K.S. Breuer. 2012. A bird? A plane? No, it's a bat: An introduction to the biomechanics of bat flight. *In* G.F. Gunnell and N.B. Simmons, eds., Evolutionary history of bats: Fossils, molecules and morphology, pp. 317–352. Cambridge, United Kingdom: Cambridge University Press.

Tavares, V. da C., and C.A. Mancina. 2008. *Phyllops falcatus* (Chiroptera: Phyllostomidae). Mammalian Species 811:1–7.

Tavares, V. da C., O.M. Warsi, F. Balseiro, C.A. Mancina, and L.M. Dávalos. 2018. Out of the Antilles: Fossil phylogenies support reverse colonization of bats to South America. Journal of Biogeography 45:859–873.

Taylor, T.K., B.K. Lim, M. Pennay, P. Soisook, L.O. Loureiro, and L.M. Moras. 2019. Family Molossidae (free-tailed bats). *In* D.E. Wilson and R.A. Mittermeier, eds., Handbook of the mammals of the world. Volume 9, Bats, pp. 598–672. Barcelona, Spain: Lynx Edicions.

Tejedor, A. 2009. The type locality of *Natalus stramineus* (Chiroptera: Natalidae): Implications for the taxonomy and biogeography of the genus *Natalus*. Acta Chiropterologica 8:361–380.

Tejedor, A. 2011. Systematics of funnel-eared bats (Chiroptera: Natalidae). Bulletin of the American Museum of Natural History 353:1–140.

Tejedor, A. 2019. Family Natalidae (funnel-eared bats). *In* D.E. Wilson and R.A. Mittermeier, eds., Handbook of the mammals of the world. Volume 9, Bats, pp. 584–596. Barcelona, Spain: Lynx Edicions.

Tejedor, A., G. Silva-Taboada, and D. Rodríguez-Hernández. 2004.

Discovery of extant *Natalus major* (Chiroptera: Natalidae) in Cuba. Mammalian Biology 69:153–162.

Tejedor, A., V. da C. Tavares, and D. Rodríguez-Hernández. 2005a. New records of hot-cave bats from Cuba and Dominican Republic. Boletín de la Sociedad Venezolana de Espeleología 39:10–15.

Tejedor, A., V. da C. Tavares, and G. Silva-Taboada. 2005b. A revision of extant Greater Antillean bats of the genus *Natalus* (Chiroptera: Natalidae). American Museum Novitates 3493:1–22.

Thiagavel, J., S. Brinkløv, I. Geipel, and J.M. Ratcliffe. 2020. Sensory and cognitive ecology. *In* T.H. Fleming, L.M. Dávalos, and M.A.R. Mello, eds., Phyllostomid bats: A unique mammalian radiation, pp. 187–204. Chicago: University of Chicago Press.

Timm, R.M., and H.H. Genoways. 2003. West Indian mammals from the Albert Schwartz Collection: Biological and historical information. Scientific Papers, Natural History Museum, University of Kansas 29:1–47.

Turvey, S.T., R.J. Kennerley, J.M. Nuñez-Mino, and R.P. Young. 2017. The last survivors: Current status and conservation of the non-volant land mammals of the insular Caribbean. Journal of Mammalogy 98:918–936.

Übernickel, K., M. Tschapka, and E.K.V. Kalko. 2013. Selective eavesdropping behaviour in three Neotropical bat species. Ethology 119:66–76.

Uyehara, K., and G. Wiles. 2009. Bats of the U.S. Pacific Islands. U.S.D.A., Natural Resources Conservation Service, Pacific Islands Area, Biology Technical Note 20:1–34.

Vareschi, E., and W. Janetzky. 1998. Bat predation by the yellow snake or Jamaican boa, *Epicrates subflavus*. Jamaica Naturalist 5:34–36.

Vásquez-Parra, O., F.J. García, D. Araujo-Reyes, H. Brito, and M. Machado. 2015. Dinámica poblacional de *Pteronotus parnellii* y *Anoura geoffroyi* (Mammalia: Chiroptera) en Venezuela [Population dynamics of *Pteronotus parnellii* and *Anoura geoffroyi* (Mammalia: Chiroptera) in Venezuela]. Ecotrópicos 28:27–37.

Vaughan, N. 1995. New records of bats on St. Vincent. Bulletin of the British Ecological Society 26:102–104.

Vaughan, N., and J.H. Hill. 1996. Bat (Chiroptera) diversity and abundance in banana plantations and rain forest, and three new records for St. Vincent, Lesser Antilles. Mammalia 60:441–447.

Vaughan Jennings, N., S. Parsons, K.E. Barlow, and M.R. Gannon. 2004. Echolocation calls and wing morphology of bats from the West Indies. Acta Chiropterologica 6:75–90.

Vela Rodríguez, H., and C.A. Mancina. 2020. Alta longevidad en el murciélago mariposa, *Nyctiellus lepidus* (Gervais, 1837) (Natalidae), uno de los murciélagos más pequeños del mundo [High longevity in the butterfly bat, *Nyctiellus lepidus* (Gervais, 1837) (Natalidae), one of the smallest bats in the world]. Journal of Bat Research and Conservation 13:100–103.

Vela Rodríguez, H., F. Amador Hernández, M. Núñez Rodríguez, and J. Pérez Paret. 2019. Aspectos de la conducta reproductiva del murciélago mariposa, *Nyctiellus lepidus* (Chiroptera: Natalidae) en cuevas al norte de la región central de Cuba [Some aspects of the reproductive behavior of the butterfly bat, *Nyctiellus lepidus* (Chiroptera: Natalidae) from caves north of Cuba's central region]. Poeyana 507:54–68.

Villalobos-Chaves, D., and B. Rodríguez-Herrera. 2021. Frugivorous bats promote epizoochoric seed dispersal and seedling survival in a disturbed Neotropical forest. Journal of Mammalogy 102:1507–1513.

Westbrook, J.K., and R.S. Eyster. 2017. Atmospheric environment associated with animal flight. *In* P.B. Chilson, W.F. Frick, J.F. Kelly, and F. Liechti, eds., Aeroecology, pp. 13–45. Cham, Switzerland: Springer.

Whitaker, J.O., Jr., and P.A. Frank. 2012. Foods of little free-tailed bats, *Molossus molossus*, from Boca Chica Key, Monroe County, Florida. Florida Scientist 75:249–252.

Whitaker, J.O., Jr., and A. Rodríguez-Durán. 1999. Seasonal variation in the diet of Mexican free-tailed bats, *Tadarida brasiliensis antillularum* (Miller) from a colony in Puerto Rico. Caribbean Journal of Science 35:23–29.

Wilburn, D. 2011. Wind energy in the United States and materials required for the land-based wind turbine industry from 2010 through 2030. U.S.G.S. Scientific Investigations Report 2011-5036:1–22.

Wiley, J.W. 2010. Food habits of the endemic ashy-faced owl (*Tyto glaucops*) and recently arrived barn owl (*T. alba*) in Hispaniola. Journal of Raptor Research 44:87–100.

Wiley, J.W., and J.M. Wunderle, Jr. 1993. The effects of hurricanes on birds, with special reference to Caribbean islands. Bird Conservation International 3:319–349.

Wilkinson, G.S., and T.H. Fleming. 1996. Migration and evolution of lesser long-nosed bats *Leptonycteris curasoae*, inferred from mitochondrial DNA. Molecular Ecolgy 5:329–339.

Willig, M.R., S.J. Presley, C.P. Bloch, and H.H. Genoways. 2009. Macroecology of Caribbean bats: Effect of area, elevation, latitude, and hurricane induced disturbance. *In* T.H. Fleming and P.A. Racey, eds., Island bats: Evolution, ecology, and conservation, pp. 216–264. Chicago: University of Chicago Press.

Wilson, D.A., and R.A. Mittermeier, eds. 2019. Handbook of the mammals of the world. Volume 9, Bats. Barcelona, Spain: Lynx Edicions.

Yi, X., and E.K. Latch. 2022. Nuclear phylogeography reveals strong impact of gene flow in big brown bats. Journal of Biogeography 49:1061–1074.

LIST OF CONTRIBUTORS

Camilo A. Calderón-Acevedo, Department of Earth and Environmental Sciences, Rutgers University, Newark, NJ, USA

Vinícius C. Cláudio, Fundação Oswaldo Cruz, Rio de Janeiro, RJ, Brazil

Liliana M. Dávalos, Department of Ecology and Evolution, Stony Brook University, Stony Brook, NY, USA

José M. De la Cruz Mora, Museo de Historia Natural "Tranquilino Sandalio de Noda," Centro de Investigaciones y Servicios Ambientales, Pinar del Río, Cuba

Héctor M. Díaz Perdomo, Departamento de Zoología, Instituto de Ecología y Sistemática, La Habana, Cuba

Andrea Donaldson, National Environment and Planning Agency, Kingston, Jamaica

Jon Flanders, Bat Conservation International, Austin, TX, USA

Hugh H. Genoways, University of Nebraska State Museum, Lincoln, NE, USA

Justin D. Hoffman, Department of Biology, McNeese State University, Lake Charles, LA, USA

Allen Kurta, Department of Biology, Eastern Michigan University, Ypsilanti, MI, USA

Gary Kwiecinski, Department of Biology, University of Scranton, Scranton, PA, USA

Burton K. Lim, Royal Ontario Museum, Toronto, Ontario, Canada

Livia Loureiro, Illumina, San Diego, California, USA

Ariel E. Lugo, International Institute of Tropical Forestry, USDA Forest Service, Río Piedras, PR, USA

Carlos A. Mancina, Centro Nacional de Biodiversidad, Instituto de Ecología y Sistemática, La Habana, Cuba

Ricardo Moratelli, Fundação Oswaldo Cruz, Rio de Janeiro, RJ, Brazil

Jafet M. Nassar, Venezuelan Institute of Scientific Research, Miranda, Venezuela

Miguel S. Núñez-Novas, Museo Nacional de Historia Natural "Prof. Eugenio de Jesús Marcano," Santo Domingo, Dominican Republic

Johanset Orihuela, Department of Earth and Environment, Florida International University, Miami, FL, USA

Ana C. Pavan, University of São Paulo, Piracicaba, SP, Brazil

Melissa E. Rodríguez, Programa de Conservación de Murciélagos de El Salvador de la Asociación Territorios Vivos, Colonia Vista Hermosa, El Salvador

Armando Rodríguez-Durán, Departamento de Ciencias Naturales, Universidad Interamericana, Recinto de Bayamón, Bayamón, Puerto Rico, USA

Bernal Rodríguez-Herrera, Universidad de Costa Rica, San José, Costa Rica

Margarita Sánchez-Lozada, Centro Oriental de Ecosistemas y Biodiversidad, Santiago de Cuba, Cuba

J. Angel Soto-Centeno, Department of Earth and Environmental Sciences, Rutgers University, Newark, NJ, and American Museum of Natural History, New York, NY, USA

Kelly A. Speer, Smithsonian's National Museum of Natural History and National Zoological Park and Conservation Biology Institute, Smithsonian Institution, Washington D.C., USA

Valéria da Cunha Tavares, Instituto Tecnológico Vale, PA, and Universidade Federal da Paraíba, PB, Brazil

Humberto Vela Rodríguez, Grupo Espeleológico Cayo-Barién, Sancti Spiritus, Cuba

Ashley K. Wilson, Department of Biology, Eastern Michigan University, Ypsilanti, MI, USA

Eumops auripendulus (black bonneted bat), Burton K. Lim
Eumops ferox (fierce bonneted bat), ©MerlinTuttle.org
Molossus milleri (pug-nosed mastiff bat), Burton K. Lim
Molossus molossus (Pallas's mastiff bat), Gary Kwiecinski
Molossus verrilli (Hispaniolan mastiff bat), Burton K. Lim
Mormopterus minutus (little goblin bat), Carlos A. Mancina
Nyctinomops laticaudatus (broad-tailed bat), R. D. Lord and
    the Mammal Images Library of the American Society of
    Mammalogists
Nyctinomops macrotis (big free-tailed bat), Brock Fenton
Tadarida brasiliensis (Brazilian free-tailed bat), Burton K. Lim
Mormoops blainvillei (Antillean ghost-faced bat), Burton K. Lim
Pteronotus fulvus (Thomas' naked-backed bat), substituting for *Ptero-
    notus davyi* (Davy's naked-backed bat), Gerardo Ceballos
Pteronotus fuscus (Allen's mustached bat), Gary Kwiecinski
Pteronotus macleayii (Macleay's mustached bat), Burton K. Lim
Pteronotus parnellii (Parnell's mustached bat), Carlos A. Mancina
Pteronotus portoricensis (Puerto Rican mustached bat), Ashley K.
    Wilson
Pteronotus pusillus (Hispaniolan mustached bat), Miguel S.
    Núñez-Novas
Pteronotus quadridens (sooty mustached bat), Carlos A. Mancina
Chilonatalus macer (Cuban lesser funnel-eared bat), Carlos A.
    Mancina
Chilonatalus micropus (Caribbean lesser funnel-eared bat), Miguel S.
    Núñez-Novas

Chilonatalus tumidifrons (Bahamian lesser funnel-eared bat), J. Angel
Soto-Centeno
Natalus jamaicensis (Jamaican greater funnel-eared bat),
©MerlinTuttle.org
Natalus major (Hispaniolan greater funnel-eared bat), J. Angel
Soto-Centeno
Natalus primus (Cuban greater funnel-eared bat), Carlos A. Mancina
Natalus stramineus (Lesser Antillean funnel-eared bat), Béatrice
Ibéné
Nyctiellus lepidus (Gervais's funnel-eared bat), J. Angel Soto-Centeno
Noctilio leporinus (greater bulldog bat), Gary Kwiecinski
Ardops nichollsi (Lesser Antillean tree bat), Béatrice Ibéné
Ariteus flavescens (Jamaican fig-eating bat), Burton K. Lim
Artibeus jamaicensis (Jamaican fruit-eating bat), Allen Kurta
Artibeus lituratus (great fruit-eating bat), Burton K. Lim
Artibeus schwartzi (Schwartz's fruit-eating bat), Gary Kwiecinski
Brachyphylla cavernarum (Antillean fruit-eating bat), Burton K. Lim
Brachyphylla nana (Cuban fruit-eating bat), J. Angel Soto-Centeno
Chiroderma improvisum (Guadeloupean big-eyed bat), Béatrice Ibéné
Erophylla bombifrons (brown flower bat), Burton K. Lim
Erophylla sezekorni (buffy flower bat), Burton K. Lim
Glossophaga longirostris (Pallas's long-tongued bat), Burton K. Lim
Glossophaga soricina (Miller's long-tongued bat), Burton K. Lim
Macrotus waterhousii (Waterhouse's leaf-nosed bat), J. Angel
Soto-Centeno
Micronycteris buriri (Saint Vincent big-eared bat), Gary Kwiecinski
Monophyllus plethodon (insular single-leaf bat), Burton K. Lim
Monophyllus redmani (Greater Antillean long-tongued bat), Allen
Kurta
Phyllonycteris aphylla (Jamaican flower bat), J. Angel Soto-Centeno
Phyllonycteris poeyi (Cuban flower bat), Carlos A. Mancina
Phyllops falcatus (Cuban fig-eating bat), J. Angel Soto-Centeno
Stenoderma rufum (red fig-eating bat), Allen Kurta
Sturnira angeli (Angel's yellow-shouldered bat), Béatrice Ibéné
Sturnira paulsoni (Paulson's yellow-shouldered bat), Gary Kwiecinski
Antrozous pallidus (North American pallid bat), substituting for
Antrozous koopmani (Koopman's pallid bat), J. Scott Altenbach
Eptesicus fuscus (big brown bat), Allen Kurta

Eptesicus guadeloupensis (Guadeloupean big brown bat), Michel and Anne Breuil

Eptesicus lynni (Jamaican brown bat), Andrea Donaldson, National Environment and Planning Agency

Lasiurus cinereus (northern hoary bat), Allen Kurta

Lasiurus borealis (eastern red bat), substituting for *Lasiurus degelidus* (Jamaican red bat), Robin M. Slider

Lasiurus intermedius (northern yellow bat), substituting for *Lasiurus insularis* (Cuban yellow bat), J. Scott Altenbach

Lasiurus minor (minor red bat), Burton K. Lim

Lasiurus pfeifferi (Pfeiffer's red bat), Carlos A. Mancina

Myotis dominicensis (Dominican myotis), Gérard Issartel

Myotis martiniquensis (Schwartz's myotis), Gérard Issartel

Myotis nyctor (Barbadian myotis), Gary Kwiecinski

Nycticeius cubanus (Cuban evening bat), Héctor M. Díaz Perdomo

INDEX